MATHEMATICS NCERT SYLLABUS 6th GRADE

R C Yadav

Ex Principal
SBMN Public School, Civil Road - Rohtak
Maharishi Vidyamandir, Gurgaon.

INDIA • SINGAPORE • MALAYSIA

Notion Press

Old No. 38, New No. 6
McNichols Road, Chetpet
Chennai - 600 031

First Published by Notion Press 2018
Republished by Notion Press 2019
Copyright © R C Yadav 2019
All Rights Reserved.

ISBN 978-1-64733-992-0

This book has been published with all reasonable efforts taken to make the material error-free after the consent of the author. No part of this book shall be used, reproduced in any manner whatsoever without written permission from the author, except in the case of brief quotations embodied in critical articles and reviews.

The Author of this book is solely responsible and liable for its content including but not limited to the views, representations, descriptions, statements, information, opinions and references ["Content"]. The Content of this book shall not constitute or be construed or deemed to reflect the opinion or expression of the Publisher or Editor. Neither the Publisher nor Editor endorse or approve the Content of this book or guarantee the reliability, accuracy or completeness of the Content published herein and do not make any representations or warranties of any kind, express or implied, including but not limited to the implied warranties of merchantability, fitness for a particular purpose. The Publisher and Editor shall not be liable whatsoever for any errors, omissions, whether such errors or omissions result from negligence, accident, or any other cause or claims for loss or damages of any kind, including without limitation, indirect or consequential loss or damage arising out of use, inability to use, or about the reliability, accuracy or sufficiency of the information contained in this book.

Contents

Chapter 1: Numbers	1
Chapter 2: Natural Numbers and Whole Numbers	17
Chapter 3: Negative Numbers and Integers	37
Chapter 4: Playing with Numbers	63
Chapter 5: Fractions	99
Chapter 6: Decimals	141
Chapter 7: Ratio, Proportion and Unitary Method	169
Chapter 8: Fundamental Concept of Algebra	197
Chapter 9: Basic Geometrical Ideas	229
Chapter 10: Understanding Elementary Shapes	271
Chapter 11: Symmetry	315
Chapter 12: Geometrical Constructions	337
Chapter 13: Mensuration	353
Chapter 14: Data Handling	381

Chapter 1
Numbers, Digits, and Numerals

Number

A number is something that represents a quantity of an item.

For example: Quantity of days in a year

= Number of days in a year

= 365

Here, 365 is a number.

Digits

To write various numbers – small or big, we use the following ten symbols:

0, 1, 2, 3, 4, 5, 6, 7, 8 and 9.

These symbols are called digits or figures.

Numeral

A numeral is a symbol or name that stands for a number.

A numeral is made of a single-digit or more than one digit.

For example '3' is a numeral,

'16' is a numeral,

3402 is a numeral.

Remember

- Digits make numerals, which is a representation of number.
- A number has numerical value while digit is just a representation.
- Numbers are made up of digits.

For example: There are thirty-six students in a class. This means the count or measurement of the class strength is **Thirty-six** (i.e. the number is 36). Its representation using symbols is 36 (i.e. the numeral is 36). Here, the symbols (digits) used are 3 and 6. Thus,

Numeral → 36

Number → Thirty – Six

Digits → 3 and 6

NOTATION AND NUMERATION

Writing a number using digits (figures) is called its notation. Whereas writing a number in words is called numeration.

DIFFERENCE BETWEEN ARITHMETIC AND MATHEMATICS

The branch of mathematics that deals with basic operations of only numbers involving +, –, × and ÷, is called *'Arithmetic.'* Whereas *'Mathematics'* is the study of various relationships among the numbers quantities and shapes. Mathematics includes **arithmetic**, and other branches like **algebra, geometry, trigonometry** etc.

READING AND WRITING NUMBERS

To express various numbers – small or big we make the use of 10 digits (**i.e. 0, 1, 2, 3, 4, 5, 6, 7, 8 and 9**). The digits are limited but numbers are unlimited. Therefore, we need a system. There are various number systems, such as **Arabic, Babylonian, Chinese, Egyptian, Greek, Roman, Decimal** etc.

Here we will study two systems:

- Hindu – Arabic Number system.
- International Number system.

To express a number, we use 'place-value system.'

The *place value* of a digit in a number depends on its place (position) in the number.

Place value of a digit = $\boxed{\text{Face value of the digit}} \times \boxed{\text{Value of the place occupied by the digit}}$

HINDU – ARABIC NUMBER SYSTEM [OR *INDIAN PLACE VALUE SYSTEM*]

In the Indian – Place value System, the places are grouped in periods.

The places ones (or units), tens and hundreds taken together form **Ones-period**.

The places thousands and ten-thousands taken together form **thousands-period**.

The places lakhs and ten-lakhs taken together form **lakhs-period**.

The places crores and ten-crores taken together form **crores-period**.

The places arabs and ten-arabs taken together form **arabs-period**.

INDIAN – PLACE-VALUE CHART

Periods →	ARABS		CRORES		LAKHS		THOUSANDS		ONES		
Places →	Ten – Arabs (TA) (10 00 00 000)	Arabs (A) (1 00 00 00 000)	Ten Crores (C) (10 00 00 000)	Crores (C) (1 00 00 000)	Ten Lakhs (TL) (10 00 000)	Lakhs (L) (1 00 000)	Ten Thousands (T-th) (10 000)	Thousands (Th) (1000)	Hundreds (H) (100)	Tens (T) (10)	Ones (O) (1)

2-Places   2-Places   2-Places   2-Places   3-Places

Each place has a value of 10 times the value of the place to its right.

The value of the 1st place (extreme right)	=	1
The value of the 2nd place	=	10 = 10 × 1
The value of the 3rd place	=	100 = 10 × 10
The value of the 4th place	=	1000 = 10 × 100
The value of the 5th place	=	10000 = 10 × 1000

and so on

For reading and writing the numbers, we use above table.

EXAMPLE

Read the numbers in the following table:

	CRORES		LAKHS		THOUSANDS		ONES		
	T-C	C	T-L	L	T-Th	Th	H	T	O
(i)	6	5	1	3	5	6	8	5	1
(ii)		2	6	5	4	0	7	4	5
(iii)	1	8	5	3	0	0	0	8	1
(iv)	4	4	4	4	4	4	4	4	4

Solution

Beginning from the right, divide the digits into groups period wise. Next, begin at the left with the largest group, and proceed to the right, reading the digits of a period together along with the name of the period (except ones).

i. Since 6 5 1 3 5 6 8 5 1 → 65, 13, 56, 851

∴ 65, 13, 56, 851 is read as:

'Sixty-five crore Thirteen lakh fifty-six thousand eight-hundred fifty one.'

ii. Since 26540745 → 2, 65, 40, 745 ∴ 2, 65, 40, 745 is read as:

Two crore sixty-five lakh **forty thousand seven hundred forty-five**

iii. Since 185300081 → 18,53,00,081

∴ 18,53,00,081 is read as:

Eighteen crore fifty-three lakh eighty-one.

iv. Since 4 4 4 4 4 4 4 4 4 → 44,44,44,444

∴ 44,44,44,444 is read as:

Forty-four crore forty-four lakh forty-four thousand four hundred forty-four.

Note

These days to express very large numbers, the period Arab is not used. For example,

723,47,98,605 is expressed as:

Crores	Lakhs	Thousands	Ones
723	47	98	605

or 'Seven hundred twenty-three crore forty-seven lakh ninety-eight thousand six hundred five.'

International Number System

In international system each period has three-places. Here also, each place has a value ten times the value of the place to its right.

INTERNATIONAL PLACE VALUE CHART

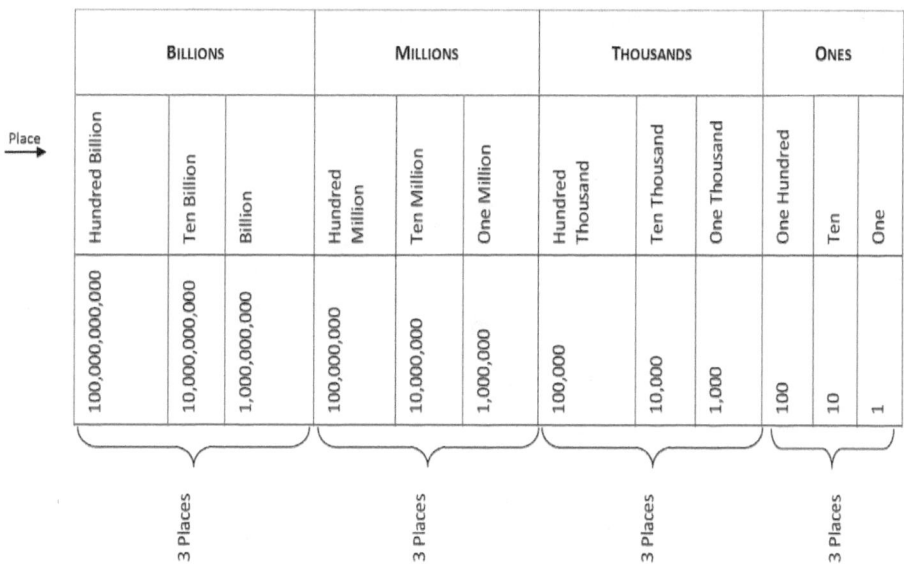

On comparing the Indian and International systems:

Indian Place-Value Chart   ⇨ **1C = 1 crore = 1,00,00,000**

Crores		Lakhs		Thousands		Ones		
TC	C	TL	L	TTh	Th	H	T	O
	1	0	0	0	0	0	0	0

⇨ **1TM = 10 million = 10,000,000**

Millions			Thousands			Ones		
HM	TM	M	HTh	TTh	Th	H	T	O
	1	0	0	0	0	0	0	0

INTERNATIONAL PLACE-VALUE CHART

From above place – value charts, we observe that:

1 ten	has	1 zero	⎫
1 hundred	has	2 zeros	
1 thousand	has	3 zeros	
1 ten-thousand	has	4 zeros	⎬ Indian System
1 lakh	has	5 zeros	
1 ten lakh	has	6 zeros	
1 crore	has	7 zeros	
1 ten-crore	has	8 zeros	⎭

1 ten	has	1 zero	⎫
1 hundred	has	2 zeros	
1 Thousand	has	3 zeros	
1 ten-thousand	has	4 zeros	
1 hundred – thousand	has	5 zeros	
1 million	has	6 zeros	⎬ International system
1 ten-million	has	7 zeros	
1 hundred million	has	8 zeros	
1 billion	has	9 zeros	
1 ten billion	has	10 zeros	
1 hundred billion	has	11 zeros	⎭

We find that:

Indian System	International System
1 lakh	= 1 hundred thousand
10 lakhs	= 1 million
1 crore	= 10 millions

Numbers, Digits, and Numerals | 7

EXAMPLE

From the following table write the number In words:

BILLIONS			MILLIONS			THOUSANDS			ONES		
Hundred Billions	Ten-Billions	Billions	Hundred Millions	Ten-Millions	Millions	Hundred Thousand	Ten-Thousands	Thousands	Hundreds	Ten	Ones
3	0	2	7	2	2	1	6	5	3	6	

SOLUTION

From the table, we have 30272216536

$\qquad$ = 30,272,216,536

= { Thirty *billion* two hundred seventy-two *million* two hundred sixteen *thousand* five hundred thirty-six

NOTE

To express a number in words, we do not use the plural of periods i.e. We use 'Thousand' and not 'Thousands' we use lakh and not 'lakhs' we use million and not 'millions'.

EXAMPLE

Write the numeral for each of the following numbers:
  i. Thirty-seven lakh thirty-eight thousand four hundred twenty.
  ii. Two crore two lakh twenty thousand four hundred twenty-five.
  iii. Thirty-four crore forty-three thousand nine hundred ninety-eight.

SOLUTION

  i. Thirty-seven lakh thirty-eight thousand four hundred twenty has '37 lakhs 38 thousands 420' or

LAKHS	THOUSANDS	ONES	
37	38	420	} = 37,38,420

  ii. Two crore two lakh twenty thousand four hundred twenty-five has 02 crores 02 lakhs 20 thousands 425 or

CRORES	LAKHS	THOUSANDS	ONES	
02	02	20	425	} = 2,02,20,425

Note

(i) **We put 0 (zero) at the vacant places.**

(ii) **We do not write 0 at the extreme left of a number.**

(iii) Thirty-four crore forty-three thousand nine hundred ninety-eight: has 34 crores 43 Thousands 998 or

Crores	Lakhs	Thousands	Ones	
34	00	43	998	= 34,00,43,998

Example

Write the face value and the place-value of the encircled digit in each of the following numbers:

(i) 8 0 0 ⑥ 8 7 2 1   (ii) 2 5 7 1 8 3 2 1 ⑦

(iii) 7 2 7 2 3 9 ⓪ 5 6   (iv) ③ 8 1 2 0 8 4 5 5

Solution

NUMBER	ENCIRCLED DIGIT	FACE-VALUE	PLACE-VALUE	
TC C TL L TTH TH H T O				
(i) 8 0 0 6 8 7 2 1	6	6	6 × 10,000	= 60,000
(ii) 2 5 7 1 8 3 2 1 7	7	7	7 × 1	= 7
(iii) 7 2 7 2 3 9 0 5 6	0	0	0 × 100	= 0
(iv) 3 8 1 2 0 8 4 5 5	3	3	3 × 10 00 00 000	= 30 00 00 000

Example

Find the difference of the place values of two 6 in 3 8 1 6 7 8 6 4 1.

Solution

The given number can be written as:

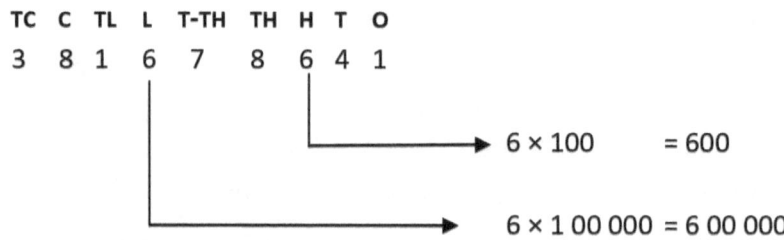

∴ The place value of 6 at Hundred's place = 600

The place value of 6 at Lakh's place = 600 000

⇒ The required difference between two place values = 600 000 – 600

= 599,400

REMEMBER

- **The face-value of a digit is the digit itself irrespective of its position in the numeral.**
- **The face-value of a digit is also called its true-value.**
- **The place value of '0' is always '0' regardless of the places it occupies in the numeral.**

EXPANDED FORM OF A NUMERAL

When a numeral expressed as the sum of the place values of the digits, it is called in its *expanded-form*.

For example, The expanded form of 980671234

98,06,71,234 = 9 × 10 00 00 000 + 8 × 100 00 000 + 0 × 10 00 000
+ 6 × 100 000 + 7 × 10 000 + 1 × 1000 + 2 × 100
+ 3 × 10 + 4 × 1

= 90 00 00 000 + 8 00 00 000 + 0 + 6 00 000 + 70 000 + 1000 + 200 + 30 + 4

= 90 00 00 00 + 8 00 00 000 + 6 00 000 + 70000 + 1000 + 200 + 30

EXAMPLE

Express each of the following numbers in expanded form:

(i) 3 6 5 6 7 0 9 2 (ii) 4 5 6 1 2 3 4 3 2

(iii) 3 9 3 9 3 9 9 0 5 (iv) 7 0 0 1 8 3 2 9

SOLUTION

i. Expanded form of 3, 65, 67, 092

= 3 × 1 00 00 000 + 6 × 10 00 000 + 5 × 100 000 + 6 × 10 000
+ 7 × 1000 + 0 × 100 + 9 × 10 + 2 × 1

= 3 00 00 000 + 60 00 000 + 5 00 000 + 60 000 + 7 000 + 0 + 90 + 2

= 3 00 00 000 + 60 00 000 + 5 00 000 + 60 000 + 7000 + 90 + 2

ii. Expanded form of 45, 61, 23, 432

= 4 × 10 00 00 000 + 5 × 100 00 000 + 6 × 10 00 000
+ 1 × 1 00 000 + 2 × 10 000 + 3 × 1000 + 4 × 100
+ 3 × 10 + 2 × 1

= 40 00 00 000 + 5 00 00 000 + 60 00 000 + 100 000 + 20 000
+ 3000 + 400 + 30 + 2

iii. The Expanded from of 39, 39, 39, 905

= 3 × 1000 00 000 + 9 × 100 00 000 + 3 × 10 00 000
+ 9 × 100 000 + 3 × 10 000 + 9 × 1000 + 9 × 100
+ 0 × 10 + 5 × 1

= 30 00 00 000 + 900 00 000 + 30 00 000 + 9 00 000 + 30 000
+ 9000 + 900 + 0 + 5

= 30 00 00 000 + 9 00 00 000 + 30 00 000 + 9 00 000 + 30 000
+ 9000 + 900 + 5

iv. The Expanded form of 700, 18, 329

= 7 × 100 00 000 + 0 × 1000 000 + 0 × 100 000 + 1 × 10 000
+ 8 × 1000 + 3 × 100 + 2 × 10 + 9 × 1

= 7 00 00 000 + 0 + 0 + 10000 + 8 000 + 300 + 20 + 9

= 7 00 00 000 + 10000 + 8000 + 300 + 20 + 9

EXERCISE – 1

1. Complete the following statements:
   a. The numeral for nine is _____
   b. Hindu-Arabic digits are _____
   c. Different places in the period crores are_____
   d. Different places in the period million are _____
   e. The face-value of 5 in 185012 is _____
   f. The face-value of 0 in 181,301,256 is _____
   g. The place-value of 0 in 26,01,236 is _____
   h. The place-value of 7 in 1,23,47,652 is _____

2. a. How many thousands make 1 lakh?
   b. How many lakhs make 1 million?
   c. How many millions make 1 crore?

d. How many crores make 1 billion?

e. How many lakhs make 1 Arab?

3. Express these numbers in words:

(a) 22,06,702   (b) 7,18,22,156

(c) 30,00,00,312   (d) 34,567,891

(e) 123,456,112,189

4. Express these numbers in figures:

a. Sixty-one Thousand three hundred thirteen.

b. Two crore two lakh twenty thousand four hundred twenty-five

c. Thirty four crore forty three thousand nine hundred ninety-eight.

d. Nine hundred million seven hundred ninety-three thousand five hundred twenty-two

e. Four hundred eleven million one hundred nine thousand nine hundred ninety-nine.

5. Use commas to express the following numerals according to Indian and international system:

Numeral	Indian System	International System
a) 67089125		
b) 10020609		
c) 786123456		
d) 471820251		
e) 554403101		
f) 345678901		
g) 234567897		

6. Write number of zeros in:

(a) 1 Ten-lakh   (b) 1 crore

(c) 1 ten thousand   (d) 1 hundred-thousand

(e) 1 ten-million   (f) 1 billion

7. Write True or False:

a. $9 \times 1000 + 5 \times 100 + 5 \times 10 + 9 \times 1 = 95{,}59$

b. $3 \times 10000 + 4 \times 1000 + 5 \times 100 + 6 \times 10 + 7 = 34{,}567$

c. 4 × 1000 + 2 × 1000 + 3 × 100 + 4 × 10 + 5 = 42,345

d. 2 × 100000 + 3 × 10000 + 4 × 1000 + 5 × 100 + 6 × 10 + 7 = 2,34,567

Answers

Exercise – 1

1. (a) 9 (b) 1, 2, 3, 4, 5, 6, 7, 8, 9 and 0
   (c) crores, ten-crores (d) millions, ten-millions, hundred-millions
   (e) 5 (f) 0
   (g) 0 (h) 7, 000

2. (a) 100 (b) 10 (c) 10 (d) 100 (e) 1000

3. (a) Twenty-two lakh six thousand seven hundred two.
   (b) Seven crore eighteen lakh twenty-two thousand one hundred fifty-six.
   (c) Thirty crore three hundred twelve.
   (d) Thirty-four million five hundred sixty-seven thousand eight hundred ninety-one.
   (e) One hundred twenty-three billion four hundred fifty-six million one hundred twelve thousand one hundred eighty-nine.

4. (a) 61,313 (b) 2,02,20,425 (c) 34,00,43,998 (d) 900,793,522
   (e) 411,109,999

5.

	Indian-system	International-system
(a)	6,70,89,125	67,089,125
(b)	1,00,20,609	10,020,609
(c)	78, 61,23,456	786,123,456
(d)	47,18,20,251	471,820,251
(e)	55,44,03,101	554,403,101
(f)	34,56,78,901	345,678,901
(g)	23,45,67,897	234,567,897

6. (a) 6 (b) 7 (c) 4 (d) 5
   (e) 7 (f) 9

7. (a) True (b) True (c) False (d) True

Fun with Numbers

    I   Think of any number.

    II   Subtract 1 from it.

    III   Multiply the difference by 3.

    IV   Add 15 to the product.

    V   Divide the answer by 3

    VI   Add 5 to result obtained in step V.

    VII   Subtract the number (thought by you).

    VIII   The ANSWER is 9

Forming Numbers

A number is an arranged group of digits. Numbers can be formed with or without the repetition of digits

Forming Smallest and Greatest Number

[*Digits are not allowed to repeat*]

To form the *greatest* number using the given digits, we arrange them in descending order.

To form the *smallest* number using the given digit, we arrange them in ascending order.

Ascending Order

Placing the things from smallest to greatest i.e. in an increasing order.

Descending Order

Placing the things from greatest to smallest i.e. in a decreasing order.

Example

Using the digits 9, 1, 3, 5 and 0, Write the greatest and the smallest numbers.

Solution

Given digits are: 9, 1, 3, 5 and 0 Ascending order of given digits:

$$0, 1, 3, 5, 9$$

Any number cannot begin with 0.

∴ Smallest number = 10, 359

Descending order of given digits:

$$9, 5, 3, 1, 0$$

Greatest umber = 95, 310

[**Digits are Allowed To Repeat**]

When repetition of digits is allowed, the greatest digit is repeated on the left most places to form the greatest number. In case of writing smallest number, the smallest digit is repeated on the left most places.

NOTE

As '0' is never written on the left most place. In case '0' is to be repeated then it is repeated on the places just after the first place from the left for the smallest number and on the last places for the greatest number.

EXAMPLE

Write the greatest and smallest 6-digit number using the digits 1, 4, 0 and 5.

SOLUTION

To form the greatest-number.

Descending order of the given digits is: 5, 4, 1, 0

∴ The greatest 6-digit number (formed by the given digits) = 5,5 5, 4 1 0

To form the smallest number

Ascending order of the given digits is 0, 1, 4, 5

∴ The smallest 6-digit number (formed by the given digits) = 1, 00,045

EXAMPLES

Find the difference between the greatest and smallest 6-digit number having three different digits.

SOLUTION

The smallest three different digits are 0, 1, 2

∴ The smallest six digit number = 1,00,002

The greatest three different digits are: 9, 8, and 7

∴ The greatest six digit number = 9,99,987

Now, the required difference is = 9,99,987 − 1,00,002 = <u>8,99,985</u>

Example

Using the digits 3, 4, 5 and 0, how many different 4-digit numbers can be formed (without repeating any digit)?

SOLUTION

A four digit number has four places:

| **THOUSAND** | **HUNDREDS** | **TENS** | **ONES** |

'0' cannot be placed in extreme left place (i.e. in the thousands place)

∴ Remaining three digits 3, 4 and 5 can be part in thousands place.

Now,

Numbers with 3 in thousands place	**Numbers with 4 in thousands place**	**Numbers with 5 in thousands place**
3 0 4 5	4 0 5 3	5 0 4 3
3 0 5 4	4 0 3 5	5 0 3 4
3 4 0 5	4 5 0 3	5 3 0 4
3 4 5 0	4 5 3 0	5 3 4 0
3 5 0 4	4 3 0 5	5 4 0 3
3 5 4 0	4 3 5 0	5 4 3 0

Thus, we can form 18 different 4-digit numbers using the digits 3, 4, 5 and 0.

EXAMPLE

Make the smallest 4-digit number by using four different digits such that the digit 2 is always in the hundred place.

SOLUTION

Four different smallest digits are 0, 1, 2 and 3

To write a smallest number using given digits, we write the digits in ascending order.

Ascending order of digits is 0, 1, 2, 3

Thousands	*Hundreds*	*Tens*	*Ones*
0	1	2	3

The digit 0 cannot be put in the extreme left (thousands) place.

Thousands	Hundreds	Tens	Ones
1	0	2	3

But the digit 2 is always in the hundreds place.

Thousands	Hundreds	Tens	Ones
1	2	0	3

Hence, the required smallest number is 1203.

Exercise – 2

1. Write the smallest 5-digit number using two different digits.
2. Write the greatest 6-digit number using three different digits.
3. How many times does the digit 6 appear in the ten's place when we write all the numbers from:

    (i) 0 to 59 (ii) 10 to 100 (iii) 60 to 900

4. In 10087092, rearrange the digits to get the

    (i) largest number (ii) smallest number

5. Write all the 3-digit numbers that can be formed, using 8, 5 and 0 **[each digit used once in a number]**
6. Write the (i) smallest 5-digit number (ii) greatest 5-digit number, using 4 different digits such that the digit 2 is always in the tens place.
7. Which of the following statements are true and which are false?

    (i) The smallest 4-digit number that can be formed using the digits 9, 8, 0 and 1 is 0189.

    (ii) The greatest 6-digit number, using only 2-digits is 9,99,998.

    (iii) The smallest eight-digit number using only 3 digits is 1,00,00,002.

Answers

Exercise – 2

1. 10,000   **2.** 9,99,987

3. (i) 0 (ii) 10 (iii) 90   **4.** (i) 9,87,21,000 (ii) 1,00,02,789

5. 850, 805, 580 and 508   **6.** (i) 10,023   (ii) 99,827

7. (i) False (ii) True (iii) True

Chapter 2
Natural Numbers and Whole Numbers

Natural Numbers

For counting objects, we need counting numbers 1, 2, 3, 4, 5, ... etc. These numbers come naturally when we start counting. So, we call the counting numbers as **Natural Numbers**.

The smallest number of objects to be counted can be **'one.'** Therefore, the first or the smallest counting number (or the natural number) is 1.

Because the objects to be counted increase as **two, three, four,** ... etc, the counting numbers associated with them also increase as 2, 3, 4, ... etc.

The number of objects to be counted can be endless, so there is no *last* or *largest* natural number, i.e. the natural numbers are 1, 2, 3, 4, ... ∞.

Remember

i. **Every natural number (except 1) can be obtained by adding 1 to the previous number (called its predecessor). There is no previous (predecessor) natural number for the number 1.**

ii. One more than a given whole number is called its successor.

The Concept of Zero

Let us consider a situation that Mrs Saleena has five flowers and she gives all the five flowers to her daughter. Now, Mrs Saleena is left with no flower. Or Mrs Saleena is left with zero flowers. So 'zero' is a number associated with '**no objects**.' The symbol for zero is '0.'

When we start counting the objects, we do not start counting with 'no-object.' Therefore, '0' is not a counting number (or natural number).

Whole Numbers

When the number '0' is included in the group of natural-numbers, then the new group is called a collection of **whole-numbers**. Thus, the numbers 0, 1, 2, 3, 4, ... are called whole-numbers.

Remember

i. **The number '0' is the smallest whole-number. There is no last or the largest whole number.**

ii. **Every whole-number (except 0) can be obtained by adding '1' to the previous whole number. There is no 'previous whole' number for the whole number '0.'**

iii. **Every 'natural number' is a whole number. But every 'whole number' is not a natural number ('0' is not a natural number).**

Number Line

A line on which we can represent numbers is called a 'number line.' We can represent whole numbers on a number line.

Representation of Whole-Numbers on a Number Line

Draw a straight line and mark point 0 on it. Let this point represents the number 0 (zero).

Mark points P, Q, R, S, T, ... to the right of 0 at equal distances. Label each of the points

O as 0 (zero)
P as 1 (one)
Q as 2 (two)
R as 3 (three)...

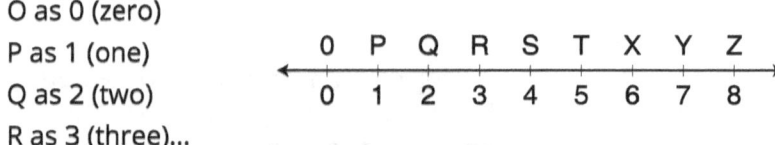

Properties of Whole Numbers

Basic operations – **addition, subtraction, multiplication** and **division** can be performed on whole numbers.

Closure Property

Consider, 5 + 2 = 7,

11 + 36 = 47 } Sum of whole numbers is a whole number

i.e. Whole number + Whole number = Whole number

In general, if 'a' and 'b' are whole numbers and their sum is 'c' then c is also a whole number, i.e.

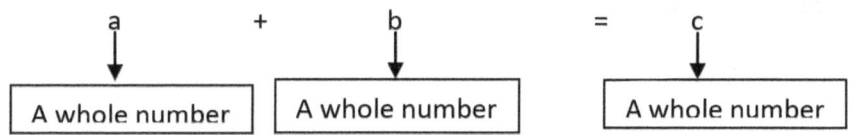

This property is called the **Closure Property of addition.**

If we add two whole numbers, we will get a whole-number.

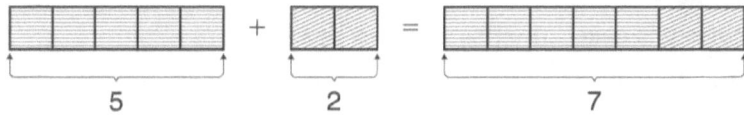

Commutative Property

Consider, $\quad 6 + 3 = 9$

and $\quad 3 + 6 = 9$

i.e. two whole numbers can be added in any order, their sum remains the same.

In general, if a and b are two whole-numbers, then a + b = b + a

This property of addition is called **commutative property of addition.**

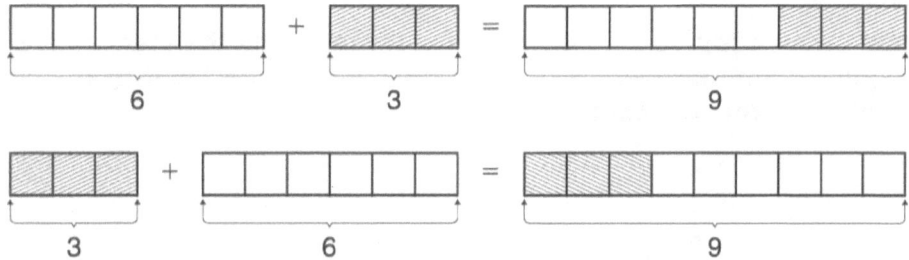

This can be represented on a number line as:

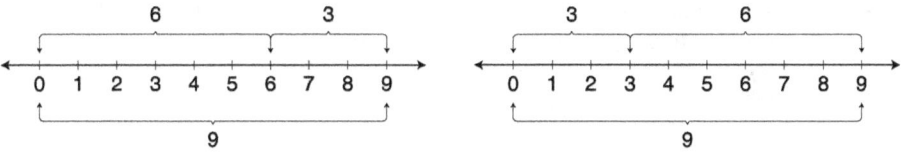

Associative Property

Consider 2 + (3 + 4) = 9
(2 + 3) + 4 = 9

i.e. in the addition of whole numbers, the sum does not change even if the grouping is changed.

In general, if a, b and c are three whole numbers, then

$$(a + b) + c = a + (b + c)$$

This property is called the **associative property** of addition.

Property of zero

Consider, 4 + 0 = 4
16 + 0 = 16
0 + 27 = 27

i.e. if zero is added to a whole number, then the result is the number itself.

In general if 'a' is a whole number then

a + 0 = a   and   0 + a = a

Here, 0 (zero) is called the *additive identity* of a whole number.

NOTE

0 is called additive identity because it maintains the identity (i.e. value) of the number i.e. it does not change the identity of the number during the operation – addition.

Closure Property

Consider, 5 ⟵ Minuend ⟶ 36
-2 ⟵ Subtrahend ⟶ -11
3 ⟵ Remainder ⟶ 25

i.e. The difference of two whole numbers is a whole number only if minuend is greater than subtrahend. If the subtrahend and minuend are equal then their difference is 0 (a whole number).

Now, consider

7 ⟵ Minuend
-10 ⟵ Subtrahend
-3 ⟵ Remainder (not a whole number)

RECALL

Minuend: **A number from which another number is subtracted.**

Subtrahend: **A number that is subtracted from another.**

Thus, for subtraction between two whole numbers, to get a whole number the *subtrahend* must be smaller than the minuend.

In general, for whole numbers 'a' and 'b,' 'a–b' is not always a whole-number. Because remainder is not always a whole number, therefore, *subtraction is not closed in whole numbers.*

Commutative Property

Consider 9 – 6 = 3 and 6 – 9 = –3 that is 9 – 6 is not equal to 6 – 9

In general, for whole numbers 'a' and 'b,' a – b and b – a are not equal.

Thus, the commutative property is not true for subtraction in whole numbers.

Associative Property

Consider, (18 – 4) – 3 = 14 – 3 = 11

18 – (4 – 3) = 18 – 1 = 17

i.e. (18 – 4) – 3 is equal to 18 – (4 – 3)

In general, if a, b and c are whole numbers, then (a – b) – c is not equal to a – (b – c)

Thus, the Associative property is not true for subtraction in whole numbers.

Property of zero

Consider, 5 – 0 = 5 and 0 – 5 = –5 (not a whole number)

18 – 0 = 18 and 0 – 18 = –18 (not a whole number)

i.e., if zero is subtracted from a whole number, then the result is the whole number itself, but a whole number subtracted from zero is not a whole number (0 – 0 = 0, which is a whole number).

In general if 'a' is whole number, then a – 0 = a, but 0 – a = –a is not a whole number.

NOTE

Property	For addition in whole numbers	For subtraction in whole numbers
Closure	is true	is not true
Commutative	is true	is not true
Associative	is true	is not true

Closure Property

Consider, $3 \times 7 = 21$ (a whole number)

$9 \times 8 = 72$ (a whole number)

$5 \times 0 = 0$ (a whole number)

In general, if 'a' and 'b' are whole number. Then, a × b is always a whole number.

i.e., the product of whole numbers

Thus, whole numbers are closed under multiplication.

Commutative Property

Consider, $6 \times 8 = 48$

$8 \times 6 = 48$

So, $6 \times 8 = 8 \times 6$

In general, for all whole numbers 'a' and 'b,'

$a \times b = b \times a$

So, regardless of the order in which whole numbers are multiplied, we get the same product. Thus, the commutative property of multiplication is true in whole numbers.

Associative Property

Consider,

$(2 \times 5) \times 7 = 10 \times 7 = 70$

$2 \times (5 \times 7) = 2 \times 35 = 70$

i.e. $(2 \times 5) \times 7 = 2 \times (5 \times 7)$

Thus, the product of any three whole numbers, in any order, is the same.

In general, if 'a,' 'b' and 'c' are any whole numbers, then (a × b) × c = a × (b × c)

Hence, associative property is true for multiplication in whole numbers.

Property of 1

The product of a whole number and 1, is the number itself. For example:

or $\left.\begin{array}{l} 5 \times 1 = 5 \\ 1 \times 5 = 5 \end{array}\right\}$ and $\left\{\begin{array}{l} 12 \times 1 = 12 \\ \text{or } 1 \times 12 = 12 \end{array}\right.$

If 'a' is whole number, then a × 1 = 1 × a = a.

Here, 1 is called *the identity element in multiplication* (or *multiplicative identity*).

Property of Zero

Consider: 6 × 0 = 0, 0 × 6 = 0

3 × 0 = 0, 0 × 3 = 0

0 × 0 = 0

In general, if 'a' is whole number, then a × 0 = 0 × a = 0

Thus, when a whole number 'a' is multiplied by zero, then product is always zero.

Closure Property

Consider, 8 ÷ 4 = 2

$10 \div 4 = 2\dfrac{1}{4}$

$12 \div 13 = \dfrac{12}{13}$

$2\dfrac{1}{4}$ and $\dfrac{12}{13}$ are not whole numbers.

In general, if 'a' and 'b' are whole numbers, then the quotient a ÷ b may or may not be a whole number.

Thus, division in whole numbers is not closed.

Recall

Dividend: **A number that is to be divided by another number.**

Divisor: **A number by which the dividend is divided.**

Quotient: **The result of division.**

Remainder: **The amount left over when we divide two numbers.**

Commutative Property

Consider, $8 \div 4 = 2$

$4 \div 8 = \dfrac{1}{2}$

i.e., $8 \div 4$ and $4 \div 8$ are not same.

In general, if 'a' and 'b' are whole numbers, then $a \div b$ is not equal to $b \div a$

Thus, the commutative property does not hold True in whole numbers.

Associative Property

Consider, $(64 \div 8) \div 4 = 8 \div 4 = 2$

and $64 \div (8 \div 4) = 64 \div 2 = 32$

That is, $(64 \div 8) \div 4$ is not equal to $64 \div (8 \div 2)$

In general, if 'a,' 'b' and 'c' are whole numbers, then $(a \div b) \div c$ is not equal to $a \div (b \div c)$

Thus, the associative property does not hold true in whole numbers.

Special Properties of division

Division by 1

Consider $9 \div 1 = 9$, $5 \div 1 = 5$, $20 \div 1 = 20$

In general, for a whole number 'a' we have:

$$a \div 1 = a$$

Thus, whenever a whole-number is divided by 1, we get the same whole number as the answer.

Division by the number itself

Consider, $9 \div 9 = 1, 5 \div 5 = 1, 20 \div 20 = 1$

In general, for a whole-number 'a,' we have:

$$a \div a = 1$$

Thus, whenever a whole number is divided by itself, we get the answer as 1.

Division of Zero by whole numbers

Consider, $0 \div 9 = 0$, $0 \div 5 = 0$, $0 \div 20 = 0$

In general, for a whole-number 'a,' we have:

$$0 \div a = 0$$

Thus, whenever 0 (zero) is divided by any whole number, the result is always 0.

Division by Zero

Division of any whole number by zero is meaningless and is not allowed.

DISTRIBUTIVE PROPERTY OF MULTIPLICATION OVER ADDITION

If 'a,' 'b' and 'c' be any whole numbers then

$$a \times (b + c) = a \times b + a \times c$$

i.e. in whole numbers, multiplication distributes over addition.

For example:

$5 \times (2 + 6)$	$5 \times (2 + 6)$
$= 5 \times 2 + 5 \times 6$	$= 5 \times (8)$
$= 10 + 30 = 40$	$= 5 \times 8 = 40$

We can say that in whole numbers multiplication distributes over addition.

NOTE

In whole numbers, multiplication also, distributes over subtraction, $a \times (b - c) = a \times b - a \times c$ provided 'b' is greater or equal to 'c'

For example,

(i) $5 \times (6 - 2) = 5 \times 6 - 5 \times 2$
$= 30 - 10 = 20$

(ii) $5 \times (6 - 6) = 5 \times 6 - 5 \times 6$
$= 30 - 30$
$= 0$ or $5 \times (6 - 6) = 5 \times 0 = 0$

EXAMPLES

EXAMPLE-1

Add the following by grouping the numbers in different ways:

$$447 + 253 + 587$$

SOLUTION

$447 + 253 + 587$	$447 + 253 + 587$
$= (447 + 253) + 587$	$= 447 + (253 + 587)$
$= (700) + 587 = 1287$	$= 447 + (840) = 1287$

EXAMPLE-2

Show that:

$$(53 + 69) + 47 = 53 + (69 + 47)$$

SOLUTION

$(53 + 69) + 47$	$53 + (69 + 47)$
$= 122 + (47)$	$= 53 + (116)$
$= 169$	$= 169$
$\therefore (53 + 69) + 47$	$= 53 + (69 + 47)$

EXAMPLES-3

Solve the following using suitable rearrangement:

$$132 \times 125$$

SOLUTION

We know that $125 \times 4 = 500$

Also, $132 \div 4 = 33 \Rightarrow 132 = 4 \times 33$

So, $132 \times 125 = 33 \times 4 \times 125$

$$= 33 \times (4 \times 125)$$

$$= 33 \times 500 = 16500$$

EXAMPLE-4

Using properties, simplify:

$$4 \times 24 \times 125$$

Solution

$$4 \times 24 \times 125$$

Rearranging the numbers:

$4 \times 24 \times 125 = 24 \times 4 \times 125$
$ = 24 \times (4 \times 125)$
$ = 24 \times 500 = 12000$

Example-5

Find the Value of:

(i) $43 \times 2571 + 357 \times 2571$

(ii) $425 \times 42 - 34 \times 425$

Solution

(i) $43 \times 2571 + 357 \times 2571$
$= (43 + 357) \times 2571$
$= 400 \times 2571$
$= 4 \times 100 \times 2571$
$= 10284 \times 100 = 10,28,400$

(ii) $425 \times 42 - 34 \times 425$
$= (42 - 34) \times 425$
$= 8 \times 425 = 3400$

More About Division

The number to be divided is called '*dividend*'

The number which divides is called '*divisor*'

The result of the operation of division is called '*quotient*'

The number which is left over after division, is called remainder.

Relation between the Dividend, Divisor, Quotient and Remainder

If 'a' is any whole number and 'b' another smaller non-zero whole number then there exist unique whole numbers q and r (r < b) such that

$a = (b \times q) + r$ or   Dividend = Divisor × Quotient + remainder

We call it *Division Rule* or *Division Algorithm*

EXAMPLE - 6

Divide 6888 by 123

SOLUTION

Here dividend = 6888 and divisor is 123.

$$6888 \div 123$$

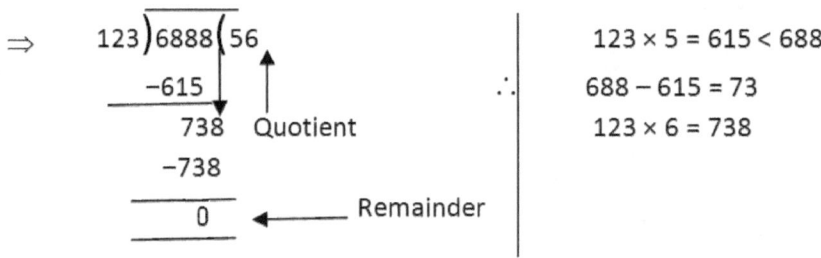

We get: Quotient = 56 and Remainder = 0

EXAMPLE -7

Find the quotient and remainder when 143193 is divided by 456. Also prove the rule of division.

SOLUTION

We have:

```
      456 ) 143193 ( 314
           - 1368
             ‾‾‾‾
              639
            - 456      Quotient
              ‾‾‾
             1833      Remainder
            -1824
             ‾‾‾‾
                9
```

∴ Quotient = 314 and Remainder = 9

To prove the division-rule, let us substitute the given values in the relation:

Dividend = Divisor × Quotient + Remainder,

We have: 143193 = 456 × 314 + 9

⇒   143193 = 143184 + 9

⇒   143193 = 143193   which is true.

Thus, the division rule is verified.

Example - 8

Find the largest four digit number which is exactly divisible by 98.

Solution

∴ The largest 4-digit number = 9999

Let us divide 9999 by 98 and find the remainder (if any):

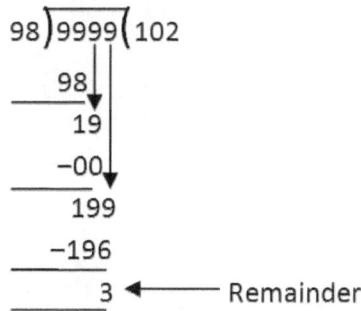

For exact divisibility, the remainder must be '0.' Thus, the 4-digit number nearest to 9999, which is exactly divisible by 98

= (9999 – Remainder)

= 9999 – 3 = 9996

Thus, the required number = 9996

Exercise - 1

1. Find the sum of:

    463 + 354 + 237 + 112 + 3046

2. Using properties of addition, fill in the blanks:

    (i)   442 + 1049 = _____ + 442

    (ii) 145 + (_____ + 213) = (145 + 112) + 213

    (iii) _____ + 1890 = 2121 + 1890

3. Add the following in two ways to verify communicative property of addition:
   (a) 4516 + 1267
   (b) 2156 + 2139
   (c) 1790 + 4035
   (d) 4460 + 119

4. By suitable arrangements, find the sum of:
   (i) 389, 10528 and 11311
   (ii) 1723, 1834 and 1277
   (iii) 1078, 8197, 1422 and 2003
   (iv) 75108, 191, 1192 and 209

5. What should be subtracted from 345678 to get the largest 5-digit number?

6. Subtract the sum of 3492 and 1008 from the smallest 6-digit number.

7. Multiply the smallest number formed by the digits 1, 0, 7 and 9 and the greatest number formed by these digits.

8. Fill in the blanks using the properties of multiplication:
   (i) $318 \times 91 = 91 \times$ _____
   (ii) $(42 \times 16) \times 48 =$ _____ $\times (16 \times 48)$
   (iii) $(413 + 25) \times 15 = 15 \times 413 + ($_____$) \times 15$
   (iv) $(213 - 78) \times 18 = 213 \times ($_____$) - 78 \times 18$

9. Determine the product of the largest 3-digit number and the greatest number formed by the digit 7, 0 and 2.

10. By using suitable arrangements find the product of:
    (i) $125 \times 18 \times 40$
    (ii) $16 \times 12 \times 125$
    (iii) $42 \times 25 \times 10 \times 4$
    (iv) $41 \times 40 \times 2 \times 25$

11. Divide the following and check the result:
    (i) $31531 \div 101$
    (ii) $25371 \div 79$

12. Which least number should be subtracted from 234567 such that the remainder is exactly divisible by 234?

13. What should be added to the smallest 4-digit number such that it is exactly divisible by 82?

14. What should be subtracted from the least 5 digit numbers such that it is exactly divisible by the smallest number formed by the digits 2, 0 and 1?

15. Which number should be divided by 312 to obtain a quotient 213 and remainder 18?

ANSWERS

1. 4212   2. (i) 1049   (ii) 112   (iii) 2121

4. (i) 22228   (ii) 4834   (iii) 12700   (iv) 76700

5. 245679   6. 95500   7. 10477090

8. (i) 318   (ii) 42   (iii) 25   (iv) 18

9. 719280

10. (i) 90,000   (ii) 24000   (iii) 42000   (iv) 82,000

11. (i) Q = 132; R = 19   (ii) Q = 321; R = 12

12. 99   13. 66   14. 4   15. 66474

Miscellaneous Exercise

1. Find the difference of the place value of two 9's in 2836978951.
2. Express 46578093 in expanded form.
3. Find the difference between the place value and the face value of 3 in 938782469
4. Write all even numbers between 10 and 30.
5. Write all odd numbers from 5 to 15.
6. Find the smallest 5-digit and greatest 5-digit numbers formed from the digits 9, 1 and 0.
7. Find the sum of 98765 and the smallest 4 digit number formed by the digits 8, 0 and 7.
8. By suitable arrangements, find the sum of: 3161, 103, 839 and 997
9. Find the value of (1234 × 12 − 1234 × 10) + (156 × 18 + 156 × 12) using the property of multiplication over whole numbers.

10. What least number be added to 10000 such that 75 divides the sum exactly.
11. Find the largest 5-digit number which is exactly divisible by 1218?

Answers

1. 8,99,100

2. 4,00,00, 000 + 60,00,000 + 5,00,000 + 70,000 + 8000 + 0 + 90 + 3

3. 2,99,99,9997   4. 12, 14, 16, 18, 20, 22, 24, 26, 28

5. 5, 7, 9, 11, 13, 15   6. 10009, 99910   7. 1,05,773

8. 5100   9. 7148   10. 50   11. 99876

HOTS

The greatest and the smallest numbers of 7-digits each were formed by using four different digit such that the digit 4 always occupied the thousands place. Find the sum of these numbers.

Answer

1,09,98,989

Mental Maths

1. What is the place value of 9 in 900021?
2. What is the difference between the place value and face value of 0 in 12304567?
3. What is the short form of:
   8,00,000 + 4,000 + 20 + 9?
4. What is the product of the smallest prime number and the smallest composite number? What is the relation between dividend, divisor, quotient and remainder.
5. What is identity element of addition?
6. Which number is called the identity element with respect to multiplication?
7. What is the product of 25 × 13 × 40 × 1
8. Using the distributive property of multiplication, find the value of: 125 (10 − 2) + 25 (14 − 10)

9. Using suitable arrangements, find the value of:

   18 × 25 – 14 × 25 + 4

ANSWERS

1. 9,00000
2. 0
3. 8,04,029
4. 8
5. Dividend = Divisor × Quotient + Remainder
6. 0
7. 1
8. 13,000
9. 1100
10. 104

MULTI CHOICE QUESTION

1. Which of the following is the face value of 0 in 3 2 0 1 6 4?
   (i) 1000   (ii) 0   (iii) 999   (iv) 1

2. Which of the following is the largest 5-digit numbers formed two different digits?
   (i) 80009   (ii) 90008   (iii) 99998   (iv) 10,000

3. Which of the following is the smallest number formed by two different digits?
   (i) 10000   (ii) 10001   (iii) 98889   (iv) 99998

4. Which of the following is the difference between the two place values of 1 in 7910761?
   (i) 10001   (ii) 99998   (iii) 9999   (iv) 10009

5. Which of the following is number of even-numbers between 6 and 16?
   (i) 10   (ii) 9   (iii) 4   (iv) 6

6. Which of the following is the number of odd numbers from 5 to 15?
   (i) 10   (ii) 9   (iii) 4   (iv) 6

7. Which of the following is the additive-identity?
   (i) 0   (ii) 1   (iii) –1   (iv) None of these

8. Which of the following can be placed in the blank of 125 × (104 – 100) × (_____) = 1000
   (i) 0   (ii) 4   (iii) 8   (iv) 2

9. Which of the following multiplicative identity?
   (i) 0   (ii) 1   (iii) –1   (iv) None of these

10. Which of the following is the value of: 100 × (26 − 25) + (26 − 25)?
   (i) 100   (ii) 101   (iii) 2500   (iv) 2600

ANSWERS

1. (ii)   2. (iii)   3. (i)   4. (iii)   5. (iii)
6. (iv)   7. (i)   8. (iv)   9. (ii)   10. (ii)

WORKSHEET

1. Match the *column A* and *column B*

	COLUMN A	COLUMN B
(i)	8 × (10 − 9) − 6 × (100 − 99)	(a) 0
(ii)	Difference between the place-value of 1 and face value of 7 in 97815	(b) 1
(iii)	Identity element corresponding to multiplication	(c) 2
(iv)	Identity element corresponding to addition	(d) 3

2. Write *'True'* or *'False'* for each of the following:

   (i) The division algorithm is

   Dividend = divisor × quotient − remainder

   (ii) The value of 16 (126 − 1) × 25 × 4 is 20100

   (iii) The value of 8 × (130 − 5) + 8 × 5 is 1040

   (iv) The smallest odd composite number is 9.

   (v) If a, b and c are any whole numbers then, a × c = b × c ⇒ a = b
   is the cancellation law of multiplication.

3. Fill in the blanks:

   (i) If 'a' and 'b' are any whole numbers then a + b = _____ + a

   (ii) If 'a' and 'b' are any whole numbers then _____ × b = a × b

   (iii) If a, b and c are any whole numbers then according to the _____ law

   (a + b) + c = a + (b + c)

(iv) If a, b and c are any whole-numbers then according to the _____ law of multiplication

a × (b + c) = a × b + a × c

(v) _____ is the identity element with respect to multiplication.

4. By suitable arrangements, find the sum of:

1109 + 1247 + 1001 + 853

5. Find the value of:

80 × (131 − 6) + 40 × (126 − 1) × 2

6. On dividing a certain whole number by 115, the quotient and remainder are 104 and 13 respectively. Find the number.

Answers

1. (i) ⟶ (c)   (ii) ⟶ (d)   (iii) ⟶ (b)   (iv) ⟶ (a)

2. (i) False   (ii) False   (iii) True   (iv) True

(v) True

3. (i) b   (ii) a   (iii) Associative   (iv) Distributive

(v) 1

4. 4210   5. 20,000   6. 11973

CHAPTER 3
NEGATIVE NUMBERS AND INTEGERS

INTEGERS

NEED FOR NEGATIVE NUMBERS

Whenever we subtract a smaller number from a greater number, we get a whole number, e.g. 9 – 5 = 4; 16 – 5 = 11; 28 – 7 = 21, etc. On the other hand, if we subtract a greater number from a smaller number, then we do not get a whole number, e.g.

$\qquad$ 5 – 9 = ? $\qquad$ 5 – 16 = ?, $\qquad$ 7 – 28 = ? etc.

Therefore, in order to find answer of such situations, we need to extend our number system. We introduce another set of numbers called *negative numbers*. Such as – 1, – 2, – 3, – 4, ... etc.

NEGATIVE NUMBERS IN DAILY LIFE

In our daily life, we come across many situations, which are of the same class but of opposite character. We consider one of them as positive and prefix the sign + before its numerical measure. We consider the other as negative and prefix the sign – before its numerical value.

EXAMPLES:

i. If a gain of Rs. 500 be represented by +500 then a loss of Rs. 500 will be represented by – 500.

ii. If an income of Rs. 900 be represented by +900 then an expense of Rs. 900 will be represented by – 900.

iii. If a height of 50 m above sea level be represented by +50 then depth of 50 m below sea level will be represented by – 50.

iv. If the temperature 10°C above freezing point be represented by +10 the temperature below freezing point will be represented by –10.

v. If moving 20 m towards east be represented by +20, then moving 20 m towards west will be represented by – 20.

REPRESENTATION NEGATIVE NUMBERS ON NUMBER-LINE

Recall that we used number line to represent whole numbers on it. We had marked a point 0 on the number line and considered it to represent 0 (zero). Starting from 0, we marked various points to the *right* of it at equal distance.

To represent negative numbers, we mark various points to the *left* of 0 at equal distance, as shown below:

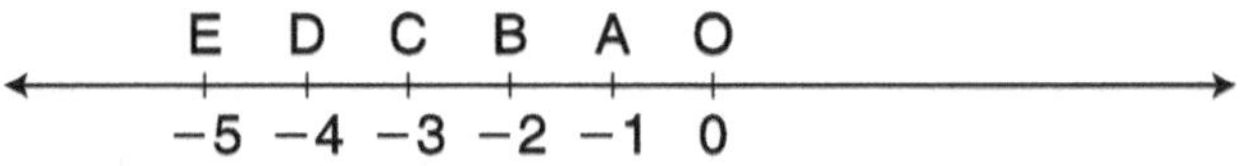

```
         E   D   C   B   A   O
    ◄────┼───┼───┼───┼───┼───┼────►
        -5  -4  -3  -2  -1   0
```

Mark these points A, B, C, D, E ......

Label each of the points starting from

O	as	0
A	as	− 1
B	as	− 2
C	as	− 3
D	as	− 4
E	as	− 5, and so on.

ORDERING NEGATIVE NUMBERS

While comparing whole numbers we have learnt that out of two given whole numbers, one on the right side of the number line is always greater. The same property applies to the negative numbers also.

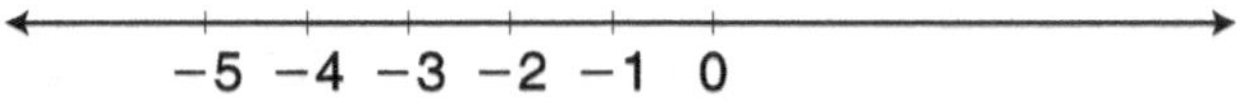

```
    ◄────┼───┼───┼───┼───┼───┼────►
        -5  -4  -3  -2  -1   0
```

Thus, − 3 > − 4, because − 3 is to the right side of − 4,

-2 > -3, because − 2 is to the right side of − 3

Similarly , − 2 < − 1, because − 2 is to the left side of − 2, and so on.

All negative numbers are less than zero.

INTEGERS

The negative numbers...... − 5, − 4, − 3, − 2, − 1 along with whole numbers 0, 1, 2, 3, 4, 5, ... are called *integers*. Thus the numbers: .... − 5, − 4, − 3, − 2, − 1, 0, 1, 2, 3, 4, 5, ... are integers.

REPRESENTATION OF INTEGERS ON THE NUMBER LINE

We know that the set of integers is:

{...... – 5, – 4, – 3, – 2, –1, 0, 1, 2, 3, 4, 5, ...}

To represent the integers on a number line we mark 0 (zero) on it.

Mark positive numbers 1, 2, 3, 4, 5, ... on the right side of 0 at equal distances.

Mark negative numbers –1, – 2, – 3, – 4, – 5...... on the left side of 0 at the same equal distances.

This representation on the number line is said to be the *representation* of integers on a number line.

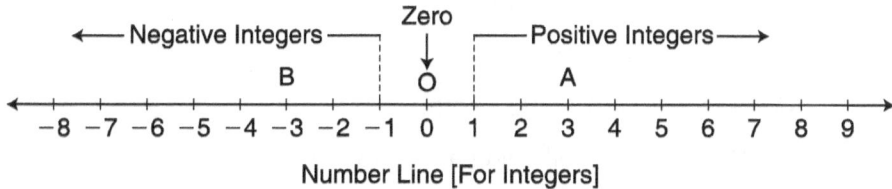

Number Line [For Integers]

NOTE

The arrow-head on both sides means that the number line extends up to infinity on both sides, so the integers continue on the positive side as well as on the negative side. That is why neither the (positive) highest integer nor the smallest (negative) integer exists.

We observe that the point A is 3 steps away from O on its right and the point B is 3 steps away from O on its left. They are at equal distance from O, but on opposite directions.

It is important to say that A is at + 3 and B is at – 3 where '+' and '–' indicate the directions. That is why the numbers – 3 and +3 etc., are called directed numbers.

DID YOU KNOW?

'0' is neither positive nor negative. It is because that '0' is the reference point from which we determine the positive and negative directions. Therefore, it is meaningless to say that 0 is positive or negative.

Exercise – 1

1. Write the following statement by appropriate integers:
   (i) Withdrawal of Rs. 5,000 from a bank.
   (ii) 150 new admissions in a school.
   (iii) Walking 10 km down the hill.
   (iv) Decrease in production by 15%.
   (v) Making a saving of Rs. 900.

2. Write the opposite of the following statements
   (i) Gain of weight by 5 kg.
   (ii) 20 meters above sea level.
   (iii) Withdrawal of Rs. 15000.
   (iv) Walking 25 km to the left.
   (v) Walking 10 km to the North.

3. (a) Which of the following are negative integers:
   $$63, \quad -72, \quad 0, \quad 72$$
   (b) Which of the following are integers:
   $$\frac{1}{2}, \quad -2, \quad -\frac{1}{2}, \quad 2$$
   (c) Write all integers between – 9 and 1

4. Represent – 7 on a number line. Is it greater than – 9?

5. Represent 12 on a number line. Is it smaller than –15?

6. Mark the successors of – 3 and 0 on a number – line.

7. Mark the predecessors of – 9 and –1 on a number line.

8. Write true or false for each of the following statements.
   i. The positive integers and non-negative integers express the same set of integers.
   ii. The greatest negative integer does not exist.
   iii. Every integer has a successor as well as a predecessor.
   iv. '0' is greater than every negative integer.
   v. The absolute value of '0' is '0' itself.

Answers

1. (i) − Rs. 5000 (ii) + 150 student
   (iii) − 10 km (iv) − 15%
   (v) + Rs. 900
2. (i) Loss of weight by 5 kg (ii) 20 m below the sea level
   (iii) Deposit of Rs. 15000 (iv) Walking 25 km to the Right
   (v) Walking 10 km to the South.
3. (a) − 72 (b) − 2, 2
8. (i) False (ii) False
   (iii) True (iv) True
   (v) True

Fundamental Operations on Integers

In previous class we studied the four fundamental operations (namely **addition, subtraction, multiplication** and **division**) on integers using their absolute values. Here, we shall discuss *Addition* and *Subtraction* of integers using the number line.

Addition

Let us add −3 to 5. We start from zero (0) and move 5 units to the right (∵ **5 > 0**) and reach at P. Then we move 3 units to the left to add −3 and reach the final position Q, denoted by 2.

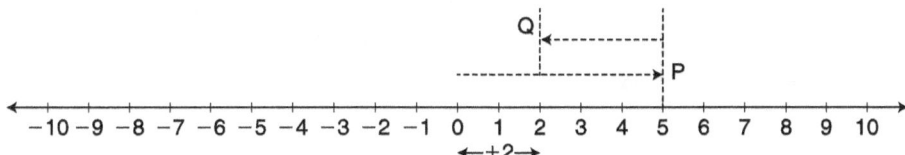

Thus $5 + (-3) = 2$

Let us take another example; to add −3 to −5. We start from zero (0) and move 5 units to the left to get −5 [∵ **−5 < 0**] 5 and reach at P.

Next, we move 3 units further to the left to add ⁻3 to ⁻5 and reach to the position Q, which is denoted by the number ⁻8.

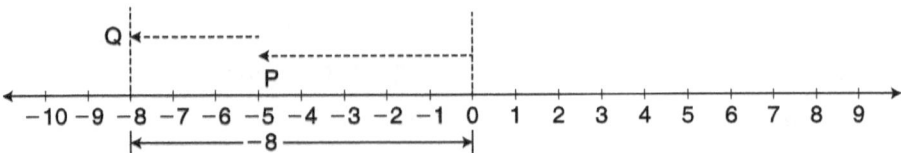

Thus, −5 + (−3) = −8

From the above discussions, we conclude that

i. To add an integers of positive sign, move to the right on the number line.

ii. To add an integer of negative sign, move to the left on the number line.

EXAMPLE – 4

Add: (i) −6 to 4   (ii) 3 to −4

SOLUTION

I.

We have to find 4 + (−6).

We start (on the number line) at 4 and move 6 units to the left. We reach to −2.

∴ 4 + (−6) = −2

ii.

We have to add (+3) to (−4), i.e. to find −4 + 3.

We start at −4 on the number line and move 3 units to the right. We reach at −1

∴ −4 + 3 = −1

EXAMPLE – 5

Find:

(i) 2 + 4   (ii) −3 + (−2)   (iii) −3 + 6

Solution

i. 2 + 4

Start from 2 on the number line and move 4 units to the right to reach at 6.

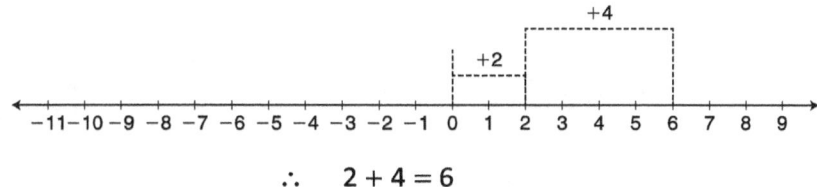

$$\therefore \quad 2 + 4 = 6$$

ii. −3 + (−2)

Start from −3 on the number line and move 2 steps to the left to reach at −5

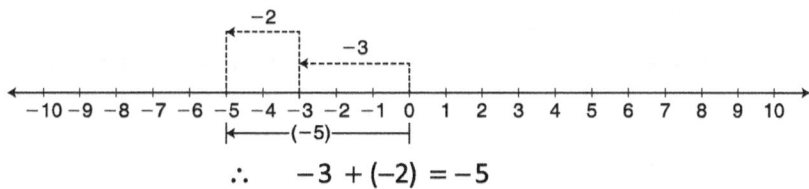

$$\therefore \quad -3 + (-2) = -5$$

iii. −3 + 6

To start from −3 on the number line and move 6 units to eight to reach at 3.

$$\therefore \quad -3 + 6 = 3$$

Addition of Zero (0)

If 0 is added to any integer then the sum equals the integer itself.

For example:

$$-5 + 0 = -5$$
$$8 + 0 = 8$$

$$0 + (-6) = -6$$
$$0 + 0 = 0$$

Since, the value of integer does not change, if 0 is added to it, therefore 0 (zero) is called the additive – identity for integers.

ADDITIVE INVERSE

Corresponding to every integer, there exists an integer such that their sum is 0. Each of such integer is called the *additive inverse* of the other.

For example:
$$2 + (-2) = 0$$

Here, -2 is the additive inverse of 2 and 2 is the additive inverse of -2.

Similarly, $-9 + 9 = 0$

So, 9 is the additive inverse of -9, and -9 is the additive inverse of 9.

NOTE

The 'additive inverse' of an integer is also called its negative.

SUBTRACTION

Subtraction is the reverse process of addition. That is, in addition we combine the integers together where as in subtraction we take away an integer from the given integer.

Therefore, we have the following rules:

   i. *To subtract a positive integer, we move to the left on the number line*

      [∵ **We had moved to the right to add positive integer.**]

      For example let us subtract (+6) from (–2).

      We start at -2 and move 6 units on the number line to the left, we reach at -8.

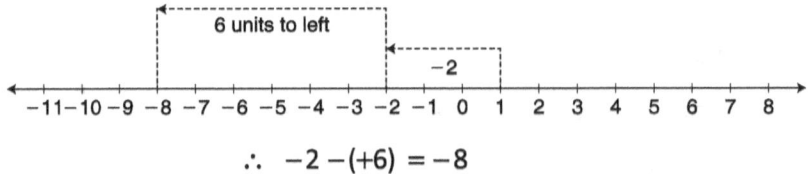

$$\therefore \quad -2 - (+6) = -8$$

   ii. *To subtract a negative integer, we move to the right on the number line*

      [∵ **We had moved to the left to add negative integer.**]

For example, let us subtract (−6) from (−2).

We start at −2 and move 6 units on the number line to the right, we reach at (+4).

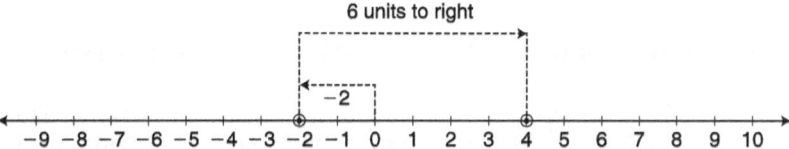

EXAMPLE − 6

Subtract

    (i) 4 from −3   (ii) −6 from 2   (iii) −5 from −4

SOLUTION

i. To subtract 4 from −3 i.e. to find −3 − (+4), we start from −3 and move 4 units to left to reach at −7.

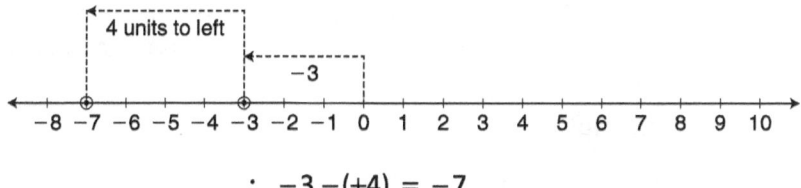

$$\therefore -3 - (+4) = -7$$

ii. To subtract (−6) from 2 i.e. to find 2 − (−6) we move 6 units to right from 2 and reach at 8.

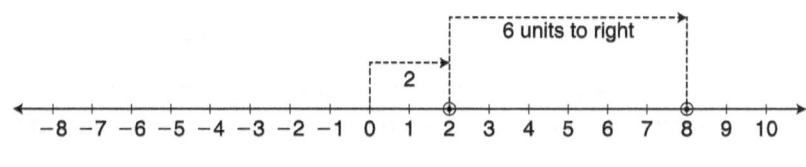

$$\therefore 2 - (-6) = 8$$

iii. To subtract −5 from −4 i.e. to find −4 − (−5)

On the number line, we start from −4 and move 5 units to right to reach at (+1).

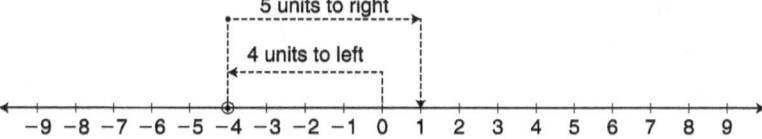

$$\therefore\ -4-(-5)=+1$$

or

$$-1-(-5)=1$$

COMPARING (ORDERING) INTEGERS ON A NUMBER LINE

Integers are compared by the same rule as we have used in case of whole numbers. Of the two integers that are marked on the same number-line, an integer indicated on the right is greater than the integer to the left.

For example,

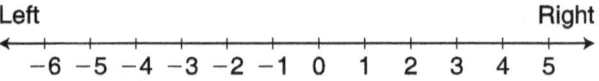

$2 > 1$; $1 > 0$; $0 > -1$; $-1 > -2$; $-2 > -3$ and so on.

From the above number line, we conclude that:

i. Zero and all negative numbers lie to the left of positive numbers. Therefore, all positive integers are greater than negative integers and zero.

   For example:

   $-3 < 3$; $-4 < 1$, $0 < 5, -5 < 4$

ii. 0 (zero) lies to the right of all negative integers, so 0 is always greater than every negative number.

   For example:

   $-2 < 0$, $-7 < 0, -1 < 0$, $-15 < 0$

iii. Among positive integers, a number with greater numerical value is greater than all those integers which are on the left of it on the number line.

   For example:

   $35 > 20$, $115 > 112$, $338 > 308$

iv. Among negative integers, a number with greater numerical value is smaller than all those integers which are on the right of it on the number line.

For example:

$-35 < -20$, $-115 < -112$, $-121 \le -99$

EXAMPLE

Arrange the following integers in ascending order:

$$-19, \quad -66, \quad 27, \quad 3, \quad -10$$

SOLUTION

Among all positive numbers,

Greatest = 27

Smallest = 3

$$3 < 27$$

Among all negative integers,

Greatest = -10

Smallest = -66

$$-66 < -19 < -10$$

Hence, the given integers in ascending order is:

$$-66 < -19 < -10 < 3 < 27$$

Write the descending order of the following integers

$$2, \quad -1, \quad 4, \quad -2$$

SOLUTION

Comparing 2 and 4, we have $4 > 2$

Comparing -1 and -2, we have $-1 > -2$

∴ The required descending order of the give integer

$$4, \quad 2, \quad -1, \quad -2$$

EXERCISE – 2

1. **Add the following integers:**
   - i. -5 and 15
   - ii. 0 and -13
   - iii. 18 and -48
   - iv. 81 and -31
   - v. -32 and 100
   - vi. -23 and -45
   - vii. -9 and 92
   - viii. 9 and -21

2. **Subtract:**
   - i. 13 from 41
   - ii. -6 from 58

iii. 19 from −26 iv. −18 from −42
v. −11 from −100 vi. 20 from −36
vii. −23 from 100 viii. 0 from −19

3. **Verify the following:**
   i. (−36) + 40 = 40 + (−36)
   ii. 2 + [(−5) + (−12)] = (−5) + [(−12) + 2]
   iii. (−15) − (12) ≠ (12) − (−15)
   iv. 5 − [(−3) − (12)] ≠ [(5) − (12)] − (−3)

4. **Fill in the blanks:**
   i. (−9) + _____ = 0   ii. 13 + (−13) = _____
   iii. (−5) + _____ = −15   iv. (−18) + (−18) = _____
   v. 12 − (−11) = _____   vi. −11 − (____) = 0

5. If the sum of two integers is −42 and one of the integers be −16 then find the other.

6. An integer is subtracted from 21. If the difference is −5, find the integer.

7. From which integer −18 be subtracted to get 10?

8. The temperature of a freezer is −10°. If the outside temperature is 10° then find the temperature difference between the two temperatures.

9. **Fill in the following blanks using the following symbols =, < or >:**
   i. 6 ___ 0   ii. 0 ___ −6   iii. −9 ___ |9 + 2|
   iv. |13| − |2| ___ |−13| − |−2| v. 8 − |−3| ___ |8 + 3|

10. **Using a number line, arrange the following integers in ascending order:**
    i. −1, −3, 0, −5, 2, −8
    ii. 0, −2, −6, −4, 5, −7

11. **Using a number line, arrange the following integers in descending**
    i. −1, 0, −3, 5

12. Compare the following pairs of integers and state which integer is greater?
    i. −10 and 0
    ii. 0 and −1
    iii. −4 and −10
    iv. −15 and 15
    v. −100 and 1

13. Write (arrange) the following integers in ascending order:
    i. −9, 0, 7, −7, 8
    ii. 5, 0, −20, 18, −19
    iii. 0, −8, −5, −12, −25
    iv. −117, −171, −711, −1
    v. 4, −44, −14, −21

14. Write (arrange) the following integers in descending order:
    i. 18, −19, 20, −21, −22
    ii. −1, 0, −38, 15, −13
    iii. 14, −41, −14, −71, −17
    iv. −100, 101, −101, −103, 102
    v. 135, −351, −531, −135

ANSWERS

1. (i) 10 (ii) −13 (iii) −30 (iv) 50
   (v) 68 (vi) −68 (vii) 83 (viii) −12

2. (i) 54 (ii) 64 (iii) −45 (iv) −24
   (v) −89 (vi) −56 (vii) 123 (viii) −19

4. (i) 9 (ii) 0 (iii) −10 (iv) −36
   (v) −1 (vi) −11

5. −26   6. 26   7. −8   8. 20°

9. (i) > (ii) > (iii) < (iv) =
   (v) <

10. (i) −8, −5, −3, −1, 0, 2   (ii) −7, −6, −4, −2, 0, 5

12. (i) 0 (ii) 0 (iii) −4 (iv) 15

(v) 1

13. (i) −9, −7, 0, 7, 8   (ii) −19, −20, 0, 5, 18

(iii) −25, −12, −8, −5, 0   (iv) −711, −171, 117, −1

(v) −44, −21, −14, 4

14. (i) 20, 18, −19, −21, −22   (ii) 15, 0, −1, −13, −38

(iii) 14, −14, −17, −41, −71   (iv) 102, 101, -100, −101, −103

(v). 135, −135, −351, −531

ABSOLUTE VALUE OF AN INTEGER

The numerical value of an integer regardless of its sign (+ or −) is called its *absolute value*.

The symbol '| |' is used to denote the absolute value of an integer.

For example, the absolute value of 9 is |9| = 9 and the absolute value of −9 is |−9| = 9

NOTE

    i. The absolute value of an integer is always non-negative.
    ii. The absolute value of 0 is 0 itself i.e. |0| = 0
    iii. The absolute value of an integer is also called its 'modulus'.

EXAMPLE

Evaluate

i. |−19| ii. |26| iii. −|−6|
iv. |−5| + |0| v. |−7| − |0| vi. |18| − |−4|
vi. |−3| + |7| viii. −|−3| − |6| ix. |−3| − |6|
x. |−3| − |−6|

REMEMBER

    i. For an integer greater than or equal to 0 (zero),
        | an integer | = the integer itself

ii. For an integer less than 0 (zero),

| an integer | = – (the integer)

SOLUTION

i. $|-19| = 19$
ii. $|26| = 26$
iii. $-|6| = -(6) = -6$ [∴ $|-6| = 6$, so $-|-6| = -6$]
iv. $|-5| + |0| = 5 + 0 = 5$
v. $|-7| - |0| = 7 - 0 = 7$
vi. $|18| - |-4| = 18 - 4 = 14$ [∴ $|-4| = 4$, so $-|-4| = -4$]
vii. $|-3| + |7| = 3 + 7 = 10$
viii. $-|-3| - |6| = -3 - 6 = -9$ [∴ $|-3| = 3$, so $-|-3| = -3$]
ix. $|-3| - |6| = 3 - 6 = -3$
x. $|-3| - |-6| = 3 - 6 = -3$

MULTIPLICATION OF INTEGERS

(a) When integers have same sign

The product of two integers of the same sign gives a positive integer because the absolute

Value of the product is equal to the product of the absolute values of the integers.

For example: $(+4) \times (+8) = +32$ or 32

$(-4) \times (-7) = +28$ or 28

∴ $|+4| \times |+8| = 4 \times 8 = 32$ and

$|-4| \times |-7| = 4 \times 7 = 28$

EXAMPLE

Find the following products:

i. $(-8) \times (-10)$
ii. $(+15) \times (+12)$
iii. $(+2) \times (+38)$
iv. $(-6) \times (-12)$

SOLUTION

i. Since (–8) and (–10) are of the same sign ( )

∴ Their product is to be positive

Since $8 \times 10 = 80$

∴ $(-8) \times (-10) = +80$ or 80

ii. (+15) × (+12) = + (15 × 12)
= + 180 or 180

Shortcut Method

To multiple two positive integers, multiply them as natural numbers.

iii. (+2) and (+38) are of positive sign

∴ Their product is also has positive sign

Now, (+2) × (+38) = +76 or 76

iv. (–6) and (–12) both have negative sign

∴ Their product is of positive sign

i.e. (–6) × (–12) = + (6 × 12)
= + (72) or (72)

(b) When Integers have unlike signs

The multiplication of two integers of opposite signs always gives a negative integer whose absolute value is equal to the product of their absolute values.

For example, the product of (–5) and (+4) is given by – {| –5| × |+4|} = – (5 × 4) = – 20

Example

Find the following products:

i. 15 × (–4)   ii. (–12) × 10   iii. (–18) × 25   iv. (–3) x 125

Solution

i. ∴ |+15| = 15 and | – 4| = 4

∴ 15 × (–4)= [15 × 4]

= –60

ii. ∴ | –12| = 12 and |10| = 10

[(–12) × 10] = – (12 × 10)

= – 120

iii. ∴ |– 18| and |25|

∴ (– 18) × 25 = – [18*25]

= – 450

∴ |−3| = 3 and |125| = 125
∴ (−3) × 125 = −(3 × 125) = −375

EXAMPLE

Fill in the blanks:

i. 3 × (−6) = (−6) × _____
ii. 10 × 7 × (−2) = 7 × (−2) × _____
iii. 3 × (7 + 9) = (_____) + (3 × 9)
iv. (−7) × _____ = 0   v. (−18) × 1 = _____

SOLUTION

i. 3 × (−6) = (−6) × **3**   ii. 10 × 7 × (−2) = 7 × (−2) × **10**
iii. 3 × (7 + 9) = (**3 × 7**) + (3 × 9)   iv. (−7) × **0** = 0
v. (−18) × 1 = **−18**

DIVISION OF INTEGERS

Division is the reverse process of multiplication. Following are the rules of division of integers:

(i) For integers of like signs, the quotient is positive.
    For example: (−10) ÷ (−2) = (−5)

(i) For integers of unlike signs, the quotient is negative.
    For example, (−10) ÷ 2 = (−5)

(i) The absolute value of the quotient is found by dividing the absolute value of the dividend by the absolute value of the divisor.

For example:

i. (−42) ÷ (−6) = (+7) | ∴ |−42| = 42, |−6| = 6 and |+7| = 7
                        | ∴ 42 ÷ 6 = 7

ii. (−48) ÷ 8 = (−6) | ∴ |−48| = 48, |8| = 8 and |−6| = 6
                        | ∴ 48 ÷ 8 = 6

NOTE

i. Above rules of division of integers apply only for division without remainder.
ii. Division of 0 by any non-zero integer gives 0.

iii. Division of an integer by 0 is not possible.

iv. Division of an integer by (+1) or 1 gives the integer itself.

v. Division of an integer by (−1) gives the integer with the opposite sign.

EXERCISE – 3

i. Find the following products:

(i) 25 × (−11)   (ii) (−10) × 7   (iii) 21 × 15   (iv) 27 × (−10)

(v) (−33) × (−5)   (vi) (−18) × 8   (vii) (315) × (−4)   (viii) (−6) × (−15)

DID YOU KNOW?

'Zero' is neither positive (+) nor negative (−), it is simply an integer. Therefore, when 'zero' is divided by any integer (positive or negative), the quotient is zero.

2. Show that:

   (i) The product of 15 and (−6) is an integer.

   (ii) (−25) × 10 = 10 × (−25)

   (iii) [5 × (−6)] × (−9) = 5 × [(−6) × (−9)]

   (iv) 28 × [1 × (−8)] = [28 × (−8)] × 1

   (v) −17 × [(−25) × 4] = [4 × (−17)] × (−25)

3. Using the Distributive property, simplify the following:

   (i) −25 × [(−10) + 18]   (ii) 18 × [(−25) + (−10)]

4. Find the quotient in each of the following:

   (i) (−10) ÷ 5   (ii) 48 ÷ (−4)

   (iii) (−75) ÷ (−5)   (iv) (−128) ÷ 16

   (v) (112) ÷ (−7)   (vi) 480 ÷ (−32)

   (vii) (−2442) ÷ (−11)   (viii) (−1260) ÷ (−6)

5. Fill in the blanks:

   i.   (−3) × (−8)   = _____

   ii. 24 × (−5)   = (−5) × _____

   iii. 6 × [(−6) × 12]   = [6 × (−6)] × _____

   iv. 315 × (−1)   = _____

   v.   1 × _____   = (−312)

   vi. _____ × 387   = 0

vii. 42 ÷ _____ = (–7)
viii. _____ ÷ 5 = 210
ix. (–311) ÷ _____ = 311
x. _____ ÷ 1 = 712

6. **Write 'True' or 'False' for each of the following:**
   i. The product two integers of same sign is a positive integer.
   ii. The product of two integers having negative signs is a negative integer.
   iii. The product of three integers _____ = (Product of any two integers) × (product of first and last integers)
   iv. The product of (–1) an integer having negative sign is the number itself with positive sign.
   v. Division of 0 by a non-zero integer is 1.
   vi. Product of 1 and an integer is 1.
   vii. [A positive integer] ÷ [A negative integer] = [An integer with – ve sign]
   viii. (3 + 5) × (–1) = (–3) + (–5)

7. If product of (–32) and an integer is 320 then find the integer.
8. When we divide (–160) by 16, then what is the quotient?
9. Which integer be divided by (–12) to get 12.
10. By which integer (–80) be multiplied to get 320?

ANSWERS

1. (i) –275   (ii) –70   (iii) 315   (iv) –270
   (v) 165   (vi) –144   (vii) –1260   (viii) 90

3. (i) –200   (ii) –630

4. (i) –2   (ii) –12   (iii) 15   (iv) –8
   (v) –16   (vi) –15   (vii) 222   (viii) 210

5. (i) 24   (ii) 24   (iii) 12   (iv) –315

(v) – 312   (vi) 0   (vii) – 6   (viii) 1050

(ix) – 1   (x) 712

6. (i) True   (ii) False   (iii) False   (iv) True

(v) False   (vi) False   (vii) True   (viii) True

7. – 10   8. – 10   9. – 144   10. – 4

HOTS

1. Complete the following square such that the numbers (integers) add up to the same sum vertically, horizontally or diagonally.

–39		–33
	–30	
		–21

2. A diver dives from the surface of seas to 249 below the surface. Then, he swims up 7 metres, down 18 metres, down another 25 metres, and then up 24 meters. Use negative and positive numbers to represent this situation. Also find the diver's depth after these swimmings and divings.

ANSWERS

1.

–39	–18	–33
–24	–30	–36
–27	–42	–21

2. 261 m

MISCELLANEOUS EXERCISE

1. Write the absolute values of:

   (i) – 15   (ii) 29   (iii) 0   (iv) – 1

2. Find the value of:

   (i) $|+9|+|-9|$ (ii) $|-18| - |9|$

3. Write the integers: 10, −5, 20, 0, −2 in ascending order
4. Write the integers: −11, 6, −5, 8, −4 in descending order.
5. If +15 stands for 15 steps forward then what is the integer which will represent 10 steps backward?
6. Add −6, 5 and −8
7. Subtract −15 from 405
8. Find the product:

    (−5) × (−10) × (−36)
9. The product of two integers is −108. If one of the numbers be 12 then find the other number.
10. Simplify:

    20 − {42 ÷ (−6) × (−3) + 7} + 8

ANSWERS

1. (i) 15 (ii) 29 (iii) 0 (iv) 1   **2.** (i) 18 (ii) 9

3. −5, −2, 0, 10, 20   **4.** 8, 6, −4, −5, −11

5. −10   **6.** −9   **7.** 420   **8.** −1800

9. −9   **10.** 0

MENTAL MATHS

1. If 36 ÷ (−9) then what is the quotient?
2. What is the value of −3 − (−8)?
3. If (−3) × 8 = −24, then what the value of −24 ÷ 8?
4. What is the successor of 0?
5. Which integer is the predecessor of -1?
6. Nine less than −1 is _____
7. What is the sum of −1 and 0?
8. Fifteen more than −2 is _____
9. Which integer is greater −1 or −5?
10. If the product of 0 and (−5) is 0 then what is the product of 0 and 5?

11. Which is greater:

    −1 − {5 ÷ 5}   or   1 − {5 ÷ 5}

12. If 0 + 0 = 0 then what is 0 − 0?

Answers

1. − 4	2. 5	3. − 3	4. 1
5. − 2	6. − 10	7. − 1	8. 13
9. − 1	10. 0	11. 1 − {5 ÷ 5}	12. 0

Multiple Choice Questions

1. If +10 means 10 units forward, then −10 means 10 units _____

    (i) Forward   (ii) Backward

    (iii) Upward   (iv) Downward

2. If a profit of Rs 1510 is written as + Rs 1510, then loss of Rs 1510 is written as _____

    (i) Rs − 1510   (ii) Rs 0

    (iii) Rs +1510   (v) Rs − 3020

3. Which of the following is the temperature difference between − 20°C and 10°C?

    (i) 30°C   (ii) 10°C   (iii) − 30°C   (iv) − 10°C

4. How many negative integers are there between − 6 and 6?

    (i) 12   (ii) 11   (iii) 6   (iv) 5

5. Which of the following is equal to −6 + (−52) − (−42)?

    (i) −58   (ii) − 16

    (iii) −100 (iv) − 6

6. Which of the following integers has the greatest value?

    (i) − 101   (ii) − 100

    (iii) − 109 (iv) − 108

7. Some salt is added to a glass of water which is at its freezing paint of 0°C such that its temperature drops to 4°C below zero. Which of the following is the present temperature of water?

    (i) 4°C   (ii) 4°C below zero

    (iii) +4°C   (iv) − 4°C

8. Which of the following is incorrect?
   (i) 18 < 19
   (ii) 18 > −19
   (iii) 5 < −4
   (iv) 1 > 0

9. Which of the following integers has the smallest value?
   (i) 0
   (ii) −1
   (iii) −2
   (iv) −3

10. Which of the following is not descending order?
    (i) −10, −9, −8, −7
    (ii) −7, −8, −9, −10
    (iii) 0, −1, −2, −3
    (iv) 1, 0, −1, −2

ANSWERS

1. (ii) 2. (i) 3. (i) 4. (iv) 5. (ii)
6. (ii) 7. (iv) 8. (iii) 9. (iv) 10. (i)

WORK SHEET

1. Match the statements of column −A with the statements of column −B:

COLUMN−A	COLUMN−B
(a) Negative of −1	(i) −1
(b) The integer which is neither negative nor positive	(ii) 2
(c) The greatest negative integer	(iii) 0
(d) 4 more than −2	(iv) 1
(e) The numerical difference between −2 and 2	(v) 4

2. **Fill in the blanks:**
   i. The smallest non-negative integer is _____
   ii. The smallest positive integer is _____
   iii. '0' plus any integer is equal to _____
   iv. _____ is the successor of −1.
   v. _____ is the predecessor of 0.
   vi. If two integers are of the same sign then the sign of their product is _____

vii. The integer _____ is greater than −5 by 5.
viii. The product of _____ and an integer is always − (the integer itself).
ix. If we divide an integer by _____ then the quotient is the integer itself.
x. To add two negative integers, we add their absolute values and give the _____ sign to the sum obtained.

3. Write 'true' or 'false' for each of the following:
   i. Zero ('0') is greater than every negative integer.
   ii. Every integer has a successor but 0 has no predecessor.
   iii. Zero is neither positive nor negative integer.
   iv. The set of non-negative integers is:
       {0, 1, 2, 3, 4, ...}
   v. The product of two negative integers is always negative.
   vi. The product of any integer and zero is always the integer itself.
   vii. The product of two integer having opposite signs is always negative.
   viii. In the process of division if the integers are of the same sign then the quotient is always positive.
   ix. When 0 is divided by any integer (negative or positive), the quotient is zero.
   x. Negative of −1 is −2.

4. The product of two integers is 1919. If one the integers is −19 then find the other integer.

5. Write the following as one directed number
   (−3) − (+4) − 5 (−5) − 6 (+6) + (−3) − (−4)

6. Fill in the blanks using the symbols; =, < or >
   (i) 3 − (−2) ☐ 5   (ii) 3 − (+4) ☐ 2   (iii) −2 − (+2) ☐ 0
   (iv) +2 − (+2) ☐ −4   (v) +1 − (+1) ☐ 0   (vi) (+2) − (−2) ☐ 3

7. Find the value of:
   (i) (−9) × 5 × (−3) × (−10)
   (ii) (−48) ÷ (−8) + (−100) ÷ (−10) − (−32) ÷ (−2)

8. Simplify:

   (−101) + [50 ÷ (−5) of 10 − 9 × 3 − (41 − 1) × 8]

9. When an integer is divided by (−9), the quotient is (−45). Find the integer.

10. Complete the following tables:

   (i)

Subtract (−)	−2	−1	0	1	2
−2					
−1					
0					
1					
2					

   1st Number (rows), 2nd Number (columns)

   (ii)

Multiply ×	−2	−1	0	1	2
−2					
−1					
0					
1					
2					

   1st Number (rows), 2nd Number (columns)

ANSWERS

1. (a) ⟶ (iv)   (b) ⟶ (iii)   (c) ⟶ (i)   (d) ⟶ (ii)

   (e) ⟶ (v)

2. (i) 0   (ii) 1   (iii) the integer itself   (iv) 0

   (v) −1   (vi) positive   (vii) 0   (viii) −1

   (ix) 1   (x) negative

4. −101   5. −17

6. (i) =   (ii) <   (iii) <   (iv) >

   (v) =   (vi) >

7. i) −1350   (ii) 0   8. −449   9. 405

CHAPTER 4

PLAYING WITH NUMBERS

SIMPLIFICATION OF BRACKETS

We know that the four fundamental operations are: *Addition, Subtraction, Multiplication* and *Division*. A group of integers connected by some or all the fundamental operations of addition (+) subtraction (−), multiplication (×) and division (÷) is called a *numerical expression*. For example:

$$5 + 12 - 7 \times (-8) \div 2 \text{ is a numerical expression.}$$

A numerical expression may also involve *brackets*.

For example:

$$(20 - 15) \times 2 + [18 - 7] \div 11$$

Simplification of a numerical expression means **"performing the operations involved and solving the brackets to get a single value of the given numerical expression, using the calculation rules."**

The *calculation-rules* tell us order in which the operations are to be done. The letters of the word '**BODMAS**' tell us the order in which the operations are to be performed. In the word BODMAS,

B stands for	**B**rackets
O stands for	**O**f
D stands for	**D**ivision
M stands for	**M**ultiplication
A stands for	**A**ddition
S stands for	**S**ubtraction

NOTE

(i) 'Of' means 'x,' but it is to be performed before ÷ and ×

(ii) 'Addition' and 'Subtraction' can be performed separately or together.

(iii) _____ is called *bar-bracket* or *Vinculum*.

(iv) ( ) are common brackets and called as *parenthesis*.

(v) { } are curly brackets and called as *braces*.

(vi) [ ] are big brackets and called as *rectangular brackets*.

We remove the brackets in the following order ____, ( ), { } and [ ]

NOTE

The vinculum '_____' is rarely used. Because it causes confusion with the subtraction symbol (–).

EXAMPLE

Simplify:

(i) –10 × 5 – 36 ÷ 9   (ii) 36 – 8 × 2 + 16 – (10 ÷ 5)

(iii) [18 ÷ 9 × 5 – (10 + 4 × 1 – 5)]   (iv) 27 – (18 + 3 ÷ 1) × 8

(v) 1 – [1 – {1 – 10 ÷ (6 + 2 × 2)} + 10]

SOLUTION

(i)   –10 × 5 – 36 ÷ 9

= –10 × 5 – 4   {36 ÷ 9 = 4

= –50 – 4   {–10 × 5 = –50

= – 54

(ii)   36 – 8 × 2 + 16 – (10 ÷ 5)

= 36 – 16 + 16 – (2)

= 36 – 16 + 16 – 2

= 52 – 18 = 34

(iii) [18 ÷ 9 × 5 – (10 + 4 × 1 – 5)]

= [18 ÷ 9 × 5 – (10 + 4 – 5)]

= [2 × 5 – (14 – 5)]

= [10 – 9]

= [1]

= 1

(iv)   27 – (18 + 3 ÷ 1) × 8

= 27 – (18 + 3) × 8

$= 27 - 21 \times 8$

$= 27 - 168 = -141$

(v) $1 - [1 - \{1 - 10 \div (6 + 2 \times 2)\} + 10]$

$= 1 - [1 - \{1 - 10 \div (6 + 4)\} + 10]$

$= 1 - [1 - \{1 - 10 \div 10\} + 10$

$= 1 - [1 - \{1 - 1\} + 10$ $\{\therefore 10 \div 10 = 1$

$= 1 - \{1 - \{0\} + 10$

$= 1 - [1 - 0 + 10]$

$= 1 - 11$

$= -10$

Exercise – 1

Simplify the following:

1. $24 \div 6 + [3 \times (-5) \times 6] - 16 \div 4$
2. $5 + \{15 - 6 \div 2 - 5 \times (-7)\} - 35 - 30$
3. $23 + \{18 \div (8 - 2) + (-3) \times 4\} - 4$
4. $26 - [17 - \{8 \div (3 \times 2 - 4)\}]$
5. $21 - \{5 \times 3 - (-6 \times 2) \times 16 \div (-8)\}$
6. $100 \div 10 - [-2 + \{-9 + (3 - 6 \text{ of } 2)\}] - 30$
7. $[12 - 16 \div 4 - \{16 \text{ of } (-2) + (5 \times 3 - 4)\}] - 28$
8. $[5 - \{7 \times 8 \div 4 + (8 - 3) \times (6 - 4)\} + 20]$
9. $[81 \div 3, \text{of } 9 \times (3 + 2) \text{ of } 6] - 10 \times 8$
10. $(16 \div 2) \times (5 \times 2) - [20 - \{(25 \times 7) \div 5 - (28 - 4) \div 6\}] - 1$

Answers

1. –90   2. –13   3. 10   4. 13   5. 30
6. 0   7. 1   8. 1   9. 10   10. 90

Divisibility Rules of 2, 3, 4, 5, 6, 8, 9 and 11

To check the divisibility of a number by another number, we have to perform actual division and see whether or not the remainder is zero. This is a time consuming process. There are certain divisibility tests to see (without actual division) whether or not a number is divisible by a certain given number.

Divisibility by 2

A given number is divisible without remainder by 2, if its last digit is either 0 or divisible by 2. For example, the numbers, 14, 18, 26, 38, 40, 152, 3405674 etc. are divisible by 2.

Divisibility by 3

If the sum of digits of a number is divisible by 3, then it is divisible by 3. For example: The number 65430 is divisible by 3, as 6 + 5 + 4 + 3 + 0 = 18, which is divisible by 3.

Divisibility by 4

A number is divisible by 4, if the number formed by its last two digits is divisible by 4 or if it ends in two zeros. For example 924, 8716, 700, 123456788 are all divisible by 4.

Divisibility by 5

A number is divisible by 5, if its last digit is either 5 or 0. For example, 925, 8720, 900, 123456895 are all divisible by 5.

Divisibility by 6

A number is divisible by 6, only if it is divisible by both 2 and 3

[i.e. its last digit should be 0 or an even number and sum of its all digits be divisible by 3]

For example: 960, 6912, 1800, 50502006 are divisible by 6.

Divisibility by 8

A number is divisible by 8, only if the number formed by its last three digits is divisible by 8 or if the number ends in three zeros. For example: 4000, 1234504, 630124936, 123456072 are all divisible by 8.

Divisibility by 9

A number is divisible by 9, only if the sum of its digits is divisible by 9. For example 63361872, 10008, 7100100, 51308172, are all divisible by 9.

Divisibility by 10

A number is divisible by 10, only if its last digit is '0'. For example, 90, 380, 4150, 71000, 1234567890 are all divisible by 10.

Divisibility by 11

A number is divisible by 11, only if the difference of the sums of its digits at alternate places are equal, or differ by 11 or by a multiple of 11. For example, the number 4802947083 is exactly divisible by 11, because

$4+0+9+7+8 = 28$ and $8+2+4+0+3 = 17$

Since, $28 - 17 = 11$,

Therefore, 4802947083 is divisible by 11.

EXAMPLE

Using the divisibility tests, determine which of the following are divisible by 4, 6, 8, 9 or 11:

(i) 231244   (ii) 316440   (iii) 840044

(iv) 10010091   (v) 12345607

SOLUTION

Test of divisibility by 4

GIVEN NUMBER	NUMBER FORMED BY LAST 2-DIGITS	DIVISION BY 4	CONCLUSION
(i) 231244	44	$44 \div 4 = 11$	231244 is divisible by 4
(ii) 316440	40	$40 \div 4 = 10$	316440 is divisible by 4
(iii) 840044	44	$44 \div 4 = 11$	840044 is divisible by 4
(iv) 10010091	91	$91 \div 4 =$ Not divisible	10010091 is not divisible by 4
(v) 12345607	07	$07 \div 4 =$ Not divisible	12345607 is not divisible by 4

Test of divisibility by 6

GIVEN NUMBER	LAST-DIGIT ODD OR EVEN	DIVISIBILITY BY 2	SUM OF DIGITS	DIVISIBILITY BY 3	CONCLUSION
(i) 231244	Even	Yes	$2+3+1+2+4+4 = 16$	$16 \div 3 =$ Not divisible	231244 Not divisible by 6
(ii) 316440	Even	Yes	$3+1+6+4+4+0 = 18$	$18 \div 3 = 6$ Divisible	316440 is divisible by 6

(Contd.)

Given Number	Last-digit Odd or Even	Divisibility by 2	Sum of Digits	Divisibility by 3	Conclusion
(iii) 840044	Even	Yes	8 + 4 + 0 + 0 + 4 + 4 = 20	20 ÷ 3 = Not divisible	840044 Not divisible by 6
(iv) 10010091	Odd	No	1 + 0 + 0 + 1 + 0 + 0 + 9 + 1 = 12	12 ÷ 3 = 4 Divisible	10010091 Not divisible by 6
(v) 12345607	Odd	No	1 + 2 + 3 + 4 + 5 + 6 + 0 + 7 = 28	28 ÷ 3 = Not divisible	12345607 Not divisible by 6

Divisibility by 8

Given Number	Number formed by last three digits is divisible by 8 or not	Conclusion
(i) 231244	244 ÷ 8 = Not divisible by 8	231244 is Not divisible by 8
(ii) 316440	440 ÷ 8 = 55 divisible by 8	316440 is divisible by 8
(iii) 840044	044 ÷ 8 Not divisible by 8	840044 is Not divisible by 8
(iv) 10010091	091 ÷ 8 = Not divisible by 8	10010091 is not divisible by 8
(v) 12345607	607 ÷ 8 = Not divisible by 8	12345607 is not divisible by 8

Divisibility by 9

Given Number	Sum of the digits	Whether sum of digits is divisible by 9 or not	Conclusion
(i) 231244	2 + 3 + 1 + 2 + 4 + 4 = 16	16 is not divisible by 9.	231244 is not divisible by 9

Given Number	Sum of the digits	Whether sum of digits is divisible by 9 or not	Conclusion
(ii) 316440	3 + 1 + 6 + 4 + 4 + 0 = 18	18 is Divisible by 9	316440 is divisible by 9
(iii) 840044	8 + 4 + 0 + 0 + 4 + 4 = 20	20 is not divisible 9	840044 is not divisible by 9
(iv) 10010091	1 + 0 + 0 + 1 + 0 + 0 + 9 + 1 = 12	12 is not divisible by 9	10010091 is not divisible by 9
(v) 12345607	1 + 2 + 3 + 4 + 5 + 6 + 0 + 7 = 28	28 is not divisible by 9	12345607 is not divisible by 9

Divisibility by 11

Given Number	Sum of digits at odd places from the right	Sum of the digits at even places from the right	Difference of the two sums	Conclusion
231244	2 + 1 + 4 = 7	3 + 2 + 4 = 9	9 − 7 = 2	231244 is not divisible by 11
316440	3 + 6 + 4 = 13	1 + 4 + 0 = 5	13 − 5 = 8	316440 is not divisible by 11
840044	8 + 0 + 4 = 12	4 + 0 + 4 = 8	12 − 8 = 4	84044 is not divisible by 11
10010091	1 + 0 + 0 + 9 = 10	0 + 1 + 0 + 1 = 2	10 − 2 = 8	10010091 is not divisible by 11
12345607	1 + 3 + 5 + 0 = 9	2 + 4 + 6 + 7 = 19	19 − 9 = 10	12345607 is not divisible by 11

Did you Observe

   i. All number divisible by 2 are even numbers.
   ii. Numbers divisible by 3 may be even or odd.
   iii. Numbers which are divisible by 4 are also divisible by 2.
   iv. Numbers which are divisible by 8 are also divisible by 4 and 2
   v. Numbers which are divisible by 9 are also divisible by 3.
   vi. Numbers which are divisible by 10 are also divisible by 2 and 5.

DIVISION OF INTEGERS

i. *When the integers have the same sign.*

When the signs of dividend integer and divisor are same, then the quotient is always positive. For example

(+32) ÷ (+4)   = (+8) or 8

(–24) ÷ (–6)   = (+4) or 4

EXPLANATION

∴ |+32| = 32 and |+4| = 4

∴ (+32) ÷ (+4)   = |+32| ÷ |+4|

= 32 ÷ 4 = 8

SIMILARLY,

|–24| ÷ 24 and |–6| = 6

∴ (–24) ÷ (–6)   = |–24| ÷ |–6|

= 24 ÷ 6 = 4

ii. When integers have the opposite signs:

If an integer is divided by an integer of opposite sign then the quotient is a negative integer, whose absolute value is equal to the quotient of the absolute values of the given numbers.

For example:

$$(-32) \div (+4) = -8$$

Also the absolute values of |–32| = 32, |+4| = 4 and |–8| = 8, |32 ÷ 4 = 8

Giving (–ve) sign to the quotient, we have 8

Thus, (–32) ÷ (+4) = –8

Similarly,   (+32) ÷ (–4) = –8

NOTE

i. **Above rules are applicable to the division without remainder.**
ii. **Division of any integer by (+1) gives the integer itself.**
iii. **Division of any integer by (–1) gives the integer itself with different sign.**
iv. **Division of 0 by any non-zero integer gives 0.**

v. Division of any integer by '0' is not possible.

EXAMPLE

Fill in the blanks:

i. $30 \div (-5) =$ _____
ii. _____ $\div (-6) = 7$
iii. $(-56) \div (8) =$ _____
iv. $(112) \div (-16) =$ _____
iv. _____ $\div (-1) = 20$

SOLUTION

i. $30 \div (-5) = -6$
   [– 30 and (–5) are having opposite signs]

ii. $(-42) \div (-6) = 7$
   [∴ Division of integers having same sign gives positive quotient.]

iii. $(-56) + (+8) = (-7)$
   [∴ Division of integers of opposite signs]

iv. $112 \div (-16) = -7$
   [∴ Dividend and divisor are of opposite signs]

v. $(-20) \div (-1) = 20$
   [∴ When divisor is (–1), the quotient is the number itself with opposite sign]

FACTOR

When two or more numbers are multiplied then the result is called *product*. For example, in 3 × 4 = 12, 3 and 4 are multiplied to get 12, so, 12 is product of 3 and 4. We observe that 3 and 4 divide 12 exactly

i.e.

```
      4                  3
   3)12       or      4)12
    -12                 -12
     x                   x
```

To divide exactly means, no remainder

Each of the numbers which are multiplied together is called a factor of the product.

So,

$$3 \times 4 = 12$$

Factors — Product

Thus, A *factor* of a number is an exact divisor of the given number.

A number can have more than two factors.

For example, the factors of 12 are:

1, 2, 3, 4, 6 and 12

An exact-divisor divides the number completely without leaving any remainder

NOTATION

The set of factors of a number (n) is denoted by $F_{(n)}$. For example, the factors of 12 are denoted as:

$$F_{(12)} = \{1, 2, 3, 4, 6, 12\}$$

The factors of a number are finite. '1' is the smallest factor and largest factor is the number itself.

EXAMPLE

(i) Write all factors of 18. (ii) show that 6 is a factor of 42.

SOLUTION

(i) $\therefore 18 = 1 \times 18$

$18 = 2 \times 9$

$18 = 3 \times 6$

$\therefore$ Factors of 18 are 1, 2, 3, 6, 9 and 18

i.e. $F_{18} = \{1, 2, 3, 6, 9, 18\}$

(ii)
```
     7
  6) 42
    -42
     0
```
i.e. 6 divides 42 exactly,

$\therefore$ 6 is a factor of 42

To write factors of a given number, we write the given number as product of two natural numbers. Then each number is a factor of the given number.

NOTE

A factor of a number is either less than or equal to the given number.

MULTIPLES

The product of two or more numbers is called a *multiple* of each of the given numbers. For example, 8 × 6 = 48, Therefore 48 is a multiple of 8 and 6.

OR

Multiples of a given number are all those numbers which can be divided completely by the given numbers. For example,

Multiples of 3 are 3, 6, 9, 12, 15, 18, 21, ... etc.

Multiples of 4 are 4, 8, 12, 16, 20, 24, 28, ... etc.

Multiples of 12 are 12, 24, 36, 48, 60, 72, 84, ... etc.

Multiples of a number are unlimited.

Every number is a multiple of itself.

NOTATION

The set of multiples of a number n denoted as M_n

Therefore, the set of multiples of 10 is given by:

M_{10} = {10, 20, 30, 40, 50, ...}

The smallest multiple of a number is the number itself. Greatest multiple of a number does not exist.

NOTE

A multiple of a number is either greater than or equal to the given number.

Example: Write first five multiples of 12.

SOLUTION

∴   12 × 1 = 12,   12 × 2 = 24

      12 × 3 = 36,   12 × 4 = 48

      12 × 5 = 60

∴ First five multiples of 12 are 12, 24, 36, 48 and 60 or

$$M_{12} = \{12, 24, 36, 48, 60, ...\}$$

To write multiples of a given natural number, find its product with 1, 2, 3, 4, ...

PRIME AND COMPOSITE NUMBERS

PRIME NUMBER

A natural number greater than 1, which has no factors other 1 and itself is called a *prime-number*.

Some prime numbers are:

2, 3, 5, 7, 11, 13, ...

DID YOU KNOW?

A prime number has two distinct factors: 1 and the number itself: But in case of 1, its factors 1 and the number itself are not distinct. So it is not a prime number.

COMPOSITE NUMBER

Numbers having more than two factors are called composite numbers.

4, 6, 8, 10, 12, 14, ... etc. are all composite numbers.

NOTE

    i.   **All natural numbers except 1 are divided into two groups.**

Natural numbers (other than 1)

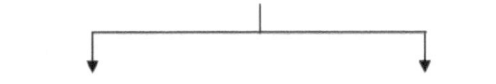

Prime-numbers   Composite-numbers

    ii.   **The natural number 1 is neither prime-number nor composite-number.**

    iii. **2 is only prime number which is even.**

    iv. **There are infinitely many prime numbers.**

CO-PRIME AND TWIN-PRIME NUMBERS

CO-PRIME NUMBERS:

Two natural numbers are called co-prime numbers, if they have no common factor other than 1.

For example:

2 and 3 are co-prime numbers.

3 and 8 are co-prime numbers,

9 and 20 are co-prime numbers and so on.

TWIN-PRIME NUMBERS

Many pairs are such that they differ by 2. Such primes are called 'twin prime numbers'.

For example, 3 and 5 $\{\therefore 5 - 3 = 2$

Some other pairs of twin-primes are:

{5, 7}, {11, 13}, {29, 31}, {41, 43} ...... etc.

EXAMPLE

i. Write five pairs of co-primes between 10 and 20.

ii. Write five pairs of twin-prime numbers between 15 and 100.

SOLUTION

i. Co-prime numbers above 10 and below 20 (between 10 and 20) are:

{11 and 13}, {13 and 15}, {13 and 14} and {17 and 20}

"Twin-prime" is a pair of prime numbers only whereas co-primes can be prime, composite or both.

ii. Twin-prime numbers between 15 and 100 are {17, 19}, {29, 31}, {41, 43}, {59, 61} and {71, 73}

PERFECT NUMBER

If the sum of all the factors of a number is two times the numbers, then it is called a *perfect number*.

For example, 6 is a perfect number.

[$\therefore F_6$ = {1, 2, 3, 6} and 1 + 2 + 3 + 6 = 12 = 2(6) = 2 (Given number)]

To Find Prime Numbers between 1 and 100 by Sieves of Eratosthenes

To find prime numbers between 1 and 100 we follow these steps:

i. Write all natural numbers up 100. Exclude 1 from the list.

ii. Mark 2 as the first prime number. Cross out all the other multiples of 2.

iii. Mark 3 as the next prime number and cross out all the other multiples of 3.

[Some of the multiples of 3 have already been crossed out as multiples of 2]

iv. Mark 5 as a prime number and cross out all other multiples of 5.

[Obviously some of them have already been crossed out as multiples of 2 and 3]

v. Mark 7 as a prime number and cross out all other multiples of 7.

[Some of them have already been crossed out as the multiples of 2, 3, or 5]

1	2	3	4	5	6	7	8	9	10
11	12	13	14	15	16	17	18	19	20
21	22	23	24	25	26	27	28	29	30
31	32	33	34	35	36	37	38	39	40
41	42	43	44	45	46	47	48	49	50
51	52	53	54	55	56	57	58	59	60
61	62	63	64	65	66	67	68	69	70
71	72	73	74	75	76	77	78	79	80
81	82	83	84	85	86	87	88	89	90
91	92	93	94	95	96	97	98	99	100

So, the sieve reveals that the prime numbers between 1 and 100 are:

2, 3, 5, 7, 11, 13, 17, 19
23, 29, 31, 37, 41, 43, 47, 53
59, 61, 67, 71, 73, 79, 83, 89 and 97

This method of sorting out prime numbers less than 100 was devised about 2000 years ago, by a Greek mathematician, His name was Eratosthenes. That is why this method is called a *"Sieve of Eratosthenes"*

NOTE

i. Pairs of prime numbers having a difference of 2, are *twin primes* are:

(3, 5), (5, 7), (11, 13), (17, 19), (41, 43), (71, 73)

ii. A set of three consecutive prime-numbers differing by 2 forms a *prime triplet* is {2, 3, 5}

EXAMPLE

How many

i. Prime numbers between 1 and 100?
ii. What is highest prime number below 100?
iii. Make a list of prime numbers between 50 and 100.

PRIME FACTORS AND FACTORIZATION

Let us make a list of all factors of 36, we have: 1, ②, ③, 4, 6, 9, 12, 18 and 36. Among these factors, Only 2 and 3 are prime numbers.

Therefore, *prime factor*, of 36 are 2 and 3. Remaining factors i.e. 4, 6, 9, 12 etc. are *composite factors*.

Factorization is a process by which a natural number (which is not a prime number) is broken down into its factors. There can be more than one factorization of a number.

For example:
$$30 = 1 \times 30$$
$$= 2 \times 15$$
$$= 3 \times 10$$
$$= 5 \times 6$$
$$= 5 \times 2 \times 3$$

DID YOU KNOW?

Prime numbers with more than 1000 digits are called *"titanic primes"*

PRIME FACTORIZATION

The process of expressing a number as the product of its prime factors only is called prime factorization.

For example, 36 = 2 × 2 × 3 × 3

30 = 2 × 3 × 5

42 = 2 × 3 × 7

NOTE

(i) **The prime factors, by convention, are listed in ascending order.**

(ii) **There is a unique prime-factorization for a number.**

NOTATION

The set of prime factors of natural number 'n' is denoted by P_n.

P_{30} = {2, 3, 5}, P_{42} = {2, 3, 7}

P_{15} = {3, 5}   etc.

METHOD [TO FIND PRIME FACTORS]

The usual method of finding the prime factors of a number is to proceed by successive division such that the first divisor being the least prime number belonging to the factors of the number.

FOR EXAMPLE, to find the prime factors of 630, we first divide by 2 and proceed as under:

2	630
3	315
3	105
5	35
	7

∴ P_{630} = {2, 3, 3, 5, 7}   or   {2, 3, 5, 7}

[Here, is set P_{630}, we not to repeat the element 3]

∴ Prime factorization of 630 = 2 × 3 × 3 × 5 × 7.

MR-5

The prime number 5 is the only prime number which can be expressed both as sum and difference of two primes: 5 = 2 + 3 = 7 − 2

Did You Know?

Every even number greater than 4 (i.e. 6, 8, 10, 12,....... etc) can be expressed as the sum of two odd prime-numbers:

6 = 3 + 3;   8 = 3 + 5
10 = 3 + 7;   12 = 5 + 7
14 = 3 + 11 etc.

Exercise − 1

1. Write all factors (in set notation) of:
   (i) 8   (ii) 12   (iii) 18   (iv) 32   (v) 75.
2. Write first 5 multiples of:
   (i) 4   (ii) 9   (iii) 19   (iv) 23   (v) 51
3. (i) List all prime numbers between 40 and 80.
   (ii) Write the smallest prime number.
   (iii) Write the greatest two-digit prime number.
   (iv) Write the smallest composite number.
   (v) Write all the prime and composite numbers up to 20.

Remember

i. A pair of prime numbers is always co-prime.
ii. A pair of consecutive numbers is always co-prime.

4.
   (i) Write 5 pairs co-prime numbers between 10 and 20.
   (ii) Write all pairs of twin primes between 50 and 80.
   (iii) Write 5 consecutive composite numbers just below 100.
   (iv) Write 5 consecutive composite numbers below 25.
   (v) Express 120 as the sum of twin primes.
5. List the elements of
   (i) P_{10} (ii) P_{12} (iii) P_{28} (iv) P_{42} (v) P_{56}

6. Write the prime factorization of:
   (i) 18   (ii) 24   (iii) 32   (iv) 63   (v) 54

ANSWERS

1. (i) $F_8 = \{1, 2, 4, 8\}$ (ii) $F_{12} = \{1, 2, 3, 4, 6, 12\}$
   (iii) $F_{18} = \{1, 2, 3, 6, 9, 18\}$ (iv) $F_{32} = \{1, 2, 4, 8, 16, 32\}$
   (v) $F_{75} = \{1, 3, 5, 15, 25, 75\}$

2. (i) 4, 8, 12, 16, 20   (ii) 9, 18, 27, 36, 45
   (iii) 19, 38, 57, 76, 95   (iv) 23, 46, 69, 92, 115
   (v) 51, 102, 153, 204, 255.

3. (i) 41, 43, 47, 53, 59, 61, 67, 71, 73 and 79.   (ii) 2
   (iii) 97   (iv) 4
   (v) Primes: 2, 3, 5, 7, 11, 13, 17 and 19; Composites: 4, 6, 8, 9, 10, 12, 14, 15, 16, 18, 20.

4. (i) (11, 12), (11, 13), (12, 13), (13, 17), (17, 19).
   (ii) (59, 61), and (71, 79)

5. (i) $P_{10} = \{2, 5\}$ (ii) $P_{12} = \{2, 3\}$
   (iii) $P_{28} = \{2, 7\}$ (iv) $P_{42} = \{2, 3, 7\}$
   (v) $P_{56} = \{2, 7\}$

6. (i) $18 = 2 \times 3 \times 3$ (ii) $24 = 2 \times 2 \times 2 \times 3$
   (iii) $32 = 2 \times 2 \times 2 \times 2 \times 2$ (iv) $63 = 3 \times 3 \times 7$
   (v) $54 = 2 \times 3 \times 3 \times 3$.

Highest Common Factor (HCF)

We know that the factors of 16 are 1, 2, 4, 8 and 16, i.e.
$$F_{16} = \{1, 2, 4, 8, 16\}$$
The factors of 24 are 1, 2, 3, 4, 6, 8, 12 and 24, i.e.
$$F_{24} = \{1, 2, 3, 4, 6, 8, 12, 24\}$$
∴ The set of common factors of 16 and 24, i.e.
$$F_{16} \cap F_{24} = \{1, 2, 4, 8, 16\} \cap \{1, 2, 3, 4, 6, 8, 12, 24\}$$
$$= \{1, 2, 4, 8\}$$
Since, the greatest element in the set of common factors = 8

Therefore, 8 is the Highest Common Factor (HCF) of 16 and 24.

Thus, *HCF of two or more natural numbers is the largest (or highest) common factor of the given numbers.*

NOTE

   i.   HCF of two or more numbers divides each number completely.
   ii. HCF is also called the HCD. i.e. Highest Common Divisor.
   iii. If the HCF of any two different number is 1, then the numbers are called co-primes.

METHOD OF FINDING HCF

1. COMMON FACTOR METHOD

We list all the possible factors of the given numbers make a set of common factors of the given numbers Identify the highest element of the set of common factors. This is the HCF of the given numbers. For example let us find HCF of 12, 18 and 24. We have:

$F_{12} = \{①, ②, ③, 4, ⑥, 12\}$, $F_{18} = \{①, ②, ③, ⑥, 9, 18\}$ and $F_{24} = \{①, ②, ③, 4, ⑥, 8, 12, 24\}$

$\therefore F_{12} \cap F_{18} \cap F_{24} = \{1, 2, 3, 6\}$

Since, the highest element of the above set is 6; So HCF = 6

2. PRIME FACTORIZATION METHOD

The HCF of two or more numbers can also be determined by prime factorization. For example, let us find the HCF of 12 and 18. First we find their prime factorization.

$\therefore$
```
2 | 12
2 |  6
  |  3
```
and
```
2 | 18
3 |  9
  |  3
```

$\therefore 12 = 2 \times 2 \times 3$ or $P_{12} = \{2, 3\}$ and $18 = 2 \times 3 \times 3$ or $P_{18} = \{2, 3\}$

Now, The set of common prime factors of 12 and 18

i.e. $P_{12} \cap P_{18} = \{2, 3\} \cap \{2, 3\}$

$= \{2, 3\}$

HCF = {The product of all the different common prime factors} = $2 \times 3 = 6$

Thus, HCF of 12 and 18 is 6

3. DIVISION METHOD

When the numbers are larger, then division method is used to find HCF of the given numbers. In this method, the greater number is divided by the smaller one. In the next step, the remainder is treated to be the divisor and the first divisor as the dividend. This process is continued till we get the remainder as '0'. The last divisor is the HCF of the given numbers.

For example, to find HCF of 90 and 324, we have

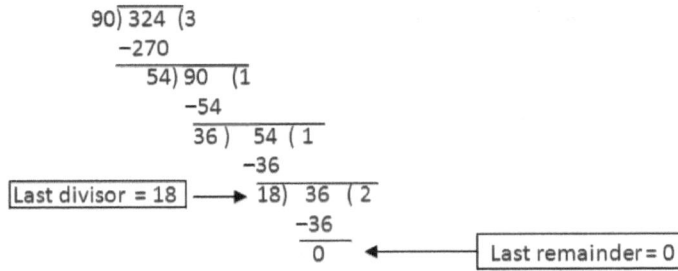

Thus, the HCF of 90 and 324 is 18

DID YOU KNOW?

The division-method of finding HCF was devised by the famous Greek *Mathematician Euclid.*

If we have to find the HCF of three or more numbers, then we have the following steps:

i. First find the HCF of any two of the given numbers.
ii. Find the HCF of Third number and the HCF of step - I.
iii. The HCF obtained in step - II is the required HCF of the three given numbers.
iv. For more than three numbers, we continue this process.

REMEMBER

i. **If a number is divisible by another number, then the former is divisible by the factors of the later.**
ii. **If a number is divisible by two or more co-prime numbers then that number is also divisible by their product.**
iii. **If a number is a factor of two given number, then it is also a factor of their difference.**
iv. **If a number is a factor of two or more given numbers, then it is also a factor of their sum.**

EXAMPLE

Find the HCF of 20 and 24

SOLUTION

$$F_{20} = \{①, ②, ④, 5, 10, 20\}$$
$$F_{24} = \{①, ②, 3, ④, 6, 8, 12, 24\}$$
$$F_{20} \cap F_{24} = \{1, 2, 4\}$$

∴ The highest element is 4

∴ HCF of 20 and 24 = 4

EXAMPLE

Use prime factorization method to find the HCF of 12 and 64.

SOLUTION

```
2 | 12         2 | 64
2 | 6    and   2 | 32
    3          2 | 16
               2 | 8
               2 | 4
                   2
```

i.e., $12 = ②×② × 3$
and $64 = ②×② × 2 × 2 × 2 × 2$
∴ HCF = 2×2 = 4

Thus, the HCF of 12 and 64 is 4

EXAMPLE

Find the HCF of 54, 162 and 144 by the division method.

SOLUTION

First let us find HCF of 54 and 144:

Since,

```
54 ) 144 ( 2
     -108
     ———
     36 ) 54 ( 1
          -36
          ———
          18 ) 36 ( 2
               -36
               ———
                 0
```

∴ HCF of 54 and 144 = 18

Now, let us find the HCF of 162 and 18:

```
    18 ) 162 ( 9           i.e., HCF of 18 and 162 is 18
       -162
       ─────
          0
```

∴ HCF of 54, 162 and 144 is 18

Exercise – 2

1. Find the HCF (using common factor method) of the following:
   (i) 27 and 36   (ii) 8, 12 and 24   (iii) 15, 18 and 32
   (iv) 36 and 48   (v) 12, 84 and 96

2. Find the HCF (using prime factorization method) of the following:
   (i) 30 and 40   (ii) 39 and 52   (iii) 70 and 84
   (iv) 36, 54 and 108   (v) 27, 36 and 45

3. Find the HCF of the following using the division method.
   (i) 132 and 156   (ii) 125, 650 and 475
   (iii) 595, 357 and 102   (iv) 75, 105 and 425   (v) 108, 180 and 324.

4. Three drums contain 54 liters, 108 liters and 144 liters of oil respectively. What is the greatest capacity of a bucket which can empty out each drum with an exact number of fillings.

5. Find the greatest number which exactly divides 1155 and 735.

6. Find the greatest length of the tape which can measure exactly 198 m and 315 m.

7. Find the greatest number which divides exactly the numbers 210, 735 and 1155.

Answers

1. (i) 9   (ii) 4   (iii) 1   (iv) 12   (v) 12
2. (i) 10   (ii) 13   (iii) 14   (iv) 18   (v) 9
3. (i) 12   (ii) 25   (iii) 17   (iv) 5   (v) 36
4. 18   5. 105   6. 9 m   7. 105.

Least Common Multiples (LCM)

A number can be multiple of many numbers For example, 12 is multiple of 1, 2, 3, 4, 6 and 12 or we can say that:

12 is a common multiple of 1, 2, 3, 4, 6 and 12

Similarly, 18 is a common multiple to 2, 3 and 6.

24 is also a common multiple of 2, 3 and 6.

Now, we can say that 12, 18 and 24 are common multiples of 2, 3 and 6.

Out of 12, 18, 24, the number 12 is the least.

Therefore, the Least common multiple of 2, 3 and 6 or LCM of 2, 3 and 6 is 12.

Thus,

LCM (Least Common Multiple) of two or more natural numbers is equal to the smallest natural number which is a multiple of all the given numbers.

For example,

Set of multiples of 9 = M_9 = {9, 18, 27, 36, 45, ...}

Set of multiples of 6 = M_6 = {6, 12, 18, 24, 30, 36........}

∴ Set of common multiples of 9 and 6 is given by

$M_9 \cap M_6$ = {18, 36, ...}

Here, the element 18 is the least.

∴ LCM of 9 and 6 is 18

METHOD OF FINDING (LCM)

1. Common Multiple Method

To find LCM of two or more given numbers, we list their individual multiples. Then we separate out all the common multiples. The least of them is the LCM the given numbers.

For example,

M_3 = {3, ⑥, 9, ⑫, 15, ⑱, 21, 24, 27, ...}

M_6 = {⑥, ⑫, ⑱, 24, 30, 36, ...}

∴ $M_3 \cap M_6$ = {6, 12, 18, 24, ...}

∴ The smallest element = 6

∴ LCM of 3 and 6 is 6

2. Prime Factorization Method

In this method:

i. We write the prime factorization of each of the given numbers.
ii. Take each prime factor as many times as the highest number of times it appears in the above prime factorization of any of the given numbers.
iii. Find the product of prime factors taken out in (ii). This product is the LCM of the given numbers.

For example, to find LCM of 12 and 18, we have

$12 = \underline{2 \times 2} \times 3$ [2 appears *2 times* and 3 appears 1 time)

$18 = 2 \times \underline{3 \times 3}$ [2 appears 1 time and 3 appears *2 times*]

$\therefore$ LCM of 12 and 18 $= (2 \times 2) \times (3 \times 3)$

$\phantom{\therefore \text{LCM of 12 and 18 }} = 4 \times 9 = 36$

3. Division Method

When it is difficult to write the prime factorization of the given numbers, then we us the simpler method known as the 'Division-method':

i. Write the given numbers in a row and separate them by commas. [They can be written in any order]
ii. Choose the first divisor a least prime number which divides exactly at least two of the given numbers.
iii. Write the quotients just below the respective numbers. Carry forward the remaining numbers as such for the next division.
iv. Repeat the steps II & III till no two numbers are available for further division.
v. Find product of all the divisors and undivided numbers left.

This product is the LCM of given numbers.

For example, let us find the LCM of 12, 20 and 24:

2	12,	20,	24
2	6,	10,	12
3	3,	5,	6
	1,	5,	2

⟵ [5 is carried forward undivided.

∴ The required LCM = 2 × 2 × 3 × 1 × 5 × 2
                          ⎵   ⎵
                       Divisors   Numbers left-undivided

= 2 × 2 × 3 × 5 × 2
= 120

EXAMPLE

Use division method to find the LCM of 18, 45 and 60.

SOLUTION

2	18,	45,	60
3	9,	45,	30
3	3,	15,	10
5	1,	5,	10
	1,	1,	2

∴ LCM = 2 × 3 × 3 × 5 × 1 × 1 × 2
      = 180

EXAMPLE

Use prime factorization to find the LCM of 20, 24 and 15

SOLUTION

∴ 20 = 2 × 2 × 5
24 = 2 × 2 × 2 × 3
15 = 3 × 5

∴ Maximum number of times of appearance of

Prime factor 2 is 3

Prime factor 5 is 1

Prime factor 3 is 1

∴ LCM = [2 taken three times] × [5 taken once] × [3 once only]
      = [2 × 2 × 2] × [5] × [3]
      = 8 × 5 × 3
      = 120

Thus, LCM of 20, 24 and 15 is 120.

EXAMPLE

Find the least number which when divided by 12, 15 and 20 leaves 5 as remainder in each case.

SOLUTION

The required number

$\quad\quad\quad$ = [Number exactly divisible by 12, 15 and 20] + 5

$\quad\quad\quad$ = [LCM of 12, 15, 20] + 5

Since,

2	12,	15,	20
2	6,	15,	10
3	3,	15,	5
5	1,	5,	5
	1,	1,	1

∴ LCM (of 12, 15 and 20) $\quad$ = 2 × 2 × 3 × 5 $\quad\quad$ = 60

∴ The required number $\quad\quad$ = 60 + 5 = 65.

EXERCISE – 3

1. Find the LCM of the following [by common multiple method]
   (i) 12 and 16 $\quad$ (ii) 3, 4 and 6
   (iii) 14 and 21 $\quad$ (iv) 24 and 36
   (v) 9, 12 and 18

2. Using prime factorization, find the LCM of the following:
   (i) 12, 16 and 20 $\quad\quad$ (ii) 48, 32 and 36
   (ii) 20, 45 and 24 $\quad\quad$ (iv) 18, 24 and 60
   (iii) 6, 11 and 18

3. Find the LCM of the following numbers, using the division method:
   (i) 100, 144 and 36 $\quad\quad$ (ii) 320, 120 and 60
   (ii) 125, 130, 115 $\quad\quad\quad$ (iv) 90, 135 and 126
   (iii) 96 and 588

4. Find the least number which when divided by 12, 16, and 24 gives 3 as remainder in each case.
5. Find the least number of 4-digits which is exactly divisible by 12, 15 and 20.
6. Find the greatest 4-digit number which is exactly divisible by 25, 15 and 20.
7. Find the greatest 4-digit number which is exactly divisible by 30, 24, 36 and 16.

ANSWERS

1. (i) 48 (ii) 12 (iii) 42 (iv) 72

   (v) 36

2. (i) 240 (ii) 288 (iii) 360 (iv) 120

   (v) 198

3. (i) 3600 (ii) 960 (iii) 74750 (iv) 1890

   (v) 4704

4. 51 5. 1020 6. 9900 7. 9360

PROPERTIES OF HCF AND LCM

1. The HCF of given numbers is not greater than any of the given numbers. At most it may be equal to the smallest given number.

   For example:

   i. HCF of 9, 15 and 24 is 3
   Here, 3 < 9;   3 < 15   3 < 24

   ii. HCF of 4, 8 and 32 is 4
   Here, 4 = 4;   4 < 8 and 4 < 32

2. L.C.M cannot be less than any of the given numbers. At the most it can be equal to the greatest.

   For example number in the given numbers.

   (i) LCM of 8, 9 and 12 is 72 obviously 8 < 72; 9 < 72 and 12 < 72

   (ii) LCM of 9, 6 and 18 is 18

i.e. LCM = Highest given number = 18.

3. The HCF of two or more co-prime numbers is equal to 1.

   For example, the HCF of 15 and 16 = 1

4. The LCM of two or more co-prime numbers is equal to the product of the given numbers.

   For example,

   LCM of 15 and 16 = 15 × 16 = 240

5. HCF of given numbers is always a factor of their LCM.

   For example,   HCF of 18, 24 and 42 is 6

   LCM of 18, 24 and 42 is 504

   ∴ 504 ÷ 6 = 84

   ∴ 6 is a factor of 504

   i.e. HCF of 18, 24 and 42 is a factor of their LCM.

6. If one number is a factor of the other then former number is their HCF and later number is then LCM.

   For example, the number 2 factor of 4

   ∴ LCM of 2 and 4 = 4 and HCF of 2 and 4 is 2.

RELATION BETWEEN HCF AND LCM

The product of HCF and LCM of two numbers is equal to the product of the numbers.

i.e. [Product of two numbers] = HCF × LCM

For example: HCF of 12 and 30 is 6

   LCM of 12 and 30 is 60

∴ 12 × 30 = 360 and 6 × 60 = 360

∴ [Product of the given two numbers] = HCF × LCM

EXAMPLE

If HCF of 27 and 36 is 9 then find their LCM.

SOLUTION

Two number (given) are: 27 and 36

   HCF = 9,   LCM =?

∴ HCF × LCM = Product of the numbers

∴ 9 × LCM = 27 × 36

Or LCM = $\dfrac{27 \times 36}{9}$ = 27 × 4 = 108

Thus, the required LCM = 108

EXAMPLE

The LCM of 40 and another number is 160. If their HCF be 8, then find the other number.

SOLUTION

HCF = 8, LCM = 160

∴ 40 × [Another number] = 8 × 160

Or Another number = $\dfrac{8 \times 160}{40}$ = 8 × 4

= 32

Thus, the required number = 32.

EXAMPLE

If the LCM of 32 and 40 is 1280. Find their HCF.

SOLUTION

We have:

1st number × 2nd number = LCM × HCF

∴ 32 × 40 = 1280 × HCF

Therefore, HCF = $\dfrac{32 \times 40}{1280}$ = 1

Thus, the required HCF is 1

EXERCISE – 4

1. Fill in the blanks:
    i. The H.C.F of given numbers is always a factor of their _____.
    ii. The L.C.M of two co-prime numbers is equal to their _____.
    iii. The HCF of two co-prime numbers is equal to _____.

iv. LCM × HCF of two numbers = _____
2. If HCF of 77 and 99 is 11, find their L.C.M.
3. IF H.C.F of two given numbers 8 and their product is 384, then find their L.C.M.
4. If the product of two numbers and their L.C.M are 54 and 18 respectively, then what their H.C.F?
5. Two numbers are such that their L.C.M and H.C.F are 240 and 6 respectively. If one of them is 48 then find the other.
6. The L.C.M of 96 and 120 is 480 find their HCF.

ANSWERS

1. (i) L.C.M   (ii) Product   (iii) 1   (iv) product of the given numbers
2. 693   3. 48   4. 3   5. 30   6. 24

BRAIN TEASER

1. Determine the two numbers nearest 10000 which are exactly divisible by 1, 2, 3, 4, 5, 6 and 7.
2. Find the greatest four-digit number which when divided by 36, 30, 24, and 16 leaves a remainder of 13 in each case.

ANSWERS

1. 9660 and 10080   2. 9373

EXERCISE – 5

1. Choose from 3960, 3825, 87360, 51840, 5720 and 91848, those numbers which are exactly divisible by all the three numbers 5, 8 and 9.
2. Which of the following are exactly divisible by 4: 92, 547, 1926, 4756 and 10428?
3. Which of the following numbers are exactly divisible by 8
   2872, 19176, 307138, 172744 and 550342?
4. Which of the following numbers are exactly divisible by 9: 1082331, 185675, 26326, 44406 and 61866?
5. Which of the following numbers are exactly divisible by 11:
   6435, 27406, 50836, 742071, 903082, 265379 and 92219281?

6. Complete the following table

	Sign of the integer as dividend	Sign of the integer as divisor	Sign of the quotient
(i)	(+)	(+)	(_____)
(ii)	(−)	(−)	(_____)
(iii)	(−)	(+)	(_____)
(iv)	(+)	(−)	(_____)

Answers

1. 3960 and 51840
2. 92, 4756 and 10428
3. 2872, 19176 and 172744
4. 1082331, 44406 and 61866
5. 6435, 742071 and 92219281
6. (i) (+) (ii) (+) (iii) (−) (iv) (−)

Miscellaneous Exercise

1. Write down all the multiples of 7 between 80 and 100.
2. Write down (i) F_{20} (ii) F_{42} (iii) F_{63}
3. Write (i) P_{18} (ii) P_{27} (iii) $P_{42} \cap P_{56}$
4. Find the HCF of 72, 108 and 144
5. List the prime numbers between 50 and 90.
6. Express each of the following odd numbers as the sum of three odd prime-numbers: (i) 35 (ii) 49
7. Write down five consecutive composite numbers just below 100.
8. Find the greatest number which can exactly divide 735 and 1890
9. Use division method to determine the HCF of 516, 1419 and 3000
10. Find the greatest number which divides 203 and 322 leaving remainder 5 and 7 respectively.
11. Find the least number which when divided by 12, 15 and 20 leaves 7 as remainder in each case.
12. Find the LCM of each set of numbers given below:

    (i) 14, 18 and 21   (ii) 15, 20 and 24   (iii) 24, 32 and 40

    (iv) 28, 35 and 42   (v) 54, 48, 72 and 36   (vi) 30, 45, 75 and 90

(vii) 49, 84, 70 and 105 (viii) 60, 72, 84 and 112.

13. Find the greatest number of 4 digits which when divided by 30, 16, 24 and 36 leaves a remainder of 9 in each case.
14. If the LCM and HCF of two numbers are 240 and 20 respectively, and one of the numbers is 80, find the other number.

Answers

1. 84, 91, 98

2. (i) {1, 2, 4, 5, 10, 20}   (ii) {1, 2, 3, 6, 7, 14, 21, 42}

(iii) {1, 3, 7, 9, 63}

3. (i) {2, 3}   (ii) {3}   (iii) {2, 7}

4. 18   5. 53, 59, 61, 67, 71, 73, 79, 83, 89

6. (i) 35 = 5 + 11 + 19   (ii) 49 = 5 + 7 + 37

7. 92, 93, 94, 95 and 96   8. 105   9. 3

10. 9   11. 67

12. (i) 126   (ii) 120   (iii) 480   (iv) 420

(v) 432   (vi) 450   (vii) 2940   (viii) 5040

13. 9369   14. 60

Mental Maths

1. Which numbers between 1 and 10 are factors of 36.
2. Which number is the smallest prime factor of 15?
3. Which number is the greatest composite number between 1 and 10.
4. What is smallest composite number between prime numbers 2 and 11?
5. Which number is neither prime nor composite.
6. What is the prime factorization of 12?
7. A number is divisible by 9. Is it also divisible by 3?

8. A number ends in two zeros. Is it always divisible by 4?
9. What is the least multiple of 8 between 15 and 20?
10. Which number are the multiples of 7 between 50 and 100?
11. What is the HCF of two co-prime numbers?
12. Which number is a perfect number between 1 and 10?
13. What is the HCF of 6 and 10?
14. What is the LCM of 90 and 91?
15. HCF and LCM of two numbers are 5 and 30 respectively. What is the product of the numbers?

Answers

1. 2, 3, 4, 6 and 9
2. 3
3. **9**
4. 4
5. 1
6. 2 × 2 × 3
7. Yes
8. Yes
9. 16
10. 56, 63, 70, 77, 84, 91 and 98
11. 1
12. 6
13. 2
14. 810
15. 150

Multiple Choice Questions

1. Which of the following is the prime factorization of 28?
   (i) 1 × 2 × 7 × 2   (ii) 7 × 4   (iii) 7 ×2 × 2   (iv) 1 × 28
2. Which of the following is HCF of 25, 50 and 160?
   (i) 5   (ii) 1   (iii) 25   (iv) 10
3. Which of the following is the LCM of 31 and 30?
   (i) 30   (ii) 31   (iii) 310   (iv) 930
4. Which of the following are common multiple of 8 and 24?
   (i) 72 and 100   (ii) 48 and 32   (iii) 48 and 72   (iv) 16 and 72
5. Which of the following the difference between largest and the smallest prime number between 6 and 18?
   (i) 8   (ii) 10   (iii) 4   (iv) 6

6. Which of the following are the prime factors of 24?
   (i) 2, 3, 4, 6, 12, 24   (ii) 1, 2, 3,
   (iii) 1, and 24   (iv) 2 and 3

7. Which of the following is the sum of the prime factors of 63?
   (i) 10   (ii) 13   (iii) 8   (iv) 14

8. Which of the following are the prime factors of 100
   (i) 4 and 5   (ii) 1, 2 and 5   (iii) 2 and 5   (iv) 5 and 20

9. The first three prime numbers greater than 50 are:
   (i) 53, 59, 61   (ii) 53, 61, 67   (iii) 51, 53, 57   (iv) 51, 53, 59

10. Which of the following is the LCM of 2, 3 and 10?
    (i) 120   (ii) 80   (iii) 60   (iv) 30

ANSWERS

1. (iii)   2. (i)   3. (iv)   4. (iii)   5. (ii)   6. (iv)   7. (i)   8. (iii)   9. (v)   10. (iv)

WORK SHEET

1. Write the factors in roster form:
   i. $F_{(8)}$ = { _____ }
   ii. $F_{(12)}$ = { _____ }
   iii. $F_{(18)}$ = { _____ }
   iv. $F_{(48)}$ = { _____ }
   v. $F_{(60)}$ = { _____ }

2. Write 'True' or 'False' for each of the following statements:
   i. The natural number 1 is not a prime number.
   ii. All prime numbers are even.
   iii. 18 is the LCM of 6 and 9.
   iv. 12 is the LCM of 4 and 6.
   v. An integer is exactly divisible by when the last digit is 0.
   vi. If a number is divisible by 4, it is also divisible by 8.
   vii. Every natural number is a multiple of itself.
   viii. 1 is a factor of every natural number.
   ix. Every natural number is either prime or composite.
   x. (LCM of two numbers) × (HCF of two numbers) = (Product of the two numbers).

3. Match the statements of column – A with the statement of the column-B:

Column – A	Column – B
(a) Twin primes between 70 and 80	(i) 3240
(b) 72 expressed as the sum of two odd primes	(ii) 5183
(c) LCM of 72, 81 and 90	(iii) 31 + 41
(d) LCM of 71 and 73	(iv) 2 × 2 × 2 × 3 × 3
(e) 72 expressed as the product of prime numbers	(v) 71, 73

4. Fill in the blanks.
   i. The number 1 is neither _____ nor composite.
   ii. The only even prime number is _____.
   iii. The smallest composite number is _____.
   iv. The largest prime number between 1 to 100 is _____.
   v. The HCF of two _____ numbers is 1
   vi. Multiple of a number is always greater than or equal to the _____.
   vii. Every number is a multiple of _____ and itself.
   viii. Factor of a number is always less than or equal to the _____.
   ix. The number _____ is a factor of every number.
   x. _____ and _____ is a pair of twin primes between 11 and 20.

5. The traffic-lights at three different road crossings change after every 35 seconds, 91 seconds and 91 seconds respectively. If they change simultaneously at 8 a. m., at what time will they change again simultaneously?

6. Two baskets contain 360 and 840 apples respectively. Find the largest number of apples that can be taken out from each basket every time to make the baskets empty.

7. Find the greatest number which divides 202 and 321 leaving remainders 4 and 6 respectively.

8. Four bells toll after intervals of 15 minutes, 12 minutes, 9 minutes and 8 minutes respectively. If they toll together at 5 p.m. When will they toll together next?

9. Find the least 5-digit number which leaves a remainder of 11 in each case when it is divided by 12, 40 and 75.
10. What is the least number which where decreased by 9 is divisible by 6, 8, and 12 respectively.

Answers

1. (i) $F_8 = \{1, 2, 4, 8\}$ (ii) $F_{12} = \{1, 2, 3, 4, 6, 12\}$

(iii) $F_{18} = \{1, 2, 3, 6, 9, 18\}$ (iv) $F_{48} = \{1, 2, 3, 4, 6, 8, 12, 16, 24, 48\}$

(v) $F_{60} = \{1, 2, 3, 4, 5, 6, 10, 12, 15, 20, 30\}$

2. (i) True (ii) False (iii) True (iv) True

(v) False (vi) False (vii) True (viii) True

(ix) False (x) True

3. (a) ⟶ (v), (b) ⟶ (iii), (c) ⟶ (i), (d) ⟶ (ii),

(e) ⟶ (iv).

4. (i) prime (ii) 2 (iii) 4 (iv) 97

(v) co-prime (vi) number itself (vii) 1 (viii) the number itself

(ix) 1 (x) 17 and 19

5. 8: 00: 40   6. 120   7. 9   8. 8.11 p.m.

9. 10211   10. 129

CHAPTER 5
FRACTIONS

Let us take a rectangular piece of paper. Fold in horizontally and vertically to divide it into 4 equal parts, as shown in the adjoining diagram.

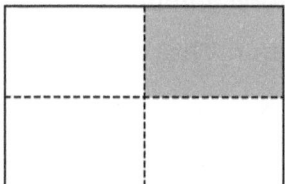

When an object is divided into equal parts then each part is called its *fraction*. Here the shaded part represents one-fourth $\left(\frac{1}{4}\right)$ of the whole piece of paper. Similarly the unshaded part represents its three-fourths $\left(\frac{3}{4}\right)$ so, $\frac{1}{4}$ and $\frac{3}{4}$ called the fractions. For example,

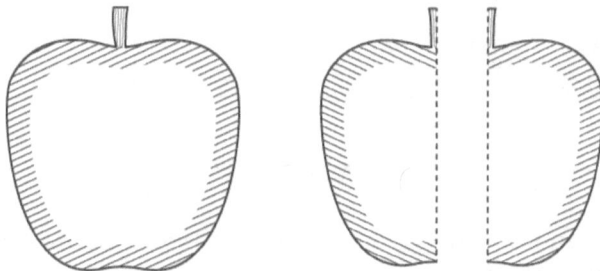

Take an apple. It is a whole object. Now cut into two equal parts. These parts of the apple are its fractions. Each of them represents $\frac{1}{2}$ of the whole.

Thus,

When a whole single object or a group of objects is divided into equal parts, then each part is called a fraction of that whole or group.

DID YOU KNOW?

The word *'fraction'* **originates from the Latin word** *"FRACTUS"* **which mean 'broken.'**

We can also say that a fraction is a way of representing a *'whole'* into equal parts and then choosing (or considering) some of them.

A fraction is in the form: $\dfrac{\text{A Top number}}{\text{A bottom number}}$ ← bar

Such that the 'top number' and the bottom number are separated by a 'bar.' The top number is named as *numerator* and bottom number is named as *denominator*. Therefore,

$$\text{Fraction} = \dfrac{\text{Numerator}}{\text{Denominator}}$$

Where, numerator = number of equal parts chosen

and denominator = total number of equal parts.

Look at the adjoining diagram. The rectangle is divided into 9 equal parts and 4 parts are shaded.

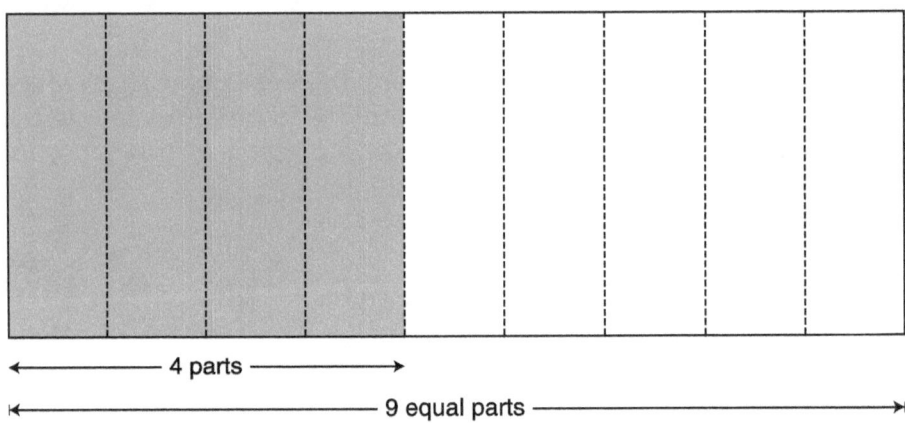

So, the shaded portion, represents 'four-ninths' of the rectangle or the shaded part is a fraction = $\dfrac{4}{9}$

In fraction, $\dfrac{4}{9}$, numerator = 4 and the denominator = 9

∴ Fraction = $\dfrac{\text{Numerator}}{\text{Denominator}}$

EXAMPLE

Write a fraction which represents the shaded part of each of the following figures:

(i)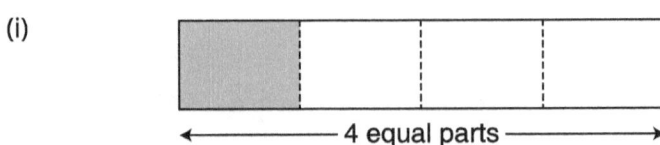

(ii)

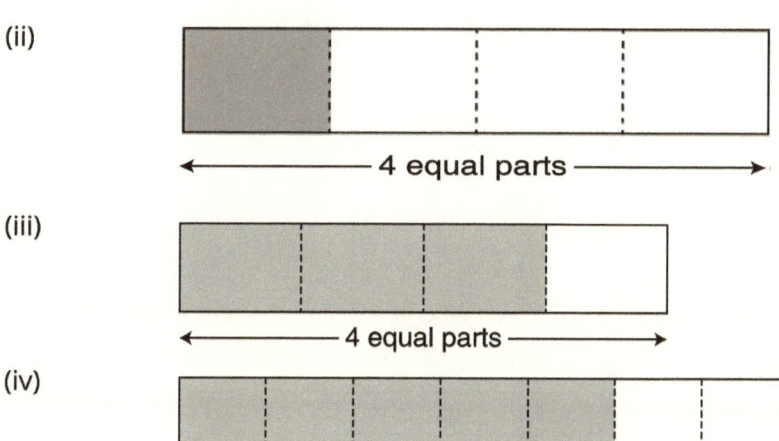

← 4 equal parts →

(iii)

← 4 equal parts →

(iv)

← 7 equal parts →

SOLUTION

i. Total number of equal parts = 4 = Denominator
Number of shaded parts = 1 = Numerator

$$\therefore \text{Fractions} = \frac{\text{Numerator}}{\text{Denominator}}$$

[represented by the shaded part] = $\frac{1}{4}$, means 'one-fourth' of the whole

Similarly,

ii. Fraction [represented by the shaded part]
= $\frac{1}{5}$, means 'One-fifth' of the whole.

iii. Fraction [represented by the shaded part]
= $\frac{3}{4}$, means 'Three-fourths' of the whole.

iv. Fraction [represented by the shaded part]
= $\frac{5}{7}$, means 'Five-Sevenths' of the whole.

EXAMPLE

Determine the numerator and denominator in each of the given fractions.

(i) $\frac{3}{8}$ (ii) $\frac{1}{6}$ (iii) $\frac{2}{3}$ (iv) $\frac{4}{5}$ (v) $\frac{7}{9}$ (vi) $\frac{5}{8}$ (vii) $\frac{1}{10}$ (viii) $\frac{2}{13}$ (ix) $\frac{6}{7}$ (x) $\frac{9}{11}$

Solution

S No	Fraction	Numerator	Denominator
1.	$\dfrac{3}{8}$	3	8
2.	$\dfrac{1}{6}$	1	6
3.	$\dfrac{2}{3}$	2	3
4.	$\dfrac{4}{5}$	4	5
5.	$\dfrac{7}{9}$	7	9
6.	$\dfrac{5}{8}$	5	8
7.	$\dfrac{1}{10}$	1	10
8.	$\dfrac{2}{13}$	2	13
9.	$\dfrac{6}{7}$	6	7
10.	$\dfrac{9}{11}$	9	11

Exercise – 1

1. Write a fraction which represents the shaded part of each of the following figures:

    (i)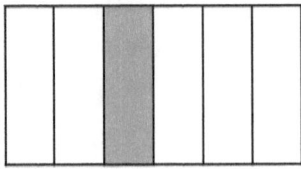

    (ii)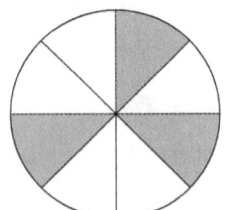

(iii) 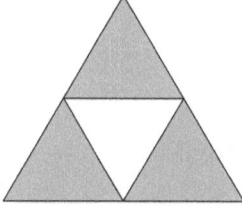   (iv)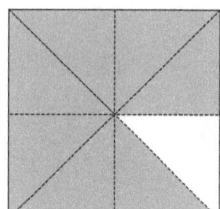

(v)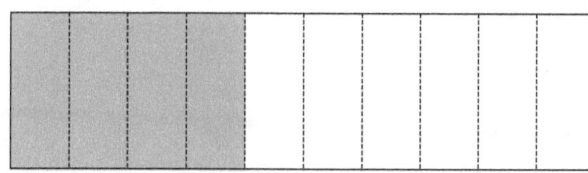

2. Write the following in fractions:

   (i) Two-fifths (ii) Four-sixths (iii) Eight-tenth,

   (iv) Seven-nineths (v) Nine-seventeenths (vi) Seven-twelfths

3. Write the following fractions is words:

   (i) $\dfrac{1}{5}$ (ii) $\dfrac{3}{10}$ (iii) $\dfrac{5}{7}$ (iv) $\dfrac{9}{13}$ (v) $\dfrac{8}{15}$ (vi) $\dfrac{6}{11}$

4. Determine the numerator and denominator in each of the given fractions.

   (i) $\dfrac{3}{19}$ (ii) $\dfrac{8}{27}$ (iii) $\dfrac{5}{21}$ (iv) $\dfrac{15}{19}$ (v) $\dfrac{38}{43}$ (vi) $\dfrac{115}{301}$

5. Find out the numerator and denominator in each of the following fractions:

   (i) Sixteen-seventeenths (ii) Five-fourteenths

   (iii) Three-thirteenths (iv) one-twelfths

   (v) Seven-twenty threes (vi) Nineteenth-thirteenths

ANSWERS

1. (i) $\dfrac{1}{6}$ (ii) $\dfrac{3}{8}$ (iii) $\dfrac{3}{4}$ (iv) $\dfrac{7}{8}$ (v) $\dfrac{4}{10}$

2. (i) $\dfrac{2}{5}$ (ii) $\dfrac{4}{6}$ (iii) $\dfrac{8}{10}$ (iv) $\dfrac{7}{9}$ (v) $\dfrac{9}{17}$ (vi) $\dfrac{7}{12}$

3. (i) One-fifth   (ii) Three-tenths   (iii) Five-sevenths
   (vi) Nine-thirteenths   (v) Eight-fifteenths   (vi) Six-elevenths

4. (i) N = 3; D = 19   (ii) N = 8; D = 27   (iii) N = 5; D = 21
   (iv) N = 15; D = 19   (v) N = 38; D = 43   (vi) N = 115; D = 301

5. (i) $\dfrac{16}{17}$ (ii) $\dfrac{5}{14}$ (iii) $\dfrac{3}{13}$ (iv) $\dfrac{1}{12}$ (v) $\dfrac{7}{23}$ (vi) $\dfrac{19}{30}$

TYPES OF FRACTIONS

1. **Proper Fraction**

   A fraction is called a 'proper-fraction,' if its numerator is smaller than the denominator.

   For example: $\dfrac{1}{3}, \dfrac{2}{5}, \dfrac{7}{9}, \dfrac{115}{312}$, etc. are proper fraction. [∴ 1 < 3, 2 < 5, 7 < 9, etc.]

2. **Improper Fraction**

   A fraction is called an 'improper-fraction,' if its numerator is greater than or equal to the denominator.

   For example: $\dfrac{5}{2}, \dfrac{9}{7}, \dfrac{12}{11}, \dfrac{310}{421}, \dfrac{5}{5}$, etc.
   [∴ 5 > 2, 7 > 9, 12 > 11 etc.]

3. **Like Fractions**

   Fractions having same denominator are called like fractions.

   For example: $\dfrac{7}{9}, \dfrac{3}{9}, \dfrac{2}{9}, \dfrac{5}{9}$, etc. are like fractions [∴ These fare fraction have a common denominator as 9]

4. **Unlike Fractions**

   Fractions having different denominators are called unlike fractions.

   For example: $\dfrac{2}{3}, \dfrac{4}{5}, \dfrac{11}{17}, \dfrac{312}{421}$, etc. are unlike fractions.
   [∴ These fraction have 3, 5, 17, etc. different denominators.]

5. **Unit Fraction**

   A fraction having its numerator as 1 is called a unit-fraction.

   For example: $\dfrac{1}{6}, \dfrac{1}{9}, \dfrac{1}{10}, \dfrac{1}{105}, \dfrac{1}{1092}$, etc, are unit fractions
   [∴ Each fraction is having numerator = 1]

6. **Mixed-Fraction**

   A combination of a natural number and a proper-fraction is called a 'mixed fraction.'

   For example; $1\frac{1}{2}, 3\frac{4}{7}, 7\frac{1}{3}, 100\frac{41}{53}$, etc. are mixed fractions.

   [∴ In $1\frac{1}{2}$, 1 is a natural number and $\frac{1}{2}$ is a proper fraction.]

   In $3\frac{4}{7}$, 3 is a natural number and $\frac{4}{7}$ is a proper fraction.

   In $7\frac{1}{3}$, 7 is a natural number and $\frac{1}{3}$ is a proper fraction.

NOTE

   i. A mixed fraction is also called a mixed number.

   ii. A mixed fraction may always be express as an improper fraction.

   iii. An improper fraction may always be expressed as a mixed fraction.

   iv. In a mixed fraction, the natural number is the "integral part" and the proper fraction is the 'fractional part'

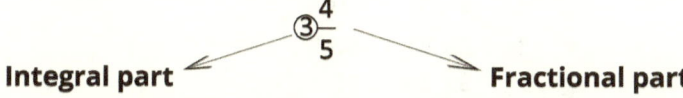

   Integral part   Fractional part

   v. **In a mixed fraction, its two parts are written together with the understanding that they are to be added to each other, i.e. $7\frac{8}{11}$ means $7 + \frac{8}{11}$.**

TO CONVERT 'UNLIKE-FRACTION' INTO 'LIKE-FRACTION'

METHOD

   i. Find the LCM of denominators of the given fractions.

   ii. Multiply the numerator denominator of each fraction by (LCM ÷ denominator).

For example: To convert $\frac{3}{4}$ and $\frac{4}{5}$ as like fraction (with common denominator), we find LCM of 4 and 5 which is 20.

∴ $\frac{3}{4} = \frac{3 \times 5}{4 \times 5} = \frac{15}{20}$ [∵ 20 ÷ 4 = 5, ∴ Multiply, the numerator as well as the denominator by 5

and $\frac{4}{5} = \frac{4 \times 4}{5 \times 4} = \frac{16}{20}$ [∵ 20 ÷ 5 = 4

$\frac{15}{20}$ and $\frac{16}{20}$ have a common denominator, so they are like-fractions.

EXAMPLE

Convert the following fractions into like-fractions:

(i) $\dfrac{3}{4}$ and $\dfrac{5}{6}$ (ii) $\dfrac{1}{2}, \dfrac{2}{3}$ and $\dfrac{3}{4}$ (iii) $\dfrac{2}{3}, \dfrac{3}{4}$ and $\dfrac{4}{5}$

SOLUTION

i. (i) LCM of 4 and 6 is 12 and 12 ÷ 4 = 3, 12 ÷ 6 = 2

$$\therefore \dfrac{3}{4} = \dfrac{3 \times 3}{4 \times 3} = \dfrac{9}{12}; \quad \dfrac{5}{6} = \dfrac{5 \times 2}{6 \times 2} = \dfrac{10}{12}$$

ii. LCM of 2, 3, and 4 = 12, Also

[12 ÷ 2 = 6, 12 ÷ 3 = 4 and 12 ÷ 4 = 3]

$$\therefore \dfrac{1}{2} = \dfrac{1}{2} \times \dfrac{6}{6} = \dfrac{6}{12}; \quad \dfrac{2}{3} = \dfrac{2 \times 4}{3 \times 4} = \dfrac{8}{12} \text{ and } \dfrac{3}{4} = \dfrac{3 \times 3}{4 \times 3} = \dfrac{9}{12}$$

So, $\dfrac{6}{12}, \dfrac{8}{12}$ and $\dfrac{9}{12}$ are like-fractions

iii. LCM of 3, 4 and 5 = 60 and 60 ÷ 3 = 20, 60 ÷ 4 = 15 and 60 ÷ 5 = 12

$$\therefore \dfrac{2}{3} = \dfrac{2 \times 20}{3 \times 20} = \dfrac{40}{60}$$

$$\dfrac{3}{4} = \dfrac{3}{4} \times \dfrac{15}{15} = \dfrac{45}{60} \quad \text{and}$$

$$\dfrac{4}{5} = \dfrac{4 \times 12}{5 \times 12} = \dfrac{48}{60}$$

So, $\dfrac{40}{60}, \dfrac{45}{60}$ and $\dfrac{48}{60}$ are like-fractions.

TO CONVERT IMPROPER FRACTIONS INTO MIXED FRACTIONS

METHOD

To change an improper fraction into a mixed number:

i. Divide the numerator with the denominator
ii. Retain the denominator.
iii. Quotient becomes the integral part (natural number part) of the mixed fraction and remainder becomes its numerator.

For example, let us change $\dfrac{16}{3}$ as a mixed fraction, we have:

```
         ┌─ 3)16(5 ─────────┐
         │   -15            └──────→ Integral part (or natural number part)
┌────────┤    1  ───────────→ Numerator
│        └──────────────────→ Denominator
```

$$\therefore \frac{16}{3} = 5\frac{1}{3}$$

TO CONVERT A MIXED-FRACTION INTO AN IMPROPER FRACTION

METHOD

To change a mixed fraction into an improper fraction:

i. Retain the denominator.
ii. Multiply the integral part (natural number) with the denominator. Add the numerator to this product. Take the sum as the numerator of the improper fraction.

$$\text{Mixed Fraction} = \frac{(\text{Integral part} \times \text{Denominator}) + \text{Numerator}}{\text{Denominator}} = \text{Improper Fraction}$$

For example, let us convert $2\frac{7}{10}$ as an improper fraction we retain the denominator 5 as such. The numerator = $(2 \times 10) + 7 = 27$

So, the improper fraction = $\frac{27}{10}$

EXAMPLE

Convert the following fractions into mixed fractions:

(i) $\frac{37}{8}$ (ii) $\frac{53}{16}$ (iii) $\frac{103}{18}$

SOLUTION

i. To convert $\frac{37}{8}$ into a mixed fraction, we divide 37 by 8:

```
      4
   8)37        Quotient = 4
    -32        Remainder = 5
    ───
      5
```

$\therefore$ mixed fraction = $4\frac{5}{8}$

ii. To convert $\frac{53}{16}$ in to a mixed fraction, we divide 53 by 16:

$$16 \overline{)53} (3$$
$$\underline{-48}$$
$$5$$

Quotient = 3
Remainder = 5

∴ The mixed fraction = $3\dfrac{5}{16}$

iii. To convert $\dfrac{103}{18}$ into a mixed number; we divide 103 by 18:

$$18 \overline{)103} (5$$
$$\underline{-90}$$
$$13$$

Quotient = 5
Remainder = 13

∴ mixed fraction = $5\dfrac{13}{18}$

EXAMPLE

Convert the following mixed fractions improper fractions:

(i) $5\dfrac{3}{8}$ (ii) $9\dfrac{1}{2}$ (iii) $12\dfrac{3}{4}$

SOLUTION

∴ An improper fraction given by

$$\dfrac{[\text{Integral part}] \times \text{Denominator} + [\text{numerator of the fractional part}]}{\text{Denominator}}$$

i. $5\dfrac{3}{8} = \dfrac{(5 \times 8) + 3}{8} = \dfrac{40 + 3}{8} = \dfrac{43}{8}$

ii. $9\dfrac{1}{2} = \dfrac{(9 \times 2) + 1}{2} = \dfrac{18 + 1}{2} = \dfrac{19}{2}$

iii. $12\dfrac{3}{4} = \dfrac{(12 \times 4) + 3}{4} = \dfrac{48 + 3}{4} = \dfrac{51}{4}$

EQUIVALENT FRACTIONS

Two or more different fractions are called 'equivalent fractions' if they represent the same part of a whole or have the same value.

Let us observe the following shaded portion in each shape. The size of each shape is the same.

Shape (i): Divided into two equal parts and one part is shaded. Shaded portion is fraction $=\frac{1}{2}$

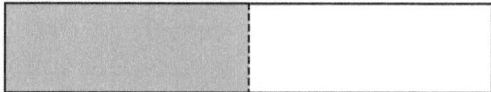

Shape (ii): Divided 4 equal parts 2 parts are shaded. Shaded portion is a fraction $=\frac{2}{4}$

Shape (iii): Divided into 8 equal parts and 4 parts are shaded. Shaded portion is a fraction $=\frac{4}{8}$

The shaded portion in each shape is identical.

So, $\frac{1}{2}, \frac{2}{4}$ and $\frac{4}{8}$ represent the identical portions, i.e., $\frac{1}{2}, \frac{2}{4}$ and $\frac{4}{8}$ have equal value.

Again, observe the following shapes:

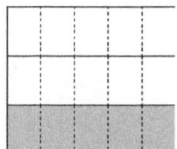 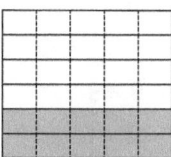 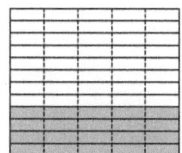

Total No. of equal parts = 15   Total No. of equal parts = 30   Total No. of equal parts = 60
No. of parts shaded = 5   No. of parts shaded = 10   No. of parts shaded = 20

Fraction $= \frac{5}{15}$ Fraction $= \frac{10}{30}$ Fraction $= \frac{20}{60}$

Obviously, fractions $\frac{5}{15}, \frac{10}{30}, \frac{20}{60}$ represent the identical (equal) shaded portions.

∴ value of $\frac{5}{15}, \frac{10}{30}$ and $\frac{20}{60}$ is the same.

Thus, (i) $\frac{1}{2}, \frac{2}{4}$ and $\frac{4}{8}$ are equivalent fractions.

(ii) $\frac{5}{15}, \frac{10}{30}$ and $\frac{20}{60}$ are equivalent fractions.

To find equivalent fractions multiply (or divide) both the numerator and denominator by the same number (except '0')

For example:

$$\frac{1}{2} = \frac{1+1}{2+1} = \frac{1}{2}$$

$$\frac{2}{4} = \frac{2+2}{4+2} = \frac{1}{2}$$

$$\frac{4}{8} = \frac{4+4}{8+4} = \frac{1}{2}$$

∵ Each of $\frac{1}{2}, \frac{2}{4}, \frac{4}{8}$ is equal to $\frac{1}{2}$

∴ $\frac{1}{2}, \frac{2}{4}$ and $\frac{4}{8}$ are equivalent fractions

i.e. $\frac{1}{2} = \frac{2}{4} = \frac{4}{8}$

EXAMPLE

a. Write two equivalent fractions for each of the following fractions:

(i) $\frac{4}{7}$ (ii) $\frac{9}{13}$ (iii) $\frac{8}{15}$

b. Write an equivalent fraction of $\frac{3}{5}$ with numerator as 12.

c. Write an equivalent fraction of $\frac{18}{35}$ with denominator as 5

SOLUTION

a. (i) $\frac{4}{7} = \frac{4 \times 2}{7 \times 2} = \frac{4 \times 3}{7 \times 3}$

or $\frac{4}{7} = \frac{8}{14} = \frac{12}{21}$ i.e. $\frac{8}{14}$ and $\frac{12}{21}$ are equivalent to $\frac{4}{7}$

(ii) $\frac{9}{13} = \frac{9 \times 2}{13 \times 2} = \frac{9 \times 3}{13 \times 3}$ i.e.

or $\frac{9}{13} = \frac{18}{26} = \frac{27}{39}$ i.e. $\frac{18}{26}$ and $\frac{27}{39}$ are equivalent to $\frac{9}{13}$

(iii) $\frac{8}{15} = \frac{8 \times 2}{15 \times 2} = \frac{8 \times 3}{15 \times 3}$

or $\frac{8}{15} = \frac{16}{30} = \frac{24}{45}$ i.e. $\frac{16}{30}$ and $\frac{24}{45}$ are equivalent to $\frac{8}{15}$

b. $\dfrac{3}{5} = \dfrac{12}{\square}$

To get 12 in the numerator we have to multiply 3 by 4

∴ multiplying the denominator also by 4, we have

$$\dfrac{3}{5} = \dfrac{3\times 4}{5\times 4} = \dfrac{12}{20}$$

∴ The fraction equivalent to $\dfrac{3}{5}$ with numerator as 12 is $\dfrac{12}{20}$

c. $\dfrac{18}{45} = \dfrac{\square}{5}$

To get 5 in the denominator, we have to divide 45 by 9,

∴ We divide the numerator 18 also by 9.

$$\therefore \dfrac{18}{45} = \dfrac{18 \div 9}{45 \div 9} = \dfrac{2}{5}$$

∴ The equivalent fraction of $\dfrac{18}{45}$ with denominator as 5 is $\dfrac{2}{5}$

EXERCISE - 2

1. Convert the following fractions into like fractions.

   (i) $\dfrac{2}{5}, \dfrac{9}{10}, \dfrac{11}{15}, \dfrac{13}{20}$

   (ii) $\dfrac{1}{3}, \dfrac{3}{4}, \dfrac{5}{6}, \dfrac{7}{8}, \dfrac{5}{12}$

   (iii) $\dfrac{1}{3}, \dfrac{3}{4}, \dfrac{5}{12}, \dfrac{7}{16}, \dfrac{17}{24}$

   (iv) $\dfrac{3}{7}, \dfrac{2}{5}, \dfrac{13}{14}, \dfrac{17}{35}, \dfrac{9}{20}$

   (v) $\dfrac{3}{5}, \dfrac{4}{10}, \dfrac{7}{12}, \dfrac{9}{20}, \dfrac{11}{15}$

2. Convert the following fraction in mixed fractions:

   (i) $\dfrac{27}{4}$ (ii) $\dfrac{61}{10}$ (iii) $\dfrac{92}{7}$ (iv) $\dfrac{103}{9}$ (v) $\dfrac{211}{12}$

3. Convert the following mixed fractions into improper fractions.

   (i) $7\dfrac{1}{4}$ (ii) $6\dfrac{5}{9}$ (iii) $11\dfrac{11}{17}$ (iv) $9\dfrac{11}{13}$ (v) $12\dfrac{5}{8}$

4. Write two equivalent fractions for each of the following.

   (i) $\dfrac{5}{6}$ (ii) $\dfrac{11}{12}$ (iii) $\dfrac{9}{13}$ (iv) $\dfrac{8}{15}$ (v) $\dfrac{7}{29}$

5. Write an equivalent fraction of:

   (i) $\dfrac{3}{8}$, with numerator 24   (ii) $\dfrac{27}{45}$, with denominator 5

   (iii) $\dfrac{30}{80}$, with numerator 3   (iv) $\dfrac{66}{132}$, with denominator 6

   (v) $\dfrac{24}{40}$, with numerator 3.

ANSWERS

1. (i) $\dfrac{24}{60}, \dfrac{54}{60}, \dfrac{44}{60}, \dfrac{39}{60}$ (ii) $\dfrac{8}{24}, \dfrac{18}{24}, \dfrac{20}{24}, \dfrac{21}{24}, \dfrac{10}{24}$ (iii) $\dfrac{16}{48}, \dfrac{36}{48}, \dfrac{20}{48}, \dfrac{21}{48}, \dfrac{34}{48}$

   (iv) $\dfrac{60}{140}, \dfrac{56}{140}, \dfrac{130}{140}, \dfrac{68}{140}, \dfrac{63}{140}$ (v) $\dfrac{36}{60}, \dfrac{24}{60}, \dfrac{35}{60}, \dfrac{27}{60}, \dfrac{44}{60}$

2. (i) $6\dfrac{3}{4}$ (ii) $6\dfrac{1}{10}$ (iii) $13\dfrac{1}{7}$ (iv) $11\dfrac{4}{9}$ (v) $17\dfrac{7}{12}$

3. (i) $\dfrac{29}{4}$ (ii) $\dfrac{59}{9}$ (iii) $\dfrac{198}{17}$ (iv) $\dfrac{128}{13}$ (v) $\dfrac{101}{8}$

4. (i) $\dfrac{10}{12}$ and $\dfrac{15}{18}$ (ii) $\dfrac{22}{24}$ and $\dfrac{33}{36}$ (iii) $\dfrac{18}{26}$ and $\dfrac{27}{39}$ (iv) $\dfrac{16}{30}$ and $\dfrac{24}{45}$ (v) $\dfrac{14}{58}$ and $\dfrac{21}{87}$

5. (i) $\dfrac{24}{64}$ (ii) $\dfrac{3}{5}$ (iii) $\dfrac{3}{8}$ (iv) $\dfrac{3}{6}$ (v) $\dfrac{3}{5}$

COMPARISON OF FRACTIONS

For comparing two fraction, we have the following situations

   i. **Situation of Like Fractions**

      When both fractions have a common denominator, then compare the values of the denominators. The fraction with the larger numerator has the larger value.

      For example, $\dfrac{3}{5}$ and $\dfrac{4}{5}$ have the common denominator as 5 and $4 > 3$

      $$\therefore \dfrac{4}{5} > \dfrac{3}{5}$$

   ii. **Situation of Unlike Fractions**

      In this situation we can have two types of fractions:

   a. *When both fractions have a common numerator:*

      We compare the values of denominators. The fraction with the smaller denominator has the larger value.

For example:

In $\frac{5}{9}$ and $\frac{5}{7}$, the common numerator as 5 and 7 < 9.

$$7 < 9$$
$$\therefore \frac{5}{7} > \frac{5}{9}$$

b. **When have both fractions different numerator and denominator;**

Change both the fractions into their respective equivalent fractions with the same denominator, i.e. by finding LCM for both the denominators, change them as the like fractions. Now compare them as like fractions.

For example: To compare $\frac{2}{5}$ and $\frac{3}{4}$; we have LCM of 5 and 4 as 20

Now, $\frac{2}{5} = \frac{2 \times 4}{5 \times 4} = \frac{8}{20}$

$$\frac{3}{4} = \frac{3 \times 5}{4 \times 5} = \frac{15}{20}$$

$$\therefore 15 > 4$$

$$\therefore \frac{15}{20} > \frac{8}{20} \text{ or } \frac{3}{4} > \frac{2}{5}$$

SHORTCUT METHOD

To compare two fractions, simply cross multiply them. The fraction having the greater cross-product overhead is greater.

For example: To compare $\frac{7}{9}$ and $\frac{4}{5}$, we have

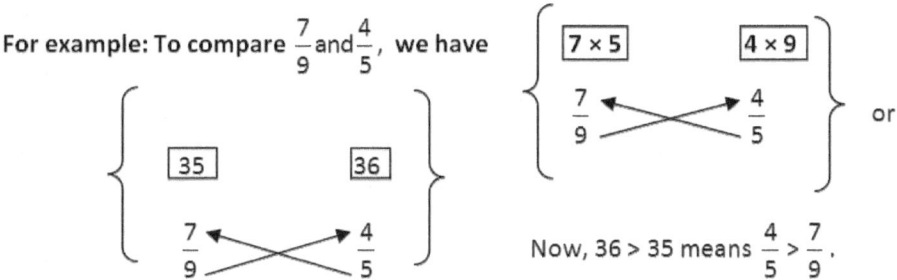

Now, 36 > 35 means $\frac{4}{5} > \frac{7}{9}$.

EXAMPLE

Which fraction is greater?

(a) (i) $\dfrac{3}{7}$ or $\dfrac{5}{7}$ (ii) $\dfrac{7}{9}$ or $\dfrac{4}{9}$ (iii) $\dfrac{8}{11}$ or $\dfrac{5}{11}$ (iv) $\dfrac{21}{101}$ or $\dfrac{92}{101}$

(b) (i) $\dfrac{2}{9}$ or $\dfrac{2}{7}$ (ii) $\dfrac{6}{13}$ or $\dfrac{6}{43}$ (iii) $\dfrac{3}{4}$ or $\dfrac{3}{5}$ (iv) $\dfrac{8}{11}$ or $\dfrac{8}{21}$

(c) (i) $\dfrac{1}{4}$ or $\dfrac{3}{5}$ (ii) $\dfrac{3}{5}$ or $\dfrac{4}{7}$ (iii) $\dfrac{7}{9}$ or $\dfrac{4}{27}$ (iv) $\dfrac{1}{12}$ or $\dfrac{2}{9}$

SOLUTION

a. When denominator is same, the fraction with greater numerator is greater.

(i) ∴ 5 > 3   ∴ $\dfrac{5}{7} > \dfrac{3}{7}$

(ii) ∴ 7 > 4   ∴ $\dfrac{7}{9} > \dfrac{4}{9}$

(iii) ∴ 8 > 5   ∴ $\dfrac{8}{11} > \dfrac{5}{11}$

(iv) ∴ 92 > 21   ∴ $\dfrac{92}{101} > \dfrac{21}{101}$

b. When numerator is same, the fraction with smaller denominator is greater.

(i) ∴ 7 < 9   ∴ $\dfrac{2}{7} > \dfrac{2}{9}$

(ii) ∴ 13 < 43   ∴ $\dfrac{6}{13} > \dfrac{6}{43}$

(iii) ∴ 4 < 5   ∴ $\dfrac{3}{4} > \dfrac{3}{5}$

(iv) ∴ 11 < 21   ∴ $\dfrac{8}{11} > \dfrac{8}{21}$

c. When the given fractions have different numerators and denominators, then they are first changed as like fraction with common denominator equal to the LCM of their denominators.

(i) ∵ LCM of 4 and 5 is 20

$$\therefore \frac{1}{4} = \frac{1 \times 5}{4 \times 5} = \frac{5}{20}$$

$$\frac{2}{5} = \frac{2 \times 4}{5 \times 4} = \frac{8}{20}$$

∴ 8 > 5 $\therefore \frac{8}{20} > \frac{5}{20}$

or $\frac{2}{5} > \frac{1}{4}$

(ii) ∵ LCM of 5 and 7 is 35

$$\therefore \frac{3}{5} = \frac{3 \times 7}{5 \times 7} = \frac{21}{35}$$

$$\frac{4}{7} = \frac{4 \times 5}{7 \times 5} = \frac{20}{35}$$

Since, 21 > 20, $\therefore \frac{21}{35} > \frac{20}{35}$

or $\frac{3}{5} > \frac{4}{7}$

(iii) ∵ LCM of 9 and 27 is 27

$$\therefore \frac{7}{9} = \frac{7 \times 3}{9 \times 3} = \frac{21}{27}$$

$$\frac{4}{27} = \frac{4 \times 1}{27 \times 1} = \frac{4}{27}$$

Since, 21 > 4, $\therefore \frac{21}{27} > \frac{4}{27}$

or $\frac{7}{9} > \frac{4}{27}$

(iv) ∵ LCM of 12 and 9 is 36

$$\therefore \frac{1}{12} = \frac{1 \times 3}{12 \times 3} = \frac{3}{36}$$

$$\frac{2}{9} = \frac{2 \times 4}{9 \times 4} = \frac{8}{36}$$

Since, 8 > 3, $\therefore \frac{8}{36} > \frac{3}{36}$

or $\frac{2}{9} > \frac{3}{36}$

EXAMPLE

Arrange the following fractions in ascending order:

$\frac{7}{9}, \frac{5}{6}, \frac{1}{2}$

SOLUTION

2	2,	9,	6
3	1,	9,	3
	1,	3,	1

LCM = 2 × 3 × 3 = 18

∴ LCM of 2, 9 and 6 = 18

$$\therefore \frac{1}{2} = \frac{1 \times 9}{2 \times 9} = \frac{9}{18}$$

$$\frac{7}{9} = \frac{7 \times 2}{9 \times 2} = \frac{14}{18}$$

$$\frac{5}{6} = \frac{5 \times 3}{6 \times 3} = \frac{15}{18}$$

Since, $\frac{9}{18} < \frac{14}{18} < \frac{15}{18}$ or $\frac{1}{2} < \frac{7}{9} < \frac{5}{6}$

∴ The ascending order of the given fractions is

$$\frac{1}{2}, \frac{7}{9}, \frac{5}{6}$$

EXERCISE - 3

1. Which of the following fractions is smaller?

   (i) $\frac{30}{80}, \frac{23}{80}$ (ii) $\frac{24}{32}, \frac{23}{32}$ (iii) $\frac{14}{15}, \frac{48}{60}$ (iv) $\frac{99}{108}, \frac{11}{36}$

   (v) $\frac{77}{121}, \frac{2}{11}$ (vi) $\frac{118}{144}, \frac{29}{36}$

2. Compare the following fractions:

   (i) $\frac{2}{5}$ and $\frac{3}{4}$ (ii) $\frac{2}{7}$ and $\frac{2}{9}$ (iii) $\frac{1}{4}$ and $\frac{3}{5}$ (iv) $\frac{5}{12}$ and $\frac{3}{4}$

   (v) $\frac{3}{19}$ and $\frac{6}{11}$ (vi) $\frac{10}{11}$ and $\frac{11}{19}$

3. Which fraction is greater in each of the following:

   (i) $\frac{6}{21}$ and $\frac{21}{23}$ (ii) $\frac{3}{7}$ and $\frac{5}{6}$ (iii) $\frac{3}{13}$ and $\frac{5}{39}$ (iv) $\frac{3}{10}$ and $\frac{4}{17}$

   (v) $\frac{15}{35}$ and $\frac{18}{23}$ (vi) $\frac{105}{107}$ and $\frac{107}{109}$

4. Arrange the following in ascending order:

   (i) $\frac{5}{6}, \frac{5}{8}, \frac{3}{4}, \frac{7}{12}, \frac{2}{3}$ (ii) $\frac{3}{4}, \frac{5}{6}, \frac{7}{10}, \frac{4}{5}, \frac{2}{3}$

5. Arrange the following in descending order:
   (i) $\dfrac{1}{2}, \dfrac{17}{20}, \dfrac{4}{5}, \dfrac{9}{10}, \dfrac{3}{4}$
   (ii) $\dfrac{5}{6}, \dfrac{2}{3}, \dfrac{20}{27}, \dfrac{13}{18}, \dfrac{7}{9}$

Answers

1. (i) $\dfrac{23}{80}$ (ii) $\dfrac{23}{32}$ (iii) $\dfrac{48}{60}$ (iv) $\dfrac{11}{36}$ (v) $\dfrac{2}{11}$ (vi) $\dfrac{29}{36}$

2. (i) $\dfrac{2}{5} < \dfrac{3}{4}$ (ii) $\dfrac{2}{7} < \dfrac{2}{9}$ (iii) $\dfrac{1}{4} < \dfrac{3}{5}$ (iv) $\dfrac{5}{12} < \dfrac{3}{4}$
   (v) $\dfrac{3}{19} < \dfrac{6}{11}$ (vi) $\dfrac{10}{11} > \dfrac{11}{19}$

3. (i) $\dfrac{21}{23}$ (ii) $\dfrac{5}{6}$ (iii) $\dfrac{3}{13}$ (iv) $\dfrac{3}{10}$ (v) $\dfrac{18}{23}$ (vi) $\dfrac{107}{109}$

4. (i) $\dfrac{7}{12}, \dfrac{5}{8}, \dfrac{2}{3}, \dfrac{3}{4}, \dfrac{5}{6}$ (ii) $\dfrac{2}{3}, \dfrac{7}{10}, \dfrac{3}{4}, \dfrac{4}{5}, \dfrac{5}{6}$

5. (i) $\dfrac{9}{10}, \dfrac{17}{20}, \dfrac{4}{5}, \dfrac{3}{4}, \dfrac{1}{2}$ (ii) $\dfrac{5}{6}, \dfrac{7}{9}, \dfrac{20}{27}, \dfrac{13}{18}, \dfrac{2}{3}$

Operations on Fractions

Addition of Fraction

1. *Addition for Like Fraction:*

   When fractions with the same denominator are added, then the result is a fraction with the same denominator. For example:
   Sum of $\dfrac{4}{7}$ and $\dfrac{2}{7}$ is given by $\dfrac{4+2}{7}$ or $\dfrac{6}{7}$

2. *Addition for Unlike Fractions:*

   When the denominators are different, the fractions must first be expressed with the same denominator.

For example;
To add $\dfrac{5}{6}$ and $\dfrac{3}{8}$ we have to first convert them as like fractions.

∴ LCM of 6 and 8 is 24

$$\therefore \frac{5}{6} = \frac{5 \times 4}{6 \times 4} = \frac{20}{24} \text{ and } \frac{3}{8} = \frac{3 \times 3}{8 \times 3} = \frac{9}{24}$$

$$\text{Now, } \frac{5}{6} + \frac{3}{8} = \frac{20}{24} + \frac{9}{24} = \frac{20+9}{24} = \frac{29}{24}$$

Since, $\frac{29}{24}$ is an improper fraction, so it can be expressed as a mixed fraction.

i.e. $\frac{29}{24} = 1\frac{5}{24}$

Thus, $\frac{5}{6} + \frac{3}{8} = 1\frac{5}{24}$

EXAMPLE

Add the following fractions:

(i) $\frac{2}{11} + \frac{7}{11}$ (ii) $\frac{7}{9} + \frac{5}{6}$ (iii) $1\frac{1}{2} + \frac{3}{4} + 2\frac{1}{2}$

SOLUTION

(i) $\frac{2}{11} + \frac{7}{11} = \frac{2+7}{11} = \frac{9}{11}$

(ii) $\frac{7}{9} + \frac{5}{6} = \frac{7 \times 2}{9 \times 2} + \frac{5 \times 3}{6 \times 3}$ ∴ LCM of 9 and 6 is 18

$$= \frac{14}{18} + \frac{15}{18} = \frac{14+15}{18} = \frac{29}{18}$$

∴ $\frac{29}{18}$ is an improper fraction, so changing it into a mixed fraction, we get $1\frac{11}{18}$

$$\therefore \frac{7}{9} + \frac{5}{6} = 1\frac{11}{18}$$

(iii) $1\frac{1}{2}+\frac{3}{4}+2\frac{1}{2} = \frac{3}{2}+\frac{3}{4}+\frac{5}{2}$

∴ LCM of 2, 4, and 2 is 4

∴ $\frac{3}{2} = \frac{3\times 2}{2\times 2} = \frac{6}{4}$; $\frac{3}{4} = \frac{3\times 1}{4\times 1} = \frac{3}{4}$ and $\frac{5}{2} = \frac{5\times 2}{2\times 2} = \frac{10}{4}$

Now, $\frac{3}{2}+\frac{3}{4}+\frac{5}{2} = \frac{6}{4}+\frac{3}{4}+\frac{10}{4}$

$= \frac{6+3+10}{4} = \frac{19}{4} = 4\frac{3}{4}$

Thus, $1\frac{1}{2}+\frac{3}{4}+2\frac{1}{2} = 4\frac{3}{4}$

SUBTRACTION OF FRACTIONS

1. **Subtraction for like fraction:**

   When fractions with the same denominators are subtracted, then the result is a fraction with the same denominator.

   For example, to subtract $\frac{1}{4}$ from $\frac{3}{4}$, we have $\frac{3}{4}-\frac{1}{4} = \frac{3-1}{4}$

   $= \frac{2}{4}$

   Since, the answer should be expressed by a fraction in the simplest form,

   ∴ $\frac{2}{4} = \frac{\cancel{2}^{1}}{\cancel{4}_{2}} = \frac{1}{2}$

   Thus, $\frac{3}{4}-\frac{1}{4} = \frac{1}{2}$

2. **Subtraction for unlike fractions:**

   When denominators are different, the fractions must first be expressed with the same denominator.

   For example, to subtract $\frac{7}{9}$ from $\frac{4}{5}$, we have to change the given fractions as like fractions.

∴ LCM of 9 and 5 is 45

$$\therefore \frac{7}{9} = \frac{7 \times 5}{9 \times 5} = \frac{35}{45} \text{ and } \frac{4}{5} = \frac{4 \times 9}{5 \times 9} = \frac{36}{45}$$

$$\text{Now, } \frac{4}{5} - \frac{7}{9} = \frac{36}{45} - \frac{35}{45}$$

$$= \frac{36 - 35}{45} = \frac{1}{45}$$

EXAMPLE

Subtract the following fractions:

(i) $\frac{3}{5} - \frac{2}{5}$ (ii) $\frac{7}{10} - \frac{4}{15}$ (iii) $7\frac{4}{9} - 3\frac{5}{8}$

SOLUTION:

(i) $\frac{3}{5} - \frac{2}{5} = \frac{3-2}{5} = \frac{1}{5}$

(ii) $\frac{7}{10} - \frac{4}{15}$

$$= \frac{7 \times 3}{10 \times 3} - \frac{4 \times 2}{15 \times 2}$$

$$= \frac{21}{30} - \frac{8}{30}$$

$$= \frac{21-8}{30} = \frac{13}{30}$$ LCM of 10 and 15 is 30

(iii) $7\frac{4}{9} - 3\frac{5}{8} = \frac{67}{9} - \frac{29}{8}$

Now, LCM of 9 and 8 is 72

$$\therefore \frac{67}{9} = \frac{67}{9} \times \frac{8}{8} = \frac{536}{72}$$

$$\frac{29}{8} = \frac{29 \times 9}{8 \times 9} = \frac{261}{72}$$

$$\therefore 7\frac{4}{9} = \frac{(7 \times 9) + 4}{8} = \frac{67}{9}$$

$$3\frac{5}{8} = \frac{(3 \times 8) + 5}{8} = \frac{29}{8}$$

Therefore,

$$\frac{67}{9} - \frac{29}{8} = \frac{536}{72} - \frac{261}{72}$$

$$= \frac{536-261}{72} = \frac{275}{72} = 3\frac{59}{72}$$

Thus, $7\frac{4}{9} - 3\frac{5}{8} = 3\frac{59}{72}$

EXAMPLE

A fruit grower uses $\frac{1}{3}$ of his land for peers $\frac{3}{8}$ for apples and $\frac{1}{6}$ for plums. Find the total area of land used to grow the fruit plants?

SOLUTION

Area of land used for peers = $\frac{1}{3}$

Area of land used for apples = $\frac{3}{8}$

Area of land used for plums = $\frac{1}{6}$

∴ Total area used $= \frac{1}{3} + \frac{3}{8} + \frac{1}{6}$ ∵ LCM of 3, 8 and 6 is 24

$$= \left(\frac{1\times 8}{3\times 8}\right) + \left(\frac{3\times 3}{8\times 3}\right) + \left(\frac{1\times 4}{6\times 4}\right)$$

$$= \frac{8}{24} + \frac{9}{24} + \frac{4}{24}$$

$$= \frac{8+9+4}{24} = \frac{21}{24} = \frac{\overset{7}{\cancel{21}}}{\underset{8}{\cancel{24}}} = \frac{7}{8}$$

Thus, the fruit grower used $\frac{7}{8}$ of land to grow fruit plants.

EXAMPLE

Ramita ate $\frac{3}{7}$ of a chocolate chip cookie. If Anjum ate $\frac{1}{3}$ of it, then how much cookie is left?

SOLUTION

Total cookie eaten

$$= \frac{3}{7} + \frac{1}{3}$$

$$= \left(\frac{3\times 3}{7\times 3}\right) + \left(\frac{1}{3}\times\frac{7}{7}\right) \qquad \therefore \text{LCM of 7 and 3 is 21}$$

$$= \frac{9}{21} + \frac{7}{21} = \frac{16}{21}$$

So, the left out part cookie

$$= 1 - \frac{16}{21}$$

$$= \frac{21-16}{21} \qquad = \frac{5}{21}$$

EXAMPLE

What must be subtracted from $6\frac{1}{12}$ to get $3\frac{5}{8}$?

SOLUTION

The required number is the difference obtained by subtracting $3\frac{5}{8}$ from $6\frac{1}{12}$

$$\therefore 6\frac{1}{12} - 3\frac{5}{8} = \frac{73}{12} - \frac{29}{8}$$

2	12,	8
2	6,	4
	3,	2

LCM = 2 × 2 × 3 × 2 = 24

$$= \left[\frac{73\times 2}{12\times 2}\right] - \left[\frac{29\times 3}{8\times 3}\right]$$

$$= \frac{146}{24} - \frac{87}{24} = \frac{146-87}{24}$$

$$= \frac{59}{24} \text{ or } 2\frac{11}{24}$$

So, the required fraction $= 2\frac{11}{24}$

Exercise - 4

1. Add the following fractions:

    (i) $\dfrac{3}{7}+\dfrac{2}{7}$ (ii) $\dfrac{5}{9}+\dfrac{8}{9}$ (iii) $\dfrac{3}{8}+\dfrac{5}{6}$ (iv) $\dfrac{17}{30}+\dfrac{2}{15}$

    (v) $2\dfrac{5}{6}+4\dfrac{11}{18}$ (vi) $4\dfrac{7}{10}+3\dfrac{2}{5}+2\dfrac{3}{4}$

2. Subtract:

    (i) $\dfrac{5}{8}$ from $\dfrac{13}{8}$ (ii) $\dfrac{32}{41}$ from $\dfrac{63}{41}$ (iii) $\dfrac{1}{12}$ from $\dfrac{3}{4}$

    (iv) $\dfrac{11}{18}$ from $\dfrac{5}{6}$ (v) $\dfrac{5}{24}$ from $\dfrac{7}{12}$ (vi) $2\dfrac{3}{4}$ from $6\dfrac{5}{8}$

3. Solve the following:

    (i) $\dfrac{7}{10}+\dfrac{1}{10}-\dfrac{2}{15}-\dfrac{1}{3}$ (ii) $4\dfrac{1}{6}+2\dfrac{5}{8}+6\dfrac{5}{12}-5\dfrac{3}{4}$

    (iii) $7\dfrac{1}{3}-5\dfrac{1}{2}-1\dfrac{2}{9}+4\dfrac{5}{6}$

4. Rambir ran $3\dfrac{5}{8}$ km on Monday, $4\dfrac{1}{8}$ km on Tuesday and $5\dfrac{1}{4}$ km on Wednesday. What distance did he run for the three days altogether?

5. Ganeshi bought $10\dfrac{1}{2}$ kg of milk from Ram singh and $8\dfrac{3}{4}$ kg from Ramji Dass. He mixed the two quantities and sold $12\dfrac{2}{3}$ kg to a tea stall and remaining to a sweets seller. How much milk did he sell to the sweet maker?

6. What must be added to the sum of $3\dfrac{1}{2}$ and $4\dfrac{5}{6}$ to make $15\dfrac{1}{2}$?

7. How much less than 7 is the sum of $2\dfrac{4}{5}$ and $2\dfrac{3}{4}$?

Answers

1. (i) $\dfrac{5}{7}$ (ii) $1\dfrac{4}{9}$ (iii) $1\dfrac{5}{24}$ (iv) $\dfrac{7}{10}$ (v) $7\dfrac{4}{9}$ (vi) $10\dfrac{17}{20}$

2. (i) 1 (ii) $\dfrac{31}{41}$ (iii) $\dfrac{2}{3}$ (iv) $\dfrac{2}{9}$ (v) $\dfrac{3}{8}$ (vi) $3\dfrac{7}{8}$

3. (i) $\dfrac{1}{3}$ (ii) $2\dfrac{5}{24}$ (iii) $5\dfrac{4}{9}$ 4. 13 km   5. $6\dfrac{7}{12}$ kg

6. $7\dfrac{1}{6}$ 7. $\dfrac{29}{20}$

MULTIPLICATION OF FRACTIONS

To multiply a fraction by a fraction, multiply the two numerators together to make the numerator of the product, and the two denominators to make its denominator

For example:

$$\dfrac{3}{4} \times \dfrac{5}{7} = \dfrac{3 \times 5}{4 \times 7} = \dfrac{15}{28}$$

When the numerator and the denominator of the product have common factors, we cancel them both.

For example:

$$\dfrac{7}{16} \times \dfrac{20}{21} = \dfrac{\cancel{7}^{1}}{\cancel{16}_{4}} \times \dfrac{\cancel{20}^{5}}{\cancel{21}_{3}}$$

7 and 21 are both divided by 7 and 16 and 20 both by 4.

If mixed numbers occur, they must be converted into improper fractions before any multiplication is done.

For example:

$$3\dfrac{1}{2} \times 2\dfrac{1}{7} \times \dfrac{14}{25}$$

$$= \dfrac{7}{2} \times \dfrac{15}{7} \times \dfrac{14}{25} = \dfrac{\cancel{7}^{1}}{\cancel{2}_{1}} \times \dfrac{\cancel{15}^{3}}{\cancel{7}_{1}} \times \dfrac{\cancel{14}^{7}}{\cancel{25}_{5}}$$

$$= \dfrac{1 \times 3 \times 7}{1 \times 1 \times 5} = \dfrac{21}{5} \text{ or } 4\dfrac{1}{5}$$

7 and 7 are divided by 7, 15 and 25 are divided by 5, 2 and 14 are divided by 2.

Example

Find the product of the following fractions:

(i) $\dfrac{3}{4} \times \dfrac{6}{7}$ (ii) $\dfrac{5}{6} \times \dfrac{7}{25} \times \dfrac{18}{35}$ (iii) $7\dfrac{1}{2} \times 14\dfrac{1}{2}$ (iv) $2\dfrac{1}{2} \times 3\dfrac{2}{5} \times 1\dfrac{7}{8}$

Solution

(i) $\dfrac{3}{4} \times \dfrac{6}{7}$

$= \dfrac{3}{\cancel{4}_{2}} \times \dfrac{\cancel{6}^{3}}{7}$

$= \dfrac{3 \times 3}{2 \times 7} = \dfrac{9}{14}$

(ii) $\dfrac{5}{6} \times \dfrac{7}{25} \times \dfrac{18}{35}$

$= \dfrac{\cancel{5}^{1}}{\cancel{6}_{1}} \times \dfrac{\cancel{7}^{1}}{\cancel{25}_{5}} \times \dfrac{\cancel{18}^{3}}{\cancel{35}_{5}}$

$= \dfrac{1 \times 1 \times 3}{1 \times 5 \times 5} = \dfrac{3}{25}$

(iii) $7\dfrac{1}{2} \times 14\dfrac{1}{2}$

$= \dfrac{15}{2} \times \dfrac{29}{2}$

$= \dfrac{15 \times 29}{2 \times 2} = \dfrac{435}{4} = 108\dfrac{3}{4}$

(iv) $2\dfrac{1}{2} \times 3\dfrac{2}{5} \times 1\dfrac{7}{8}$

$= \dfrac{5}{2} \times \dfrac{17}{5} \times \dfrac{15}{8}$

$= \dfrac{\cancel{5}^{1}}{2} \times \dfrac{17}{\cancel{5}_{1}} \times \dfrac{15}{8} = \dfrac{1 \times 17 \times 15}{2 \times 1 \times 8}$

$= \dfrac{255}{16} = 15\dfrac{15}{16}$

Reciprocal or Multiplicative Inverse of a Fraction

Two numbers are called reciprocal of one another, if their product is 1.

FOR EXAMPLE

The reciprocal of $\frac{5}{7}$ is $\frac{7}{5}$

And reciprocal of $\frac{7}{5}$ is $\frac{5}{7}$

$$\left\{ \therefore \frac{7}{5} \times \frac{5}{7} = \frac{\cancel{7}^1}{\cancel{5}_1} \times \frac{\cancel{5}^1}{\cancel{7}_1} = 1 \right.$$

The reciprocal 7 is $\frac{1}{7}$, and the reciprocal of $\frac{1}{7}$ is 7 $\left\{ \therefore 7 \times \frac{1}{7} = 1 \right\}$

Thus, if the numerator and denominator of a fraction are interchanged, then the new fraction formed is called the *reciprocal* of the given fraction.

For example,

The reciprocal of 3 is $\frac{1}{3}$ $\left[\therefore \text{'3' mean } \frac{3}{1} \text{ and } 3 \times \frac{1}{3} = 1 \right]$

The reciprocal of $1\frac{3}{5}$ i.e. $\frac{8}{5}$ is $\frac{5}{8}$

The reciprocal of $\frac{7}{11}$ is $\frac{11}{7}\left(\text{or } 1\frac{4}{7}\right)$

REMEMBER

i. **The reciprocal of a natural number is always a unit fraction.**
ii. **The reciprocal of unit fraction is always a natural number**
iii. **The reciprocal of a mixed number (or an improper fraction) is a proper fraction and vice-versa**
iv. **'1' is called the identity element of multiplication of fractions.**

EXAMPLE

Find the reciprocal of following:

(i) 5 (ii) $3\frac{2}{5}$ (iii) $\frac{11}{13}$

SOLUTION

(i) Reciprocal of 5 is $\frac{1}{5}$

(ii) Reciprocal of $3\frac{2}{5}$ i.e. $\frac{17}{5}$ is $\frac{5}{17}$

$$\left[\therefore 3\frac{2}{5} = \frac{(3 \times 5) + 2}{5} = \frac{17}{5} \right]$$

(iii) Reciprocal of $\dfrac{11}{13}$ is $\dfrac{13}{11}$ or $1\dfrac{2}{11}$

Did You Know?

'0' does not have a reciprocal, because its product with any fraction cannot be 1.

Division of Fractions

To divide a fraction by another fraction, we multiply the first fraction (dividend) by the reciprocal of the second fraction (divisor).

Method

i. Before division, convert the mixed fractions (if any) into improper fraction.
ii. Get the reciprocal of the divisor.
iii. Multiply the dividend and the reciprocal of the divisor.
iv. Reduce the product to the lowest terms and to a mixed number (if required).

Example

Find the quotient of the following:

(i) $\dfrac{4}{3} \div 1\dfrac{3}{5}$ (ii) $2\dfrac{1}{4} \div \dfrac{2}{7}$ (iii) $3\dfrac{3}{4} \div 2\dfrac{1}{4}$ (iv) $2\dfrac{5}{8} \div 8\dfrac{1}{6}$

Solution

(i) $\because$ Reciprocal of $1\dfrac{3}{5}$ or $\dfrac{8}{5}$ is $\dfrac{5}{8}$

$\therefore \dfrac{4}{3} \div \dfrac{8}{5} = \dfrac{4}{3} \times \dfrac{5}{8} = \dfrac{\cancel{4}^{1}}{3} \times \dfrac{5}{\cancel{8}_{2}} = \dfrac{1 \times 5}{3 \times 2} = \dfrac{5}{6}$

(ii) $\therefore 2\dfrac{1}{4} = \dfrac{9}{4}$ and reciprocal of $\dfrac{2}{7}$ is $\dfrac{7}{2}$

$\therefore 2\dfrac{1}{4} \div \dfrac{2}{7} = \dfrac{9}{4} \times \dfrac{7}{2}$

$= \dfrac{9 \times 7}{4 \times 2} = \dfrac{63}{8}$ or $7\dfrac{7}{8}$

(iii) $\because 3\frac{3}{4} = \frac{15}{4}$ and $2\frac{1}{4} = \frac{9}{4}$; Also reciprocal of $\frac{9}{4}$ is $\frac{4}{9}$

$\therefore 3\frac{3}{4} \div 2\frac{1}{4} = \frac{15}{4} \div \frac{9}{4}$

$= \frac{15}{4} \times \frac{4}{9} = \frac{\cancel{15}^{5}}{\cancel{4}_{1}} \times \frac{\cancel{4}^{1}}{\cancel{9}_{3}}$

$= \frac{5 \times 1}{1 \times 3} = \frac{5}{3}$ or $1\frac{2}{3}$

(iv) $\because 2\frac{5}{8} = \frac{21}{8}$, $8\frac{1}{6} = \frac{49}{6}$ and reciprocal of $\frac{49}{6}$ is $\frac{6}{49}$

$\therefore 2\frac{5}{8} \div 8\frac{1}{6} = \frac{21}{8} \div \frac{49}{6}$

$= \frac{21}{8} \times \frac{6}{49} = \frac{\cancel{21}^{3}}{\cancel{8}_{4}} \times \frac{\cancel{6}^{3}}{\cancel{49}_{7}}$

$= \frac{3 \times 3}{4 \times 7} = \frac{9}{28}$

Using 'Of' Along with Multiplication and Division

In a mathematical sentence, 'of' means a part of something. The word 'of' between two numbers or fractions may be a thought of multiplication.

For example

$\frac{3}{4}$ of $\frac{8}{15}$ means $\frac{\cancel{3}^{1}}{\cancel{4}_{1}} \times \frac{\cancel{8}^{2}}{\cancel{15}_{5}} = \frac{2}{5}$

☞**Caution**

When 'of' is involved along with other operation (×, ÷ etc), then 'of' must be simplified before '÷' and '×'

For example, In $\frac{1}{2}$ of $\frac{3}{4} \div \frac{5}{8} \times \frac{4}{3}$; the operation 'of' is performed before ÷ and ×.

i.e. $\frac{1}{2}$ of $\frac{3}{4} \div \frac{5}{8} \times \frac{4}{3}$

$= \frac{3}{8} \div \frac{5}{8} \times \frac{4}{3}$

$= \frac{\cancel{3}^{1}}{\cancel{8}_{1}} \times \frac{\cancel{8}^{1}}{5} \times \frac{4}{\cancel{3}_{1}}$

$= \frac{1 \times 1 \times 4}{1 \times 5 \times 1} = \frac{4}{5}$

SIMPLIFICATION OF FRACTIONS

To simplify the fractions involving various operations (+, −, ×, ÷, of) etc., we make use of a specific order. The order of letters in the word BODMAS help us to remember the order in which the operations are performed.

The first letter 'B' means, the bracket is to be removed first. Brackets are removed in the order as [ ], { } and ( ). Removing a bracket mean to condense all fraction inside the bracket as a single fraction.

The second letter 'O' means 'of', then 'D' means division, 'M' means multiplication, 'A' means addition and 'S' means subtraction.

Example: Simplify the following:

(i) $\dfrac{5}{6} + \dfrac{3}{2} \times \dfrac{1}{2} \div \dfrac{7}{8}$ (ii) $15 + \left(\dfrac{1}{2} \text{ of } \dfrac{3}{4} \div \dfrac{1}{4}\right) - \dfrac{17}{8}$

(iii) $\left(\dfrac{4}{7} \div \dfrac{6}{14}\right) \text{ of } \dfrac{9}{16} + \left[10 - \left\{\dfrac{1}{8} + 3\dfrac{1}{2} \times \left(\dfrac{3}{4} - \dfrac{1}{2}\right)\right\}\right]$

Solution

(i) $\dfrac{5}{6} + \dfrac{3}{2} \times \dfrac{1}{2} \div \dfrac{7}{8}$

$= \dfrac{5}{6} + \dfrac{3}{2} \times \dfrac{1}{2} \times \dfrac{8}{7}$

$= \dfrac{5}{6} + \dfrac{6}{7}$ ∴ LCM of 6 and 7 is 42

$= \dfrac{(5 \times 7) + (6 \times 6)}{42}$

$= \dfrac{35 + 36}{42} = \dfrac{71}{42} \text{ or } 1\dfrac{29}{42}$

(ii) $15 + \left(\dfrac{1}{2} \text{ of } \dfrac{3}{4} \div \dfrac{1}{4}\right) - \dfrac{17}{8}$

$= 15 + \left(\dfrac{3}{8} \div \dfrac{1}{4}\right) - \dfrac{17}{8}$

$$= 15 + \frac{3}{\cancel{8}_2} \times \frac{\cancel{4}^1}{1} - \frac{17}{8}$$

$$= 15 + \frac{3}{2} - \frac{17}{8}$$

$$= \frac{(15 \times 8) + (3 \times 4) - (17 \times 1)}{8}$$

$$= \frac{120 + 12 - 17}{8}$$

$$= \frac{132 - 17}{8} = \frac{115}{8} \text{ or } 14\frac{3}{8}$$

$\therefore$ LCM of 2 and 8 is 8 and 15 means $\frac{15}{1}$

(iii) $\left(\frac{4}{7} \div \frac{6}{14}\right)$ of $\frac{9}{16} + \left[10 - \left\{\frac{1}{8} + 3\frac{1}{2} \times \left(\frac{3}{4} - \frac{1}{2}\right)\right\}\right]$

$= \left(\frac{4}{7} \times \frac{14}{6}\right)$ of $\frac{9}{16} + \left[10 - \left\{\frac{1}{8} + \frac{7}{2} \times \left(\frac{3-2}{4}\right)\right\}\right]$

$= \left(\frac{\cancel{4}^2}{\cancel{7}_1} \times \frac{\cancel{14}^2}{\cancel{6}_3}\right)$ of $\frac{9}{16} + \left[10 - \left\{\frac{1}{8} + \frac{7}{2} \times \frac{1}{4}\right\}\right]$

$= \frac{\cancel{4}^1}{\cancel{3}_1}$ of $\frac{\cancel{9}^3}{\cancel{16}_4} + \left[10 - \left\{\frac{1}{8} + \frac{7}{8}\right\}\right]$

$= \frac{3}{4} + \left[10 - \left\{\frac{1+7}{8}\right\}\right]$

$= \frac{3}{4} + [10 - 1] \qquad = \frac{3}{4} + 9 \qquad = 9\frac{3}{4}$

EXERCISE - 5

1. Find the product of the following:

   (i) $\frac{10}{13} \times \frac{3}{5}$ (ii) $\frac{20}{21} \times \frac{9}{16} \times \frac{7}{25}$ (iii) $\frac{5}{33} \times 2\frac{2}{7} \times 5\frac{1}{2}$ (iv) $\frac{1}{14} \times 4\frac{1}{2} \times 2\frac{1}{3}$

   (v) $\frac{4}{9} \times 1\frac{1}{5} \times 3\frac{3}{4} \times \frac{5}{14}$ (vi) $\frac{8}{15} \times 1\frac{3}{46} \times \frac{9}{14} \times 2\frac{1}{7} \times 1\frac{5}{18}$

2. Find the reciprocal of each of the following fractions:

   (i) $\frac{1}{9}$ (ii) $\frac{25}{37}$ (iii) $\frac{18}{7}$ (iv) $3\frac{4}{5}$ (v) $5\frac{3}{4}$

3. Find the quotients of the following:

   (i) $4\dfrac{1}{2} \div \dfrac{3}{7}$ (ii) $4\dfrac{1}{5} \div 1\dfrac{5}{16}$ (iii) $6\dfrac{3}{10} \div 1\dfrac{5}{16}$ (iv) $3\dfrac{1}{5} \div 4\dfrac{4}{5}$

   (v) $3\dfrac{1}{5} \div 6\dfrac{2}{3}$ (vi) $9\dfrac{2}{7} \div 3\dfrac{9}{10}$

4. Simplify the following expressions:

   (i) $\left[2\dfrac{5}{7} \times \dfrac{7}{19} + 3\dfrac{1}{2} - \dfrac{1}{2} \div 2 \right] \div 12\dfrac{3}{4}$

   (ii) $1\dfrac{7}{30} + \dfrac{3}{5} - \left(\dfrac{4}{5} - \dfrac{2}{5} \right) \div \dfrac{6}{5} \times \dfrac{3}{2} - \dfrac{13}{15}$

   (iii) $\dfrac{11}{12} + \left[\dfrac{1}{3} \times 8 \left\{ \dfrac{2}{5} \text{ of } \dfrac{15}{8} - \left(3 - 2\dfrac{1}{2} \right) \right\} \right] + \dfrac{2}{3}$

   (iv) $\left[\left\{ \left(2\dfrac{3}{4} + 1\dfrac{1}{4} \right) \div \left(3\dfrac{1}{3} + 2\dfrac{2}{3} \right) \right\} \div \dfrac{2}{3} \right] - \dfrac{3}{7} \text{ of } \dfrac{14}{15}$

   (v) $\left[\left\{ \left(10\dfrac{2}{5} - 9\dfrac{1}{5} \right) \div \left(11\dfrac{3}{5} - 9\dfrac{2}{5} \right) \right\} \text{ of } 1\dfrac{5}{6} \right] - \dfrac{3}{7} \text{ of } 2\dfrac{1}{3}$

ANSWERS

1. (i) $\dfrac{6}{13}$ (ii) $\dfrac{3}{20}$ (iii) $1\dfrac{19}{21}$ (iv) $\dfrac{3}{4}$ (v) $\dfrac{5}{7}$ (vi) 1

2. (i) 9   (ii) $\dfrac{37}{25}$ (iii) $\dfrac{7}{18}$ (iv) $\dfrac{5}{19}$ (v) $\dfrac{4}{23}$

3. (i) $10\dfrac{1}{2}$ (ii) $3\dfrac{1}{5}$ (iii) $4\dfrac{4}{5}$ (iv) $\dfrac{2}{3}$ (v) $\dfrac{12}{25}$ (vi) $2\dfrac{8}{21}$

4. (i) $\dfrac{1}{3}$ (ii) $\dfrac{7}{15}$ (iii) $2\dfrac{1}{4}$ (iv) $\dfrac{3}{5}$ (v) 0

MISCELLANEOUS EXERCISE

1. Write down the following fractions in simplest form:

   (i) $\dfrac{28}{63}$ (ii) $\dfrac{108}{192}$ (iii) $\dfrac{120}{480}$ (iv) $\dfrac{99}{110}$

2. Write as mixed numbers.

   (i) $\dfrac{79}{9}$ (ii) $\dfrac{87}{8}$ (iii) $\dfrac{207}{13}$ (iv) $\dfrac{623}{24}$

3. Fill in the blanks using the symbol >, = or <

   (i) $\dfrac{4}{5}$ ----- $\dfrac{6}{7}$ (ii) $\dfrac{25}{48}$ ----- $\dfrac{31}{72}$ (iii) $\dfrac{29}{30}$ ----- $\dfrac{21}{20}$ (iv) $\dfrac{105}{106}$ ----- $\dfrac{101}{107}$

4. Arrange in ascending order:

   (i) $\dfrac{1}{3}, \dfrac{3}{5}, \dfrac{5}{12}, \dfrac{4}{7}$ (ii) $\dfrac{1}{2}, \dfrac{1}{3}, \dfrac{1}{5}, \dfrac{1}{4}$

5. Arrange in descending order:

   (i) $\dfrac{1}{8}, \dfrac{5}{16}, \dfrac{3}{8}, \dfrac{1}{4}$ (ii) $\dfrac{1}{6}, \dfrac{3}{4}, \dfrac{5}{12}, \dfrac{2}{3}$

6. Which of $\dfrac{2}{5}$ and $\dfrac{16}{47}$ is smaller?

7. Find the quotient:

   (i) $\left(6\dfrac{1}{2} - 3\dfrac{1}{2}\right) \div \left(7\dfrac{3}{4} - 2\dfrac{1}{4}\right)$ (ii) $\left(12 - 2\dfrac{3}{4}\right) \div \left(8\dfrac{1}{2} - 4\dfrac{1}{4}\right)$

8. Find the quotient:

   (i) $\left(2 - \dfrac{3}{4}\right) \div \left(3 - \dfrac{1}{2}\right)$ (ii) $\left(4 - \dfrac{4}{5}\right) \div \left(5 - \dfrac{3}{4}\right)$

9. What is $\dfrac{3}{8}$ of $\left(\dfrac{8}{9} - \dfrac{5}{6}\right)$?

10. What is $\dfrac{300}{811}$ of $\left(\dfrac{99}{100} - \dfrac{89}{90} + \dfrac{9}{10}\right)$?

11. Simplify:

    $$\dfrac{2}{3} + \left[10\dfrac{11}{12} - 11\dfrac{5}{6} + 13\dfrac{1}{3} - 12\dfrac{3}{4}\right]$$

12. Simplify:

    $$\dfrac{9}{11} + \left[\left\{\left(1\dfrac{1}{4} + 2\dfrac{3}{4}\right) \div \left(2\dfrac{2}{3} + 3\dfrac{1}{3}\right)\right\} \div \dfrac{2}{3}\right] + \dfrac{2}{11}$$

13. There 40 students in a class. Each of the students in the class contribute Rs. $10\frac{1}{2}$ to class fund. $\frac{11}{14}$ of the total fund was used to donate for relief-fund. How much money is left in the class fund?

14. In class of 40 students 15 are girls. $\frac{3}{5}$ of the girls and $\frac{2}{5}$ of the boys went to visit "world book Fair" at pragati maidan. How many students did not go to "world book Fair"?

15. What must be added to the product of $1\frac{7}{9}$ and $3\frac{3}{4}$ to make a total $19\frac{1}{3}$?

Answers

1. (i) $\frac{4}{9}$ (ii) $\frac{9}{16}$ (iii) $\frac{1}{4}$ (iv) $\frac{9}{10}$

2. (i) $8\frac{7}{9}$ (ii) $10\frac{7}{8}$ (iii) $15\frac{12}{13}$ (iv) $25\frac{23}{24}$

3. (i) < (ii) > (iii) < (iv) >

4. (i) $\frac{1}{3}, \frac{5}{12}, \frac{4}{7}, \frac{3}{5}$ (ii) $\frac{1}{5}, \frac{1}{4}, \frac{1}{3}, \frac{1}{2}$

5. (i) $\frac{3}{8}, \frac{5}{16}, \frac{1}{4}, \frac{1}{8}$ (ii) $\frac{3}{4}, \frac{2}{3}, \frac{5}{12}, \frac{1}{6}$ 6. $\frac{16}{47}$

7. (i) $\frac{6}{11}$ (ii) $2\frac{3}{17}$ 8. (i) $\frac{1}{2}$ (ii) $\frac{64}{85}$

9. $\frac{1}{48}$ 10. $\frac{1}{3}$ 11. $\frac{2}{9}$ 12. 2

13. Rs. 90   14. 21   15. $12\frac{2}{3}$

Hots

1. One-fourth of a herd of deer have gone to the forest and one-third of the total are grazing in a nearly field. The remaining 25 are drinking water on a river bank. What is the total number of deer.

2. Find the fraction which goes $4\frac{1}{2}$ times into 6 and leaves a remainder of $\frac{6}{7}$.

Answers

1. 60 deer   2. $1\frac{1}{7}$

Mental Maths

1. What is $\frac{1}{5}$ of 3 km in metres?
2. What fraction of 1 minute is 15 seconds?
3. What is $\frac{1}{7}$ of 2.1 kg in grammes?
4. What fraction of 3 km is 600 m?
5. What is the simplest form of $\frac{24}{32}$?
6. What is the lowest form of $\frac{84}{112}$?
7. Write fraction with denominator 24 equivalent to $\frac{2}{3}$.
8. Write a fraction with numerator 20 equivalent to $\frac{5}{8}$.
9. Write $\frac{113}{12}$ as a mixed fraction?
10. Write $9\frac{7}{16}$ as an improper fraction?
11. What is the sum of $\frac{7}{9}$ and $\frac{2}{9}$?
12. What is the difference between $\frac{7}{12}$ and $\frac{5}{12}$?
13. What is $\frac{2}{3}$ of $\frac{4}{5}$?
14. What is the reciprocal of $1\frac{1}{2}$?
15. Write the quotient when $1\frac{2}{3}$ is divided by 5?

Answers

1. 600 m   2. $\frac{1}{4}$ 3. 300 gm   4. $\frac{1}{5}$ 5. $\frac{3}{4}$ 6. $\frac{3}{4}$ 7. $\frac{16}{24}$ 8. $\frac{20}{32}$

9. $9\frac{5}{12}$ 10. $\frac{151}{16}$ 11. 1   12. $\frac{1}{6}$ 13. $\frac{8}{15}$ 14. $\frac{2}{3}$ 15. $\frac{1}{3}$

Multi Choice Questions

1. Shaded portion of which of the following shape represent 'one-fourth'?

   (i) 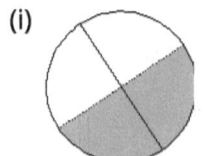   (ii) 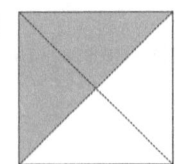   (iii)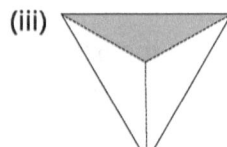

(iv)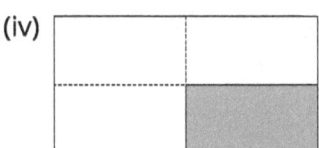

2. How much $\frac{21}{50}$ is less than $\frac{16}{35}$?

   (i) $\frac{13}{700}$ (ii) $\frac{13}{350}$ (iii) $\frac{1}{35}$ (iv) $\frac{74}{350}$

3. Which of the following is the mixed fraction for $\frac{139}{5}$?

   (i) $27\frac{1}{5}$ (ii) $27\frac{2}{5}$ (iii) $27\frac{3}{5}$ (iv) $27\frac{4}{5}$

4. Which of the following is the improper fraction for $26\frac{2}{5}$?

   (i) $\frac{132}{5}$ (ii) $\frac{131}{5}$ (iii) $\frac{130}{5}$ (iv) $\frac{129}{5}$

5. Which of the following is equivalent to $\frac{3}{5}$?

   (i) $\frac{625}{375}$ (ii) $\frac{5}{3}$ (iii) $\frac{375}{625}$ (iv) $\frac{525}{315}$

6. Which of the following in ascending order?

   (i) $\frac{1}{2},\frac{1}{3},\frac{1}{4}$ (ii) $\frac{1}{4},\frac{1}{3},\frac{1}{2}$ (iii) $\frac{1}{3},\frac{1}{2},\frac{1}{4}$ (iv) $\frac{1}{3},\frac{1}{4},\frac{1}{2}$

7. Which of the following is the sum of $\frac{3}{4}$ and $\frac{1}{4}$?

   (i) $\frac{3}{4}$ (ii) $\frac{4}{8}$ (iii) 1 (iv) $\frac{1}{2}$

8. Which of the following is the sum of $\frac{7}{10}$ and $\frac{3}{20}$?

   (i) $\frac{1}{2}$ (ii) $\frac{20}{13}$ (iii) $\frac{17}{20}$ (iv) $\frac{20}{17}$

9. Which of the following is $\frac{1}{2}+\frac{3}{4}-1\frac{1}{4}$

   (i) 1 (ii) $1\frac{1}{2}$ (iii) $\frac{3}{4}$ (iv) 0

10. Which of the following is equal to $\frac{1}{2} \times \frac{2}{3} \times \frac{3}{4}$?

   (i) $\frac{1}{2}$ (ii) $\frac{1}{3}$ (iii) $\frac{1}{4}$ (iv) 1

ANSWERS

1. (iv)   2. (ii)   3. (iv)   4. (i)   5. (iii)   6. (ii)   7. (iii)   8. (iii)   9. (iv)   10. (iii)

WORK SHEET

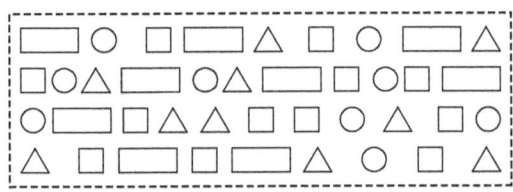

Look at the above collection of shapes and answer the following:

(i) What fraction of the collection are △?

(ii) What fraction of the collection are ○?

(iii) What fraction of the collection are □?

(iv) What fraction of the collection are △ and □ together?

(v) What fraction of the collection are □ and ○ together?

(vi) What fraction of the collection are ○, □ and △ taken together?

(vii) What fraction of the collection is □ more than △?

(viii) What fraction of the collection is ○ less than △?

(ix) What fraction of [○ + □ + □] is △?

(x) What fraction of □ is △?

2. Reduce the following fractions to the lowest terms:
   $$\frac{9}{27}, \frac{36}{48}, \frac{20}{35}, \frac{52}{78}, \frac{24}{30}$$

   Hence, arrange the above fractions in ascending order.

3. Arrange the following fractions in descending order:
   $$\frac{5}{6}, \frac{5}{8}, \frac{2}{3}, \frac{7}{12}, \frac{3}{4}$$

   and reduce them with the denominator 48.

4. Match the column A with column B

Column A		Column B
(i) The lowest terms of $\dfrac{441}{567}$:	(a)	1
(ii) The improper fraction for $3\dfrac{7}{13}$:	(b)	$\dfrac{3}{5}$
(iii) The sum of $\dfrac{1}{2}$ and product of $\dfrac{3}{4}$ with $\dfrac{2}{3}$:	(c)	$\dfrac{2}{3}$.
(iv) The product of $7\dfrac{1}{5}$ and $\left(\dfrac{3}{4}-\dfrac{2}{3}\right)$:	(d)	$\dfrac{7}{9}$
(v) $\left(\dfrac{5}{6}-\dfrac{2}{9}\right)$ of $\left[\text{The reciprocal of }\left(3\dfrac{2}{3}\div 4\right)\right]$	(e)	$\dfrac{46}{13}$

5. Write 'True' or 'False' for the following statements:

   (i) In an improper fraction, the numerator is always less than the denominator.

   (ii) A mixed number can be expressed as an improper fraction.

   (iii) The reciprocal of a mixed number is always a proper fraction.

   (iv) The product of a fraction with its reciprocal is always equal to zero.

   (v) The multiplicative inverse of $\dfrac{6}{7}$ is $1\dfrac{1}{6}$.

   (vi) $\dfrac{1}{2}\div\dfrac{3}{4}$ of $\dfrac{8}{15}\times\dfrac{2}{5}=1$

   (vii) To divide a fraction (other than 0), multiply by its equivalent fraction.

   (viii) 6 months of year is $\dfrac{1}{2}$ of a year.

   (ix) The difference of two given fractions is always less than each of the given fraction.

   (x) The sum of two given fractions is always greater than each of the given fractions.

6. Fill in the blanks:

   (i) The _____ of a unit fraction is always equal to 1.

(ii) The _____ of a whole number is always 1.

(iii) Every _____ fraction can converted into a mixed number.

(iv) _____ fractions have different denominators.

(v) All _____ fractions have equal value.

(vi) In two like fractions, the fraction having _____ numerator is greater.

(vii) Two fractions are called _____ of each other if their product is 1.

(viii) To divide fraction by another fraction, we multiply the dividend by the _____ of the divisor.

(ix) The operation of "of" is always is done before _____ and multiplication.

(x) A fraction is always _____ than 1.

7. Simplify:

(i) $5\dfrac{1}{2} - \left[\dfrac{7}{3} + \left\{ \dfrac{3}{4} - \dfrac{1}{2} \text{ of } \left(\dfrac{2}{3} - \dfrac{1}{24} \right) \right\} \right] + 5\dfrac{1}{3} + \dfrac{1}{2}$

(ii) $\dfrac{5}{7} - \dfrac{3}{16} \text{ of } \dfrac{4}{5} + \dfrac{7}{10} + \left\{ \dfrac{23}{12} - \left(\dfrac{3}{4} + \dfrac{2}{3} \right) \right\}$

8. A 45 minutes time slot on a television network contains 35 minutes of comedy 10 minutes of commercials. What fraction of the program time is devoted to commercials?

ANSWERS

1. (i) $\dfrac{1}{4}$ (ii) $\dfrac{9}{40}$ (iii) $\dfrac{3}{10}$ (iv) $\dfrac{19}{40}$ (v) $\dfrac{21}{40}$ (vi) $\dfrac{7}{10}$ (vii) $\dfrac{1}{20}$

(viii) $\dfrac{1}{40}$ (ix) $\dfrac{1}{3}$ (x) $\dfrac{5}{6}$

2. $\dfrac{1}{3}, \dfrac{3}{4}, \dfrac{4}{7}, \dfrac{2}{3}, \dfrac{4}{5}; \left(\dfrac{9}{27}, \dfrac{20}{35}, \dfrac{52}{78}, \dfrac{36}{48}, \dfrac{24}{30} \right)$

3. $\dfrac{5}{6}, \dfrac{3}{4}, \dfrac{2}{3}, \dfrac{5}{8}, \dfrac{7}{12}; \left(\dfrac{40}{48}, \dfrac{30}{48}, \dfrac{32}{48}, \dfrac{28}{48}, \dfrac{36}{48} \right)$

4. (i) ⟶ (d)   (ii) ⟶ (e)   (iii) ⟶ (a)   (iv) ⟶ (b)   (v) ⟶ (c)

5. (i) False   (ii) True   (iii) True   (iv) False   (v) True
   (vi) False   (vii) False (viii) True   (ix) False   (x) True
6. (i) Numerator   (ii) Denominator   (iii) Improper
   (iv) Unlike   (v) Equivalent   (vi) Greater
   (vii) Reciprocal (viii) Reciprocal   (ix) Division
   (x) Smaller.
7. (i) 6   (ii) 0   8. $\dfrac{2}{9}$

Chapter 6
Decimals

Decimal Fractions

We are familiar with the common fractions such as
$$\frac{1}{2}, \frac{3}{4}, \frac{9}{10}, \frac{7}{9}, \frac{13}{20}, \frac{15}{23}, \frac{18}{100}, \frac{115}{203} \text{ etc.}$$

In common fractions, numerators and denominators are any natural number. Fractions, like $\frac{9}{10}, \frac{18}{100}$ etc have a speciality. Such fractions always have their denominator as 10 or a multiple of 10.

For examples:
$$\frac{1}{10}, \frac{4}{100}, \frac{19}{1000}, \frac{22}{10000}, \frac{1}{1000}, 9\frac{7}{1000}, 26\frac{3}{1000}$$

A special name is given to this type of fractions we name them as *Decimal Fractions*.

We name common fractions as:

$\frac{3}{8}$ = Three - eights; $\frac{4}{5}$ = Four fifths, etc. Similarly, decimal-fractions are named as:

$\frac{1}{10}$ = one tenths

$\frac{2}{10}$ = two tenths

$\frac{4}{100}$ = four - hundredths, $\frac{16}{1000}$ = sixteen thousandths etc.

Did You Know?

The word 'decimal' is derived from the Latin word "DECIMA" meaning *Tenth*

DECIMAL POINT

We have a special way of representing decimal fractions. In this method denominator is not mention but a dot '.' is placed at an appropriate position in the numerator. This dot is called as *decimal-point*

For example,

$$\frac{1}{10} \text{ is written as } 0.1$$

$$\frac{2}{10} \text{ is written as } 0.2$$

$$\frac{3}{10} \text{ is written as } 0.3$$

$$\frac{14}{100} \text{ is written as } 0.14$$

$$\frac{5}{100} \text{ is written as } 0.05 \text{ etc.}$$

DECIMAL NUMBERS

The representation of a decimal-fraction without denominator but using a dot '.' gives a decimal-number. Therefore,

.1, .2, .3, .14, .05 etc are decimal numbers

NOTE

Decimal-numbers are also called as 'decimals'

READING AND WRITING A DECIMAL-NUMBER

Decimal numbers, after the decimal point, are read as separately one by one. For example:

- 346 is read as **'decimal three four six.'**
- 1178 is read as **'decimal one one seven eight.'**
- 709 is read as **'decimal seven zero nine.'**
- 0653 is read as **'decimal zero six five three.'**

To understand how to read and write decimal fractions and decimals, let us observe the following table carefully:

DECIMAL-FRACTION		DECIMAL NUMBER	
Written as	Read as	Written as	Read as
$\dfrac{1}{10}$	One tenth	.1	Decimal one
$\dfrac{2}{10}$	Two tenths	.2	Decimal two
$\dfrac{3}{10}$	Three-tenths	.3	Decimal three
$\dfrac{1}{100}$	One-hundredth	.01	Decimal zero one
$\dfrac{15}{100}$	Fifteen-hundredths	.15	Decimal one five
$\dfrac{99}{100}$	Ninety nine-hundredths	.99	Decimal nine nine
$\dfrac{1}{1000}$	One-thousandth	.001	Decimal zero zero one
$\dfrac{2}{1000}$	Two-thousandths	.002	Decimal zero zero two
$\dfrac{87}{1000}$	Eighty-seven thousandths	.087	Decimal zero eight seven
$\dfrac{429}{1000}$	Four hundred twenty nine thousandths	.429	Decimal four two nine
$\dfrac{999}{1000}$	Nine hundred ninety nine thousandths	.999	Decimal nine nine nine

MIXED DECIMALS

Look at some more decimals such as:

.27, 3.18, .0623, 25.012, 1.501 etc. some of these decimal numbers are having nothing on the left of the decimal point such as .27, .0623.

But others are having a whole number on the left of the decimal point, such decimals are called mixed decimals.

Therefore, 3.18, 25.012 and 1.501 are mixed-decimals, because each one of them has a whole-number to the left of the decimal number.

PARTS OF A DECIMAL NUMBER

A decimal number having whole number on the left of the decimal point is called a mixed decimal number. Every decimal can have two parts

i. The whole number part (also called the *Integral part*)
ii. The decimal part (also called as *Fractional part*). Both these parts are separated by the decimal point.

For example, In 123.106, the whole number part is 123 and the decimal part is .106. It can be shown in the following manner:

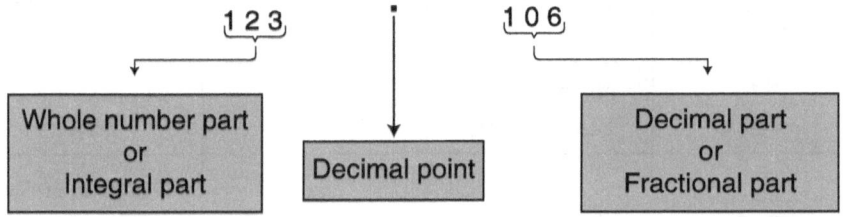

NOTE

If the 'whole number' part of a decimal-number is missing, then we write '0' to the left of decimal point. For example, .13 is written as 0.13. Similarly .0109 is written as 0.0109.

Let us examine the two parts of the following decimal numbers: 8.125; 0.172, 9.96, 14.0

Decimal Number	Whole number part	Decimal part
8.125	8	125
0.172	0	172
9.96	9	96
14.0	14	0

DECIMAL PLACES

The number digits after the decimal point gives the number of decimal places in the decimal number. For example, in 15.102 there are three digits (1, 0 and 2) after the decimal point, so its places are 3.

Let us consider some more examples:

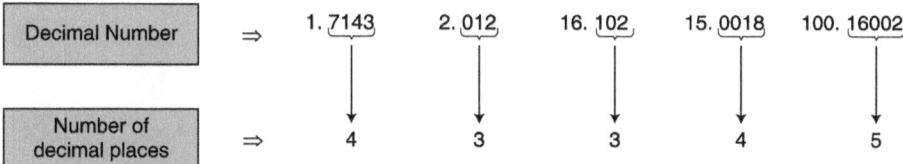

DECIMAL PLACE VALUE CHART

The position of a digit in a decimal number is very important. Each place has a specific value. Value of each place differ by a factor of 10.

If we move from right to left, place values *increase*. On moving from left to right, place values *decrease*.

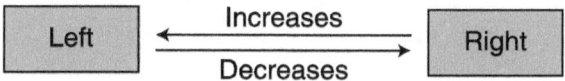

The position of a digit in a decimal number is very important. Every place has its specific value and a specific name.

The Hindu-Arabic system of numeration is 10-based. In this system, as we move from **RIGHT** to **LEFT** of unit place, we have:

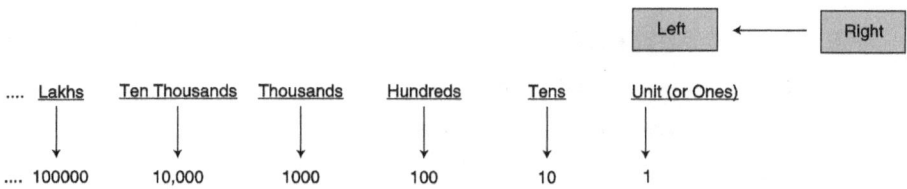

If we extend this method to the **RIGHT** of unit place, we have:

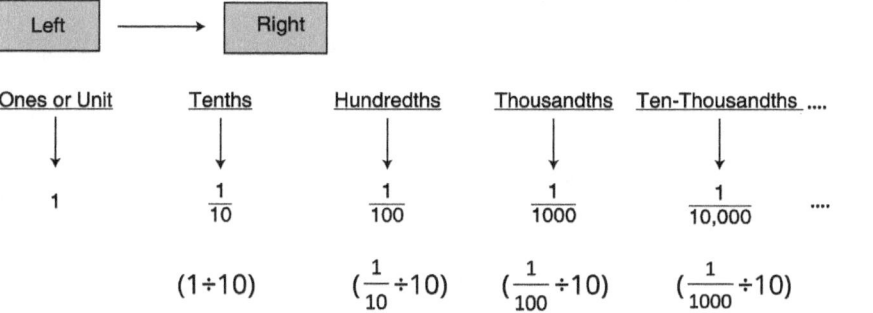

Thus, the first place to the right of unit-place is the *Tenths-place,* next is the *hundredths,* next to hundredths place is the *thousandths* and so on. We call this system of numeration as the decimal-system.

The 'decimal place value chart' is an extended form of the whole number 'place value chart' as shown under:

DECIMAL PLACE VALUE CHART

PLACES	INTEGRAL PART					Decimal Point	DECIMAL PART		
	Ten-Thousands	Thousands	Hundreds	Tens	Ones		Tenths	Hundredths	Thousandths
VALUE OF THE PLACE	10000	1000	100	10	1		$\frac{1}{10}$	$\frac{1}{100}$	$\frac{1}{1000}$

Let us place following decimal numbers into the place value chart:

15.38,   175.127,   198.073,   42.105

Numbers	Ten thousands (10,000)	Thousands (1000)	Hundreds (100)	Tens (10)	Ones (1)	Decimal point (.)	Tenths $\left(\frac{1}{10}\right)$	Hundredths $\left(\frac{1}{100}\right)$	Thousandths $\left(\frac{1}{1000}\right)$
15.38				1	5	.	3	8	
175.127			1	7	5	.	1	2	7
198.073			1	9	8	.	0	7	3
42.105				4	2	.	1	0	5

PLACE-VALUE OF DECIMALS

The place of a digit in a decimal numbers is determined exactly in the same way as we have done in case of whole numbers, i.e.

Place value of a digit} = [Face value of the digit] × [Value of the place it occupies]

For example, let us write the place value of each digit in the decimal number 4216.107

First, we fix the position of each digit in the place value chart.

Place	Thousands	Hundreds	Tens	Ones	Decimal point (.)	Tenths	Hundredths	Thousandths
Digit	4	2	1	6		1	0	7
Value of the place	1000	100	10	1		$\frac{1}{10}$	$\frac{1}{100}$	$\frac{1}{1000}$
Face value of the digit	4	2	1	6		1	0	7
Place Value of the digit	4 × 10000 = 4000	2 × 100 = 200	1 × 10 = 10	1 × 6 = 6		$1 \times \frac{1}{10}$ $= \frac{1}{10}$	$0 \times \frac{1}{100}$ $= 0$	$7 \times \frac{1}{1000}$ $= \frac{7}{1000}$

EXPANDED FORM OF DECIMALS

Writing a decimal number as sum of its place-values is called the expanded form of the decimal. Let us go back to the decimal number 4216.107, whose place values of all digits we have determined above.

Therefore, the expanded form of 4216.107 is:

$$4216.107 = 4000 + 200 + 10 + 6 + \frac{1}{10} + 0 + \frac{7}{1000}$$

LIKE AND UNLIKE DECIMALS

In a decimal number, the number of digits after the decimal point determine its decimal places.

Two or more decimal numbers having *same* number of places, are called *like decimals*.

For example, all of 3.462, 1.023, 101.123 and 3456.135 have same number of places (3-each) in their respective decimal part.

Therefore, 3.462, 1.023, 101.123 and 3456.135 are like decimals.

On the other hand decimal numbers having different number of digits in their decimal part, are called *Unlike decimals*.

For example, 1.25, 14.2051, 181.5 and 17.045 are unlike decimals, as they have different decimal places.

CONVERSION OF UNLIKE DECIMALS AS LIKE DECIMALS

We know that unlike decimals have different decimal places. For example 12.36 and 7.125 contain different number of places. To change unlike decimal as like decimals, we have the following steps.

I. Write the given decimals as decimal fractions,

$$12.36 = \frac{1236}{100}$$

$$7.125 = \frac{7125}{1000}$$

SHORT CUT METHOD

Identify the decimal having maximum number of decimal places. Make number of decimal places equal by putting zeros at the right hand end of the decimal part-in the remaining decimal numbers.

II. Make denominators of both decimal fractions equal by multiplying the denominator and numerator by 10, 100, etc. Here $\dfrac{1236}{100}$ has 2 zeros in the denominator and $\dfrac{7}{1000}$ has 3 zeros. To make the number of zeros equal we have to multiply 100 by 10

$$\text{Therefore } \dfrac{1236}{100} = \dfrac{1236 \times 10}{100 \times 10} = \dfrac{12360}{1000}$$

III. Now convert these like decimal fractions $\left(\text{here } \dfrac{1236}{1000} \text{ and } \dfrac{7}{1000}\right)$ into corresponding decimal numbers.

$$\therefore \dfrac{1236}{1000} = 12.360 \text{ and } \dfrac{7}{1000} = 0.007$$

Thus, 12.360 and 0.007 are like decimals.

NOTE

The zeros before the integral part and zeros after the decimal part of a decimal number, do not matter. For example, in 04.3500 one '0' before 4 and two 0's after 5 have no real value. If we write 045.3500 in expanded form, we get:

$$045.3500 = (0 \times 100) + (4 \times 10) + (5 \times 1) + \left(3 \times \dfrac{1}{10}\right) + \left(5 \times \dfrac{1}{100}\right) + \left(0 \times \dfrac{1}{1000}\right) + \left(0 \times \dfrac{1}{10000}\right)$$

$$= 0 + 40 + 5 + \dfrac{3}{10} + \dfrac{5}{100} + \dfrac{0}{1000} + \dfrac{0}{10000}$$

$$= 0 + 40 + \dfrac{3}{10} + \dfrac{5}{100} + 0 + 0$$

and sum of zeros has no effect. Thus, 045.3500 and 45.35 are same.

EXAMPLE - 1

Express the following decimals in figures:

  i. Fifteen hundredths
  ii. Seventy five thousandths
  iii. Fourteen decimal three nine five
  iv. Two thousand eight hundred fifty-five decimal one two nine.
  v. Zero decimal seven zero four.
  vi. One hundred decimal zero zero one

SOLUTION

  i. Fifteen hundredth $= \dfrac{15}{100}$ or 0.15

ii. Seventy five thousandths $= \dfrac{75}{1000}$ or 0.075

iii. Fourteen decimal three nine five = 14.395

iv. Two thousand eight hundred fifty-five decimal one two nine = 2855.129

v. Zero decimal seven zero four = 0.704

vi. One hundred decimal zero zero one = 100.001

EXAMPLE - 2

Write the following decimal number in place value chart and also write in words:

   i. 1.346   ii. 101.352   iii. 0.123   iv. 123.062

   v. 1768.106   vi. 25.246

	Places	Thousands	Hundreds	Tens	Ones	Decimal point	Tenths	Hundredths	Thousandths
	Place Values Numbers	(1000)	(100)	(10)	(1)	(.)	$\left(\dfrac{1}{10}\right)$	$\left(\dfrac{1}{100}\right)$	$\left(\dfrac{1}{1000}\right)$
(i)	1.346				1	.	3	4	6
(ii)	101.352		1	0	1	.	3	5	2
(iii)	0.123				0	.	1	2	3
(iv)	123.062		1	2	3	.	0	6	2
(v)	1768.106	1	7	6	8	.	1	0	6
(vi)	25.246			2	5	.	2	4	6

NUMBERS IS WORDS

   i. 1.346 = one decimal three four six

   ii. 101.352 = one hundred one decimal three five two.

   iii. 0.123 = Zero decimal one two three

   iv. 123.062 = One hundred twenty three decimal zero six two

   v. 1768.106 = One thousand seven hundred sixty eight decimal one zero six

   vi. 25.246 = Twenty-five decimal two four six

EXAMPLE - 3

Express the following decimal in expanded form:

   i. 23.156   ii. 9.601   iii. 106.123   iv. 50.005

   v. 36.43   vi. 3636.043

Solution

i. Since, the place values of all digits of 23.156 are:

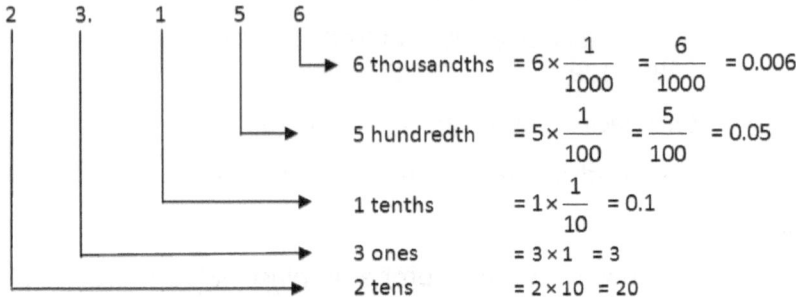

Since expanded form of a decimal is the sum of places values of its digits.

∴ 23.156 = 20 + 3 + 0.1 + 0.05 + 0.006

ii. The places of all digits are given here:

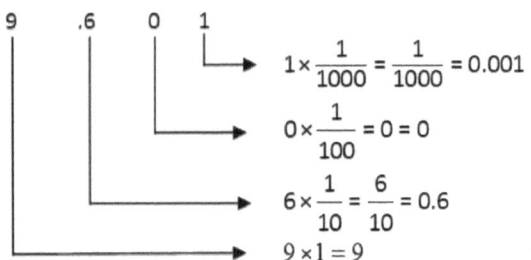

∴ Expanded form of 9.601 = 9 + 0.6 + 0 + 0.001

iii. The place values of 106.123 are:

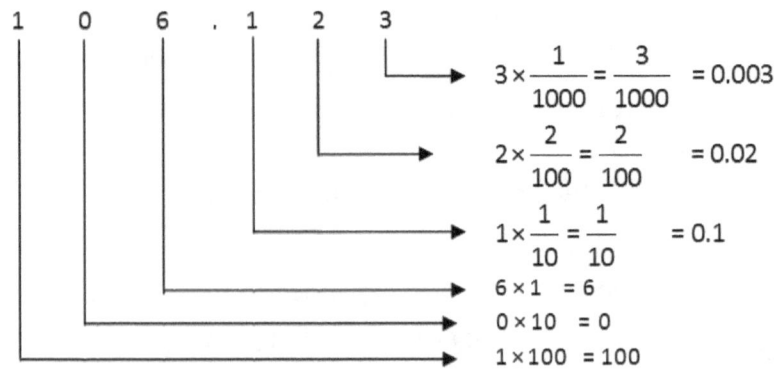

∴ Expanded form of 106.123 = 100 + 0 + 6 + 0.1 + 0.02 + 0.003

iv. Place values of all digits of 50.005 are:

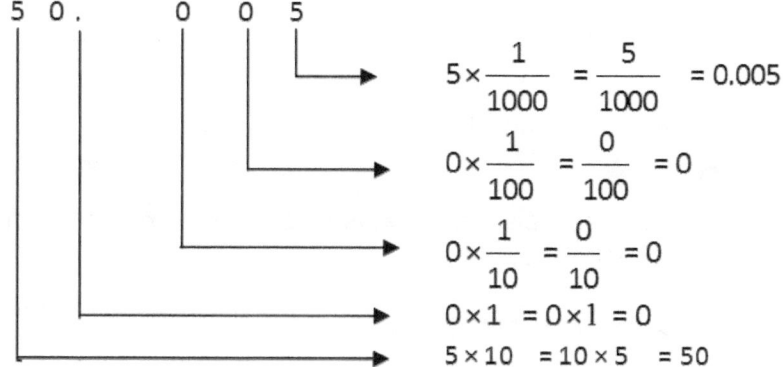

∴ Expanded form of 50.005 = 50 + 0 + 0 + 0 + 0.005

v. Place values of all digits of 36.43

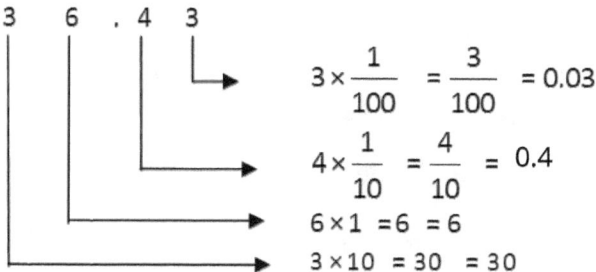

∴ Expanded form of 36.43 = 30 + 6 + 0.4 + 0.03

vi. Place values of all digits of 3636.043 are

3 6 3 6 . 0 4 3

$$3 \times \frac{1}{1000} = \frac{3}{1000} \text{ or } 0.003$$

$$4 \times \frac{1}{100} = \frac{4}{100} \text{ or } 0.04$$

$$0 \times \frac{1}{10} = 0$$

6 × 1 = 6
3 × 10 = 30
6 × 100 = 600
3 × 1000 = 3000

∴ The expanded form of 3636.043 = 3000 + 600 + 30 + 6 + 0 + 0.04 + 0.003

EXAMPLE - 4

Convert the following decimals in like decimals:

    i. 8.918 and 18.15   ii. 1.305 and 31.02
    iii. 3.912, 14.39 and 103.2   iv. 25.1, 280.123 and 13.13
    v. 0.4563, 3.3 and 21.201.   vi. 44.001, 0.1234 and 70.35

SOLUTION

i. Given decimals are 8.918, and 18.15 since, maximum number of decimal places = 3

∴ 8.918 = 8.918
    18.15 = 18.150

i.e. 8.918 and 18.150 are like decimals.

ii. Given decimals 1.305 and 31.02 maximum number of decimal places = 3

∴ 1.305 = 1.305
    31.02 = 31.020

i.e. 1.305 and 31.020 are like decimals.

iii. Given decimals are: 3.912, 14.39 and 103.2

∵ Maximum number of decimal places = 3

∴ 3.912 = 3.912
    14.39 = 14.390
    103.2 = 102.200

i.e. 3.912, 14.390 and 102.200 are like decimals.

iv. Given decimals are: 25.1, 280.123 and 13.13,

Since, maximum number of decimal places = 3

∴   25.1   = 25.100
    280.123 = 280.123
    13.13   = 13.130

i.e. 25.100, 280.123 and 13.130 are like decimals.

v. Given decimals are: 0.4563, 3.3 and 21.201, since, maximum number of decimal places are 4

∴   0.4563 = 0.4563

       3.3 = 3.3000

    21.201 = 21.2010

i.e. 0.4563, 3.3000 and 21.2010 are like decimals.

vi. Given decimals are 44.001, 0.1234 and 70.35, Since, the maximum number places are four

∴   44.001 = 44.0010

    0.1234 = 0.1234

    70.35 = 70.3500

Thus, 44.0010, 0.1234 and 70.3500 are like decimals.

Exercise - 1

1. Write the following decimals in figures:
   i. Thirty eight decimal zero six two.
   ii. Four hundred ninety-one decimal zero six three.
   iii. One thousand five hundred twenty-eight decimal six zero two four.
   iv. One thousand nine decimal seven eight three seven.
   v. Nine thousand nine hundred three decimal nine eight zero seven.

2. Write following decimals place value chart and also write in words.

   (i) 312.012   (ii) 102.321   (iii) 10.101

   (iv) 1728.12   (v) 105.501   (vi) 112.001

3. Find the place value of the encircled digit in following decimals:

   (i) 70.3①5   (ii) 117.①09   (iii) 123.③

   (iv) 3.123④   (v) 5167.31⓪4

4. Express the following in expanded form:

   (i) 3.31   (ii)   105.213   (iii)   422.0123

   (iv) 57.1234   (v)   123.456   (vi)   789.012

5. Write each of the following in short form:

   (i) $90 + 8 + \dfrac{1}{10} + \dfrac{7}{100} + \dfrac{8}{1000}$

   (ii) $6 + 0.1 + 0 + 0.004$

   (iii) $500 + 20 + 5 + 0.1 + 0 + 0.002$

   (iv) $400 + 60 + 0.2 + 0.06 + 0.002$

6. Convert the following decimals into like-decimals.

   i. 113.01, 151.002, 10.01

   ii. 121.1021, 0.5, 63.06

   iii. 71.17, 107.701, 1007.001

   iv. 463.63, 14.1278, 38.062

ANSWERS

1. (i) 38.062   (ii) 491.063   (iii) 1528.6024

   (iv) 1009.7837   (v) 9903.9807

2.

	Th	H	T	O	Decimal Point(.)	Tenths	Hundredths	Thousandths	Number-name
(i)		3	1	2	.	0	1	2	Three hundred twelve decimal zero one two
(ii)		1	0	2	.	3	2	1	One hundred two decimal three two one
(iii)			1	0	.	1	0	1	Ten decimal one zero one
(iv)	1	7	2	8	.	1	2		One thousand seven hundred twenty eight decimal one two
(v)		1	0	5	.	5	0	1	One hundred five decimal five zero one.
(vi)		1	1	2	.	0	0	1	One hundred twelve decimal zero zero one.

3. (i) .01   (ii) .1   (iii) .3   (iv) .0004   (v) 0.

4. (i) $3 + 0.3 + 0.01$ (ii) $100 + 0 + 5 + 0.2 + 0.01 + 0.003$

   (iii) $400 + 20 + 2 + 0 + 0.01 + 0.002 + 0.0003$

   (iv) $50 + 7 + 0.1 + 0.02 + 0.003 + 0.0004$

   (v) $100 + 20 + 3 + 0.4 + 0.05 + 0.006$

   (vi) $700 + 80 + 9 + 0 + 0.01 + 0.002$

5. (i) 98.178   (ii) 6.104   (iii) 525.102   (iv) 460.262

6. (i) 13.010, 151.002, 10.010

(ii) 121.1021, 0.5000, 63.0600

(iii) 71.170, 107.701, 1007.001

(iv) 463.6300, 14.1278, 38.0620

CONVERSION

CONVERSION OF FRACTIONS INTO DECIMALS

To convert a decimal fraction into decimal-number we have these steps:

I. Write the numerator only.

II. Place the decimal point in the number such that decimal part has as many digits as there are zeros in the denominator of given decimal fraction.

For example:

(i) $\dfrac{356}{10}$ = 35.6 | There is 1 zero in the denominator so, there should be 1 decimal place.

(ii) $\dfrac{1030}{1000}$ = 1.030 | There are three zero after 1, So decimal part will have 3 digits

To convert fractions like $\dfrac{3}{4}, \dfrac{2}{25}$ etc. into decimals, we first change such fractions into decimal fractions and use the above method.

For example, $\dfrac{3}{4} = \dfrac{3 \times 25}{4 \times 25} = \dfrac{75}{100} = 0.75$

EXAMPLE - 5

Convert the following fractions into decimals:

(i) $\dfrac{2}{5}$ (ii) $\dfrac{4}{25}$ (iii) $\dfrac{15}{8}$ (iv) $\dfrac{801}{125}$

SOLUTION

First convert the given fraction into an equivalent decimal-fraction, and then write the decimal-fraction so obtained as decimal number.

(i) $\dfrac{2}{5} = \dfrac{2 \times 2}{5 \times 2} = \dfrac{4}{10} = 0.4$

(ii) $\dfrac{4}{25} = \dfrac{4 \times 4}{25 \times 4} = \dfrac{16}{100} = 0.16$

(iii) $\dfrac{15}{8} = \dfrac{15 \times 125}{8 \times 125} = \dfrac{1875}{1000} = 1.875$

(iv) $\dfrac{801}{125} = \dfrac{801 \times 8}{125 \times 8} = \dfrac{6408}{1000} = 6.408$

CONVERSION OF DECIMALS IN FRACTIONS

We use following steps to convert a decimal number into a decimal fraction.

I. Remove the decimal point from the given decimal number and write this number (without the decimal point) as the numerator.

II. Write 1 in the denominator put zeros to its right such that the number of zeros is equal to the number of digits in decimal part of the given number.

For example, to convert 6.389 into a decimal-fraction. We have:

$6.389 = \dfrac{6389}{1000}$ — There are 3 digits after the decimal point. So, the denominator will have 1 followed by 3 zeros.

Similarly,

$175.25 = \dfrac{17525}{100}$ — There are 2 digits after the decimal point. So the denominator will have two digits after 1

NOTE

The decimal fraction, such as $\dfrac{17525}{100}$ can be further simplified to its lowest terms.

$\dfrac{\cancel{17525}^{701}}{\cancel{100}_{4}} = \dfrac{701}{4}$

$\dfrac{17575}{100}$ is called *decimal-fraction* where as $\dfrac{701}{4}$ is called a common-fraction. Both are equivalent fractions.

EXAMPLE - 6

Convert the following decimals into fractions:

(i) 45.9   (ii) 4.59   (iii) 0.459   (iv) 0.0459   (iv) 0.8   (v) 391.5

Solution

(i) $45.9 = \dfrac{459}{10}$ (ii) $4.59 = \dfrac{459}{100}$ (iii) $0.459 = \dfrac{459}{1000}$

(iv) $0.0459 = \dfrac{459}{10000}$ (v) $0.8 = \dfrac{8}{10}$ (vi) $391.5 = \dfrac{3915}{10}$

Note

A whole number can also be written as a decimal number by writing a decimal point after its unit (last) digit and a zero after it. Any whole number can be expressed a fraction by putting 1 in its denominator.

Exercise – 2

1. Convert the following fractions into decimals:

   (i) $\dfrac{3}{2}$ (ii) $\dfrac{5}{4}$ (iii) $\dfrac{4}{5}$ (iv) $\dfrac{5}{8}$ (v) $\dfrac{148}{125}$ (vi) $\dfrac{63}{25}$

   (vii) $\dfrac{403}{40}$ (viii) $8\dfrac{2}{5}$ (ix) $3\dfrac{7}{40}$

2. Convert the following decimals into decimal fractions:

   (i) 63.05   (ii) 312.501   (iii) 0.89   (iv) 0.048

   (v) 0.06   (vi) 18.7   (vii) 212.067   (viii) 301.103

   (ix) 35.053

Answers

1. (i) 1.5   (ii) 1.25   (iii) 0.8   (iv) 0.625   (v) 1.184   (vi) 2.52
   (vii) 10.075   (viii) 8.4   (ix) 3.175.

2. (i) $\dfrac{6305}{100}$ (ii) $\dfrac{312501}{1000}$ (iii) $\dfrac{89}{100}$ (iv) $\dfrac{48}{1000}$

   (v) $\dfrac{6}{100}$ (vi) $\dfrac{187}{10}$ (vii) $\dfrac{212067}{1000}$

   (viii) $\dfrac{301103}{1000}$ (ix) $\dfrac{35053}{1000}$

ADDITION OF DECIMALS

Addition of decimals is like whole numbers. Only case is to be taken that the place values be lined up correct by steps to add decimals:

i. Convert all the given decimals into like decimals.
ii. Write the decimals one under the other such that the decimal points are in the same column
iii. Add them from right to left as in the case of whole numbers
iv. In the sum, put the decimal point directly under the decimal point in the addends.

EXAMPLE - 9

Add the following:

(i) 38.06 and 2.178   (ii) 126.01 and 374.245

(iii) 1.78, 0.234 and 54.309   (iv) 3.001, 103.32 and 6.789

SOLUTION

i.

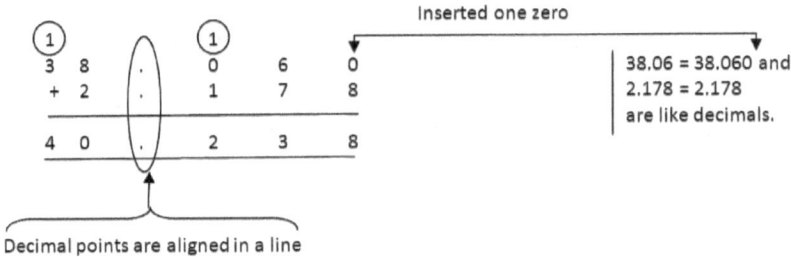

Thus, the sum of 38.06 and 2.178 is 40.238

ii.

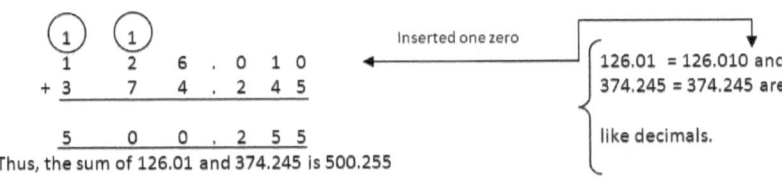

Thus, the sum of 126.01 and 374.245 is 500.255

iii.

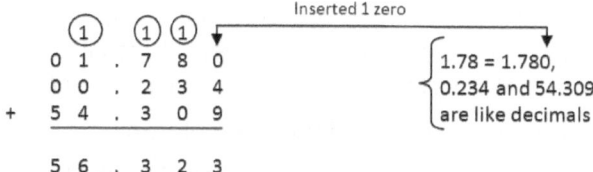

Thus, the sum of 1.78, 0.234 and 54.309 is 56.323

iv.

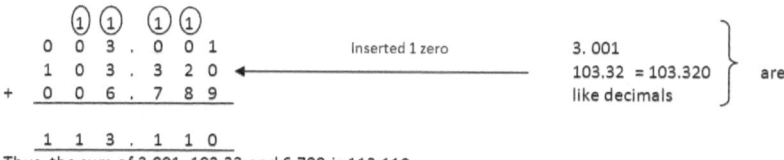

Thus, the sum of 3.001, 103.32 and 6.789 is 113.110

SUBTRACTION OF DECIMALS

Subtraction of decimals is carried out using the following steps:

i. Convert the given decimals as like decimals.
ii. Write the decimals one under the other such that the their decimal points are in the same column.
iii. Subtract as we have done the case of whole numbers.
iv. In the difference, put the decimal point directly under the decimal point of the given numbers.

EXAMPLE - 10

Subtract the following decimals:

(i) 13.17 − 1.397   (ii) 141.039 − 44.12

(iii) 100.06 − 29.345   (iv) 60.117 − 43.279

SOLUTION

(i)

```
           10 16
      2   8 6 10
   1 3 . 1 7 0
 − 0 1 . 3 9 7
 ─────────────
   1 1 . 7 7 3
```

13.17 = 13.170 and
1.397 = 1.397 are
like decimals

Thus, 13.17 − 1.397 = 11.773

EXPLANATION

i. Line up the given numbers in a column, line up the decimal points, since, 13.17 has only two decimal places whereas 1.397 has 3 decimal places, so they are converted into like decimals by putting one '0' at the right end of 13.17 making it as 13.170.

ii. Starting from right, subtract down the columns

iii. Notice that in the thousandths column 7 > 0. Therefore, we "borrow" from the hundredths column, making 0 as 10 and leaving 6 behind in the hundredths column. Here 10 thousandths − 7 thousandths = 3 thousandths.

```
          2    6 10
    1  3 . 1  7̶  0̶
          1 . 3 9 7
    _____
                    3
```

iv. In hundredths column, we have 9 > 6, so we must 1-"borrow" from the tenths column making it 16 hundredths

```
              16
          0  6̶  10
    1  3 . 1̶ 7̶ 0̶
  − 0  1 . 3  9  7
    _____
                 7  3
```

∴ 16 hundredths-9 hundredths = 7 hundredths.

v. Continue to subtract down the columns for the remaining digits from right to left.

```
           10 16
      2    6̶  6̶ 10
    1  3 . 1̶  7̶ 0̶
  − 0  1 . 3  9  7
    _____
    1  1 . 7  7  3
```

i.e.

10 tenths − 3 tenths	=	7 tenths
2 ones − 1 ones	=	1 ones
1 tens − 0 tens	=	1 tens

(ii)

$$
\begin{array}{r}
13\;10\\
0\;\;\cancel{3}\;\;\cancel{0}10\\
\cancel{1}\;\;\cancel{4}\;\;\cancel{1}\;.\;\;\cancel{0}\;\;3\;\;9\\
-\;\;\;0\;\;4\;\;4\;.\;\;1\;\;2\;\;0\\
\hline
0\;\;9\;\;6\;.\;\;9\;\;1\;\;9
\end{array}
$$

$\therefore$ 141.039 − 44.12 = 96.919

$\left\{\begin{array}{l}\text{141.039 and}\\ \text{(44.12 = 44.120) and}\\ \text{like decimals}\end{array}\right.$

(iii)

$$
\begin{array}{r}
9\;\;9\\
0\;\;10\;\;10\;\;\;10\;\;5\;\;10\\
\cancel{1}\;\;\cancel{0}\;\;\cancel{0}\;.\;\;\cancel{0}\;\;\cancel{6}\;\;\cancel{0}\\
-\;\;\;0\;\;2\;\;9\;.\;\;3\;\;4\;\;5\\
\hline
7\;\;0\;.\;\;7\;\;1\;\;5
\end{array}
$$

Thus, 100.06 − 29.345 = 70.715

$\left\{\begin{array}{l}\text{100.060 and}\\ \text{29.345 and}\\ \text{like decimals}\end{array}\right.$

(iv)

$$
\begin{array}{r}
9\;\;10\;\;10\\
5\;\;10\;\;\cancel{0}\;\;\cancel{0}\;\;17\\
\cancel{6}\;\;\cancel{0}\;.\;\;\cancel{1}\;\;\cancel{1}\;\;\cancel{7}\\
-\;\;\;4\;\;3\;.\;\;2\;\;7\;\;9\\
\hline
1\;\;6\;.\;\;8\;\;3\;\;8
\end{array}
$$

Thus, 60.117 − 43.279 = 16.838

$\left\{\begin{array}{l}\text{60.117 and}\\ \text{43.279 are already}\\ \text{like decimals}\end{array}\right.$

Example - 11

Evaluate the following:

   i.   612.48 − 511.7 + 128.193   ii. 0.818 − 36.06 + 71.11

Solution

Converting the given decimal as like decimals:

    612.48   =   612.480
    511.7   =   511.700
    128.193   =   128.193

We know that when addition and subtraction are to be solved together, then first we add and then subtract:

Adding 612.480 and 128.193, we get 740.673

$$\begin{array}{r} 612.480 \\ + 128.193 \\ \hline 740.673 \end{array}$$

Next, we subtract 511.7 from 740.673, we get:

```
            9
       3  10  16
    7  4̶  0̶ . 6̶  7  3
  - 5  1  1 . 7  0  0
    ─────────────────
    2  2  8 . 9  7  3
```

$\begin{cases} (511.7 = 511.700) \\ 740.673 \\ \text{are like decimals} \end{cases}$

∴ 612.48 − 511.7 + 128.197

= (612.48 + 128.193) − 511.700

= 740.673 − 511.700 = 228.973

Thus, 612.48 − 511.7 + 128.193 = 228.973

ii.

0.818 − 36.06 + 71.11

= 0.818 − 36.060 + 71.110

= (0.818 + 71.110) − 36.060

= 71.928 − 36.060

= 35.868

$\begin{cases} 00.818 \\ + 71.110 \\ \hline 71.928 \end{cases}$

0.818 = 0.818

36.06 = 0.060

71.11 = 71.110

are like decimal

$\begin{cases} 611812 \\ 71̶.9̶2̶8 \\ -36.060 \\ \hline 35.868 \end{cases}$

Thus, 0.818 − 36.06 + 71.11 = 35.868

Example - 12

From a grocery shop, Mrs Ranjeeta bought a chocolate pack and a biscuit pack for Rs 450.928 and Rs 38.55 respectively. How much did Mrs Ranjeeta pay altogether?

Solution

Amount paid for chocolate = Rs 450.928

Amount paid for Biscuits = Rs 38.55

Converting 38.55 as like decimals (38.55 = 38.550) and adding,

```
    4 5 0 . 9 2 8
+   0 3 8 . 5 5 0
    ─────────────
    4 8 9 . 4 7 8
```

∴ Mrs Ranjeeta paid Rs 489.478

Example - 13

There was a class collection of Rs 1128.40 for a fun-day celebration on last closing day for the summer vacations. If Rs 897.60 was spent on the activity. What amount is left unspent?

Solution

Total collection = Rs 1128.40

Amount spent on the 'fun-day' activity

= Rs 897.60

Amount left unspent = Rs 1128.40 − Rs 897.60

```
         10
    0  0  12 7   14
    1̶  1̶  2̶ 8̶ . 4̶ 0
  −    8 9 7 . 6 0
       ─────────────
       2 3 0 . 8 0
```

= Rs 230.80

Exercise - 3

1. Add the following decimals:
   (i) 34.103 and 18.2014
   (ii) 304.15 and 10.001
   (iii) 131.12, 17.102 and 109.05
   (iv) 1048.5, 0.001 and 99.0099
   (v) 21.4263, 426.213 and 44.0044
   (vi) 10.004, 20.002 and 30.008.

2. Subtract the following decimals:
   (i) 32.003 from 99.01
   (ii) 15.896 from 201.01
   (iii) 18.183 from 25.12
   (iii) 36.005 from 39.39
   (v) 109.18 from 901.81

3. Simplify the following:
   (i) 41.18 + 10.5 − 38.08
   (ii) 88.175 − 11.8 − 60.02
   (iii) 634.043 − 312.789 + 0.234
   (iv) 118.09 − 100.950 + 10.123

(v) 415.06 + 183.642 − 221.403

4. How much less the sum of 3.102 and 10.06 is than 15?
5. Shanta buys sugar for Rs 118.45, flour for Rs 500.40 and pulses for Rs 312.80. What did she pay for all these items.
6. In the above Q - 5, if shanta gave a Rs 1000 note, what amount the shopkeeper returned back?
7. What should be added to the sum of 11.05 and 25.78 to get 50?
8. How much the sum of 15.001 and 101.3 is less than 120?

ANSWERS

1. (i) 52.3044   (ii) 314.151   (iii) 257.272
   (iv) 1147.5109   (v) 491.6437   (vi) 60.014
2. (i) 67.007   (ii) 185.114   (iii) 6.937
   (iv) 3.385   (v) 792.63
3. (i) 13.6   (ii) 16.355   (iii) 321.488
   (iv) 27.263   (v) 377.299
4. 1.838   5. Rs. 931.65   6. 68.35
7. Rs. 13.17   8. 3.699

MENTAL MATHS

1. A decimal number has two parts–whole-number part and the decimal part. Which of these two parts is called integral part?
2. What is the other name of the decimal part of a decimal number?
3. A decimal is multiplied by 100. Its decimal point will shift by two places to right or left?
4. Fractions having denominator as 10, 100, 1000 etc. are called common-fractions. Is the statement true or false?
5. Write $\frac{25}{100}$ as a common fraction.
6. Write the sum of 0.21 and 2.1
7. What is difference when 0.21 is subtracted from 0.7?
8. Do (6.4 ÷ 0.8) and (0.64 ÷ 0.08) represent the same number?
9. What is the simplest form of the common fractions represented by (1.2 ÷ 14.4)?
10. Write the decimal fraction for $\frac{1}{20}$

Answers

1. Whole number part
2. Fractional Part
3. Right
4. False
5. $5\frac{1}{4}$
6. 2.31
7. 0.49
8. Yes
9. $\frac{1}{12}$
10. $\frac{5}{100}$

Multi Choice Questions

1. Which of the following decimal represents $\frac{2}{1000}$?
    (i) 0.0002   (ii) 0.200   (iii) 0.002   (iv) 0.2000

2. Which of the following is the smallest decimal?
    (i) 0.009   (ii) 0.08   (iii) 0.7   (iv) 6.0

3. Which of the following is the place value of 3 in 124.07392?
    (i) $\frac{3}{10}$ (ii) $\frac{3}{1000}$ (iii) $\frac{3}{100}$ (iv) 3

4. Which of the following pair represents like fractions?
    (i) 0.345 and 5.34   (ii) 1.0035 and 0.10035
    (iii) 123.78 and 123.078   (iv) 4507.001 and 0.963

5. Which of the following is the short form of $\frac{7}{10}+\frac{1}{1000}+\frac{3}{10000}$?
    (i) 7.13   (ii) 0.713   (iii) 71.3   (iv) 0.7013

6. "Three Thousand decimal Three Thousandths" in the number name of which of the following decimals
    (i) 3000.0003   (ii) 3000.003   (iii) 3000.03   (iv) 3000.3000

7. Which of the following are in ascending order?
    (I) 0.1, 0.06, 0.009   (ii) 0.009, 0.1, 0.06
    (iii) 0.009, 0.06, 0.1   (iv) 0.06, 0.009, 0.1

8. Which of the following is not correct?
    (i) 0.01 + 0.01 = 0.02   (ii) 0.02 − 0.01 = 0.01
    (iii) $\frac{3}{100}+\frac{2}{1000}=\frac{5}{1000}$ (iv) 0.1 − 0.01 = 0.09

9. 1 l = 1000 ml then which of following is equal to 900 ml?
    (i) 9.0l   (ii) 0.009l   (iii) 0.0009l   (iv) 0.9l

10. 1 cm = 0.01 m; then which of the following is equal to 30 cm?
    (i) 0.003 m   (ii) 0.03 m   (iii) 0.3 m   (iv) 3.0 m

11. 1 km = 1000 m then which of the following is equal to 0.001 km?

   (i) 100 m   (ii) 10 m   (iii) 1 m   (iv) 0.1 m

12. 0.7624 lies between

   (i) 0.7 and 0.76   (ii) 0.76 and 0.763

   (iii) 0.77 and 0.78   (iv) 0.76 and 0.761

13. 0.07 + 0.009 is equal to:

   (i) 0.079   (ii) 0.16   (iii) 0.016   (iv) 0.79

14. Which of the following is the sum of 0.1 and 1.001?

   (i) 1.001   (ii) 100.1   (iii) 1.001   (iv) 1.101

15. Which of the following is the difference of 0.202 − 0.02?

   (i) 0.182   (ii) 0.0182   (iii) 182.0   (iv) 0.1802

ANSWERS

1. (iii)   2. (i)   3. (ii)   4. (iv)   5. (iv)   6. (ii)
7. (iii)   8. (iii)   9. (iv)   10. (iii)   11. (iii)   12. (ii)
13. (i)   14. (iv)   15. (i)

WORKSHEET

1. Write 'true' or 'false' for the following statements

   i. 15.730 = 15.73   ii. 18.9270 = 18.0927

   iii. 20 + 0.053 = 20.0530   iv. 43.28 − 4.19 = 39.09

   v. To divide a decimal by 1000, we shift the decimal point to the left by three places.

   vi. When a decimal is multiplied by 100, the decimal point moves to the left by two places.

   vii. 2 Thousandths − 1 hundredths = 19 hundredths

   viii. 1 Ones − 1 tenths = 9 tenths

   ix. $3\frac{2}{100}$ is expressed in decimal form as 3.2

   x. $\frac{3}{4} < 0.760$

2. Match the column A and Column B

Column A	Column B
(a) 0.011, 1.001, 0.101, 0.110 arranged in ascending order are:	(i) 22.002
(b) Sum of 20.02 and 2.002 is	(ii) 1.001, 0.110, 0.101, 0.011
(c) 24.004 − 2.002 is:	(iii) 22.2
(d) The descending order of 0.110, 0.101, 1.001, 0.011 is:	(iv) 0.011, 0.101, 0.110, 1.001
(e) The difference of 28.220 − 6.02	(v) 22.02

3. Fill in the blanks:

   i. $\frac{11}{20}$ express as a decimal fraction = _____

   ii. 0.041 express as a common fraction = _____

   iii. 7.03 expressed mixed fraction = _____

   iv. 9.012 subtracted from 10, then the difference = _____

   v. 30.72 when increased by 4.880, the result is = _____

   vi. 11.001 when decreased by 5.8959, the result is = _____

   vii. The sum of 1.1, 2.02 and 3.003 is = _____

   viii. 3.001, 2.20, 3.005 and 2.1 when expressed as like decimals are: _____, _____, _____, and _____

   ix. If y = 22.1 − 0.1 then y = _____

   x. Out of 385.0762 and 385.7620 which is equal to 385.762

4. Find the total bill amount:

	Items	Rate	Amount
(i)	5 tin biscuits	Rs 37.05 per tin	_____
(ii)	2 bags of rice	Rs 64.85 per bag	_____
(iii)	10 boxes of sweets	Rs 16.05 per box	_____

Bill amount = _____

Answers

1. (i) True (ii) False (iii) True (iv) True
   (v) True (vi) False (vii) False (viii) True
   (ix) False (x) True

2. (a) → (iv) (b) → (v) (c) → (i)
   (d) → (ii) (e) → (iii)

3. (i) $\frac{55}{100}$ (ii) $\frac{41}{1000}$ (iii) $7\frac{3}{100}$ (iv) 0.998
   (v) 35.6 (vi) 5.1051 (vii) 6.123
   (viii) 3.001, 2.200, 3.005 and 2.100   (ix) 22
   (x) 385.7620

4. Rs 475.45

CHAPTER 7
RATIO AND PROPORTION

RATIO

The relation between two quantities of the same kind is called the ratio. It is the comparison of the magnitudes of two quantities. For example: Let the weights of two boxes be 3 kg and 5 kg.

If we compare the weights of these two boxes, then we get:

$$\frac{\text{weight of the first box}}{\text{weight of the second box}} = \frac{3 \text{ kg}}{5 \text{ kg}} = \frac{3}{5}$$

[Here, the same unit is cancelled from both the numerator and the denominator]

Here, $\frac{3}{5}$ is the ratio between 3 kg and 5 kg

A ratio is a pure number given by a fraction in which the numerator denotes the magnitude of the first quantity, and the denominator given the magnitude of the second quantity. A ratio has no unit. The ratio between '3 kg to 5 kg' is written as 3:5 (**read as 3 to 5 or 3 is to 5**). A ratio has two numbers. The colon (:) is inserted between them. These two numbers are called the *Terms* of the ratio. In 3:5, the first term 3 and second term is 5. Of the two numbers forming a ratio, first one is called the *antecedent* and the second one the *consequent*.

NOTE

The word '*ANTECEDENT*' literally means '**anything that goes before**'

The word '*CONSEQUENT*' literally means '**anything that goes after.**'

Some examples of ratios:

The ratio between 5 cm and 8 cm = 5:8

The ratio between Rs 4 and Rs 5 = 4:5

The ratio between 3 kg and 2 kg = 3:2

The ratio between 12 cm and 110 mm = 12:11

[∴ **110 mm = 11 cm**]

The ratio between 7 kg and 9000 gm = 7:9

[∴ **9000 gm = 9 kg**]

Important-Facts about Ratio

i. The two quantities forming a ratio must be of same kind. For example, the 'speed' of a car can be compared with speed and not with the 'height' of an object. 'Price' can be compared with 'price'. Similarly number of books can be compared with number of books and not with the number of vehicles. Thus, the ratio of two dissimilar quantities (e.g. 5 kg and 12 metres) cannot be found.

ii. Ratio is a number and it has no unit.

iii. Ratio is usually expressed in the simplest form (or in its lowest terms).

iv. In a ratio, the order of the terms is important. On changing the order of the terms, the value of the ratio changes, i.e. the ratio 5:6 is different from the ratio 6:5.

v. A ratio is smaller or greater than 1 according to whether the antecedent is smaller or greater than the consequent.

vi. Two or more ratios are equal if their corresponding fractions are equal.

vii. The second term of a ratio cannot be 0 (zero).

DIFFERENCE BETWEEN FRACTION AND RATIO

FRACTION	RATIO
A fraction is a number that represents a part of a whole.	A ratio is a comparison between two or more quantities of the same kind.
A fraction is a single number. $\frac{1}{2}$ is a number which is equal to half of something.	Ratio is a combination of two numbers, each representing an amount of something. 4:5 is a ratio such that 4 is amount of one thing and 5 is the amount of another thing.

FRACTION	RATIO
Fraction is a quotient obtained by dividing numerator by denominator. Value of the fraction $\frac{3}{4}$ is 0.75	Terms of a ratio express the relationship between the amounts of first and second quantities. Ratio of Rs 300 and Rs 400 is 3:4
A fraction may have a unit. Half of 1 kg is $\frac{1}{2}$ kg or 0.5 kg	Ratio of 4 kg and 5 kg is 4:5
Upper part in a fraction is called numerator and the lower part is denominator.	First number in a ratio is called 'first term' or 'antecedent' and the second number is called the 'second term' or 'consequent'.
Fraction is a part of the same number	Ratio is a relationship between amounts of two separate things.

ORDER OF NUMBERS IN A RATIO

In a ratio, there are two numbers. Their order is important. For example, seller have some *mangoes* and a fruit in the ratio 6:8. It means he has 8 oranges for every 6 mangoes

The ratio mangoes to oranges is 6:8

We do not write it as 8:6

Here, terms of ratio are 6 and 8. The first term is 6 and the second term is 8

Therefore, the order of numbers is very important in ratio.

EXAMPLE

Find the ratio between:

       a. ₹150 and ₹500   b. 30 bananas and 40 oranges

       c. 12.5 kg and 75 kg   d. 24 boys and 18 girls

SOLUTION

    a. The ratio between ₹150 and ₹500

       = ₹150:₹500

$$= \frac{₹150}{₹500} = \frac{150}{500} = \frac{150 \div 50}{500 \div 50} = \frac{3}{10}$$

       ∴ The required ratio = 3:10

b. The ratio between bananas and oranges

$$= \frac{30}{40} = \frac{30 \div 10}{40 \div 10} = \frac{3}{4}$$

∴ The required ratio = 3:4

c. The ratio between 12.5 kg and 75 kg

$$= \frac{12.5 \text{ kg}}{75 \text{ kg}} = \frac{\frac{125}{10}}{75} \qquad 12.5 = \frac{125}{10}$$

$$= \frac{125}{750} = \frac{125 \div 125}{750 \div 125} = \frac{1}{6}$$

∴ The required ratio = 1:6

d. The ratio between 24 boys and 18 girls

$$= \frac{24}{18} = \frac{24 \div 6}{18 \div 6} = \frac{4}{3}$$

∴ The required ratio = 4:3

EXAMPLE

Reduce 45:162 to the simplest form.

SOLUTION

HCF of 45 and 162 = 3×3 = 9

The HCF of 45 and 162 is 9

∴ $45:162 = \frac{45}{162}$

$= \frac{45 \div 9}{162 \div 9} = \frac{5}{24}$

= 5:24

3	45,	162
3	15,	54
	5,	18

So, the simplest form of 45:162 is 5:24

EXAMPLE

Find the ratio between ₹2.70 and 90 paise.

SOLUTION

Ratio between ₹2.70 and 90 paise can be obtained when both quantities are in the same units.

∴ ₹2.70 = 270 paise

∴ Ratio between 270 paise 90 paise = 270:90

HCF of 270 and 90 is 90

∴ $270:90 = \dfrac{270}{90} = \dfrac{270 \div 90}{90 \div 90} = \dfrac{3}{1}$

∴ The required ratio = 3:1

EXAMPLE

Simplify the ratio $\dfrac{2}{3} : 5$

SOLUTION

The given ratio is $\dfrac{2}{3} : 5$ or $\dfrac{2}{3} : \dfrac{5}{1}$

Multiplying both terms by 3,

$\dfrac{2}{3} \times 3 : \dfrac{5}{1} \times 3$

or 2:15

Thus, the required ratio = 2:15

EXAMPLE

A hall is 30 m long and 20 m broad what is the ratio between its length and breadth.

SOLUTION

Length = 30,   Breadth = 20 m

∴ $\dfrac{\text{Length}}{\text{Breadth}} = \dfrac{30}{20} = \dfrac{30 \div 10}{20 \div 10} = \dfrac{3}{2} = 3:2$

Thus, the required ratio = 3:2

COMPARING THE RATIOS

EXAMPLE

Compare the ratios 2:3 and 4:5

SOLUTION

Write the given ratios as fractions.

∴ $2:3 = \dfrac{2}{3}$ and $4:5 = \dfrac{4}{5}$

Now, convert $\dfrac{2}{3}$ and $\dfrac{4}{5}$ as like fractions.

$\dfrac{2}{3} = \dfrac{2 \times 5}{3 \times 5} = \dfrac{10}{15}$ and $\dfrac{4}{5} = \dfrac{4 \times 3}{5 \times 3} = \dfrac{12}{15}$

$\dfrac{10}{15}$ and $\dfrac{12}{15}$ are like fractions and $\dfrac{12}{15} > \dfrac{10}{15}$

$\therefore \dfrac{4}{5} > \dfrac{2}{3}$ or (4:5) > (2:3)

EXAMPLE

Simplify the following ratios:

(i) 3.5:7   (ii) 4.5:18

SOLUTION

(i) $3.5:7 = \dfrac{35}{10}:7$ or $\dfrac{35}{10} \div 7$

$= \dfrac{\cancel{35}^{\;5}}{\cancel{10}_{\;2}} \times \dfrac{1}{\cancel{7}^{\;1}} = \dfrac{1}{2}$ or 1:2

(ii) $4.5:18 = \dfrac{45}{10}:18$ or $\dfrac{45}{10} \div 18$

$= \dfrac{\cancel{45}^{\;5}}{\cancel{10}_{\;2}} \times \dfrac{1}{\cancel{18}_{\;2}}$

$= \dfrac{1}{2} \times \dfrac{1}{2} = \dfrac{1}{4}$ or 1:4

EXAMPLE - 3

20 students from a society building go to the same school. Of them 12 are boys and 8 are girls. Find:

   i. The ratio of boys to girls.
   ii. The ratio of girls to boys.
   iii. The ratio of boys to all students.
   iv. The ratio of girls to all students.

SOLUTION

Here, number of boys   = 12

Number of girls   = 8

Total number of students = 20

∴ (i) ∴ $\dfrac{\text{Number of boys}}{\text{Number of Girls}} = \dfrac{12}{8} = \dfrac{12 \div 4}{8 \div 4} = \dfrac{3}{2} = 3.2$

∴ Ratio boys to girls = 3:2

(ii) ∴ $\dfrac{\text{Number of girls}}{\text{Number of boys}} = \dfrac{8}{12} = \dfrac{8 \div 4}{12 \div 4} = \dfrac{2}{3} = 2:3$

∴ Ratio of girls to boys = 2:3

(iii) ∴ $\dfrac{\text{Number of boys}}{\text{Number of all students}} = \dfrac{12}{20} = \dfrac{12 \div 4}{20 \div 4} = \dfrac{3}{5} = 3:5$

∴ Ratio of boys to all students = 3:5

(iv) ∴ $\dfrac{\text{Number of girls}}{\text{Number of all students}} = \dfrac{8}{20} = \dfrac{8 \div 4}{20 \div 4} = \dfrac{2}{5} = 2:5$

∴ Ratio of girls to all students = 2:5

EXAMPLE - 4

Mr Prashant earns Rs 105000 in a month, of which saves Rs 15000. Find the ratio of his income to his expenditure.

SOLUTION

Mr Prashant's monthly income = Rs 105000

His savings = Rs 15000

∴ His expenditure = Rs 105000 − Rs 15000 = Rs 90,000

Now, Ratio of his income to his expenditure

$= \dfrac{\text{Rs } 105000}{\text{Rs } 90000} = \dfrac{105000}{90000} = \dfrac{105000 \div 15000}{90000 \div 15000}$

[HCF of 105000 and 90000 = 15000]

$= \dfrac{7}{6} = 7:6$

Thus, the required ratio = 7:6

EXAMPLE - 5

An alloy contains 4.5 gm of zinc and 13.5 g of copper. Find the ratio of weight of copper to the weight of zinc in the alloy.

Solution

∴ Weight of copper in the alloy = 13.5 g.

Weight of zinc in the alloy = 4.5 gm.

∴ Ratio of (wt. of copper) to (wt. of zinc) = $\dfrac{13.5}{4.5}$

$= \dfrac{\cancel{135}^{\,\,3}}{\cancel{45}_{\,\,1}} = \dfrac{3}{1}$ or 3:1   [∵ HCF of 135 and 45 is 45.

Thus, the required ratio is 3:1

Exercise - 1

1. (i) What is the ratio of:
   (a) balloons to kites
   (b) kites to balloons

   (ii) What is the ratio of
   (a) Laptops to cell phones
   (b) Cell phones to laptops.

   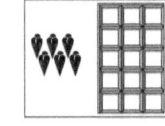

   (iii) What is the ratio of
   (a) ice-cream cups to chocolate-bars
   (b) Chocolate bars to ice cream cups.

   (iv) What is ratio of:
   (a) number of books to the total number of stationery items
   (b) number of pens to total stationery items
   (c) number of pencils to Total number of stationery items.

   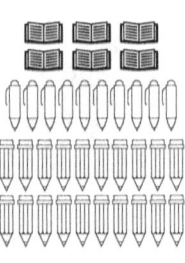

   (v) Write the antecedent and consequent of each of the following ratios:
   (a) 1:5   (b) 3:4   (c) 7:9
   (d) 10:8   (e) 15:2   (f) 13:17

   (vi) A school has 42 periods working class VI is allotted 7 periods to mathematics and 6 periods to English. Write the ratio of

(a) number of periods allotted for mathematics to the total number of periods of school working.

(b) 'number of period allotted to English' to 'the number of period allotted to mathematics'.

2. What is the ratio between:

   (i) Rs 6 and Rs 2   (ii) 5 cm and 45 cm

   (iii) Rs 7.5 and Rs 30   (iv) 5 kg and 2 kg

   (v) 12 balloons and 18 balloons (vi) 40 boys and 25 girls

3. What is the ratio between:

   (i) 60 mm and 5 cm   (ii) 3 kg and 2000 gm

   (iii) 8 cm and 2 m   (iv) 200 cm and 200 mm

   (v) 3 dozen and 2 gross   (vi) 60 paise and Rs 4.2

4. (i) What is the ratio of Rs 8 to 80 paise?

   (ii) A steel tape is 10 m long and 2.4 cm wide. What is ratio of its breadth to the length?

   (iii) What is the ratio of the number edges of a cube to the number of sides of a square?

   (iv) If sohita's annual income is Rs 252000. Her annual savings amount to Rs 1,12000. What is the ratio of her savings to the expenditure?

   (v) The mathematics text contains 350 pages. If number of chapters is 25, then find ratio of the number chapters to the total number of pages.

   (vi) In a school – library 2100 books are having hard-binding whereas the remaining 2800 books have paper-binding. Find the ratio of number of books with hard binding to the total number of books in the library.

5. Write the following ratios in the simplest form:

   (i) 13:39   (ii) 42:63   (iii) 48:64

   (iv) 36:80   (v) 27:63   (vi) 60:300

6. Look at the following diagram

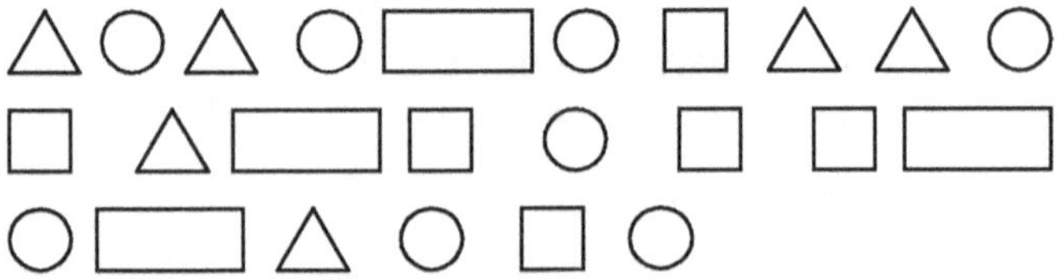

△ ○ △ ○ ▭ ▭ ○ ▭ △ △ ○
▭ △ ▭ ▭ ○ ▭ ▭ ▭
○ ▭ △ ○ ▭ ○

Find in the lowest terms the ratio of

(i) The number of △ to the number of all shapes.

(ii) The number of ○ to the number of all shapes.

(iii) The number of ▭ to the number of all shapes.

(iv) The number of ▭ to the number of ○

(v) The number of △ to the number of ▭

(vi) The number of all shapes to the number of ▭

7. Write an equivalent ratio of:

    (i) 4:5 having its consequent as 20

    (ii) 1:3 having its antecedent as 11

    (iii) 6:7 having its antecedent as 72

    (iv) 5:8 having its consequent as 40

    (v) 2:13 having its consequent as 130

8. Convert each of the following ratios in its lowest terms:

    (i) 6 months: $1\frac{1}{2}$ years   (ii) 20 kg: 1 quintal

    (iii) $12\frac{1}{2} : 6\frac{1}{4}$ (iv) 1.5: 7.5

    (v) 2 m 40 cm: 2 m 88 cm   (vi) 5 kg 200 gm: 2 kg 600 gm

9. Convert each of the following into whole number ratio:

    (i) $\frac{1}{4} : \frac{1}{5}$ (ii) $\frac{2}{7} : \frac{3}{11}$

    (iii) $\frac{4}{3} : \frac{2}{5}$ (iv) $\frac{1}{8} : \frac{2}{9}$

10. Fill in the blank by < or > in each of the following:

    (i) 2:3 ▭ 3:4   (ii) 3:5 ▭ 4:7

    (iii) 5:7 ▭ 11:23   (iv) 15:7 ▭ 7:9

11. (a) Write the following ratios in ascending order:

   2:1,   15:8,   9:13

   (b) Write the following ratios in descending order:

   4:7,   2:3,   6:7

12. Divide 144 in the ratio $1\frac{1}{2}:2\frac{1}{2}$

13. Divide 450 between A and B in the ratio 8:7

14. An alloy of zinc and copper weighs $37\frac{1}{2}$ kg. If in the alloy the ratio of copper and zinc is 4:1 find the weight of copper in it.

15. In a Bal-mela on children's day class VI students Rajat and Mohan organised a 'funny-game' corner and earned a profit of Rs 1200. Rajat got as much as twice the share mohan. Find the share of Rajat in the profit.

16. A total of 1800 students appeared in a Board's examination. The ratio of the number of students passed with 'A' grade to the number of students passed with 'B' grade is 7:5. If all students were declared pass, then how many students passed the examination with 'A' grade?

Answers

1. (i) (a) 5:6   (b) 6:5   (ii) (a) 1:4   (b) 4:1

   (iii) (a) 2:5   (b) 5:2   (iv) (a) 1:5   (b) 1:3   (c) 2:3

   (v)

	(a)	(b)	(c)	(d)	(e)	(f)
Antecedent	1	3	7	10	15	13
Consequent	5	4	9	8	2	17

   (vi) (a) 1:6   (b) 6:7

2. (i) 3:1   (ii) 1:9   (iii) 1:4   (iv) 5:2   (v) 2:3   (vi) 8:5
3. (i) 6:5   (ii) 3:2   (iii) 1:25   (iv) 10:1   (v) 1:8   (vi) 1:7
4. (i) 10:1   (ii) 3:1250   (iii) 3:1   (iv) 4:5   (v) 1:14   (vi) 3:8
5. (i) 1:3   (ii) 2:3   (iii) 3:4   (iv) 9:20   (v) 3:7   (vi) 1:5
6. (i) 1:4   (ii) 1:3   (iii) 1:4   (iv) 1:2   (v) 1:1   (vi) 6:1
7. (i) 16:20   (ii) 11:33   (iii) 72:84   (iv) 25:40   (v) 20:130
8. (i) 1:3   (ii) 1:5   (iii) 2:1   (iv) 1:5   (v) 5:6   (vi) 2:1
9. (i) 5:4   (ii) 22:21   (iii) 10:3   (iv) 9:16

10. (i) <   (ii) >   (iii) >-   (iv) <   (v) >
11. (a) (9:13),   (15:8),   (2:11)   (b) (4:7),   (2:3),   (6:7)
12. 54 and 90   13. 240 and 210   14. 30 kg
15. Rs 800   16. 1050

PROPORTION

An equality relation between two ratios is called the proportion.

For example, 3:5 = 6:10

∴ 3:5 and 6:10 form a proportion.

In a proportion, four quantities are related in a specific manner, such that the ratio between the first two is equal to the ratio of the last two.

Thus, proportion is equality of two ratios.

The symbol for proportion is a double colon "::"

∴ 3:5 = 6:10 is written as   3:5 :: 6:10

It is read as:

    "3 is to 5 as 6 is to 10"

Similarly, 4:3 and 12:9 are equal ratios,

    ∴   4:3 :: 12:9

  i.e.   "4 is to 3 as 12 is to 9"

In general, if (a:b) = (c:d) Then the two ratios form a proportion i.e.

                    a:b :: c:d

The four quantities a, b, c, d forming a proportion are said to be in proportion.

                or,

a, b, c and d are proportional.

Since, in a proportion, there are four terms.

They are named as first term, second term, third term and fourth term.

So, in a:b :: c:d

a, b, c and d are respectively the *first term, second-term, third term* and the *fourth term.*

The first and fourth terms are called the 'extremes'.

The second and third terms are called the *'means'*.

i.e. 'a' and 'd' are extremes, and   'b' and 'c' are means.

In a proportion,

> Product of Extremes = Product of Means

$$\text{Thus, a mean} = \frac{\text{Product of extremes}}{\text{other mean}}$$

$$\text{An extreme} = \frac{\text{Product of means}}{\text{other extreme}}$$

NOTE
Though in a ratio, the two quantities must be of the same kind. But in a proportion it is not necessary that all the four quantities to be of the same kind. Of course, the first two must be of the same kind and the last two must be of the same kind.

To solve different types of problems on proportion following facts are very useful.

i. A proportion is an expression which states that the two ratios are equal.
ii. Each quantity in a proportion is called its term or its proportional.
iii. If four quantities a, b, c and d are in proportion (i.e. a:b :: c:d), then $a \times d = b \times c$.
iv. If a:b :: b:d, then a, b, d are said to be in continued proportion.
v. If a, b, d are in continued proportion, then 'b' is called the mean proportional between 'a' and 'd'.
vi. If a, b, d are in continued proportion, then d is called the third proportional, to the first and the second.

In a continued proportion, the three quantities must be of the same kind.

EXAMPLE - 13

State whether the numbers 4, 12, 9 and 27 are in proportion or not.

SOLUTION

The given four numbers in order are: 4, 12, 9, 27.

∴   Product of extremes = $4 \times 27 = 108$

      Product of means   = $12 \times 9 = 108$

i.e. [Product of means] = [Product of extremes]

∴ 4, 12, 9, 27 are in proportion.

EXAMPLE - 14

Find the fourth proportion of 4, 3 and 24.

SOLUTION

Let the fourth proportional is 'x' such that 4, 3, 24 and x are proportional.

Extremes are 4 and x

∴ Means are 3 and 24

Now, Product of means = 3 × 24

Product of extremes = 4 × x

We know that, if the four quantities are proportional then

[product of extremes] = [product of means]

∴ 4 × x = 3 × 24

or x = $\dfrac{3 \times \cancel{24}^{6}}{4}$ = 3 × 6 = 18

Thus, the fourth proportional is 18.

EXAMPLE - 15

If two extremes of a proportion are 5 and 80, and one of the means is 10, then find the other mean.

SOLUTION

Let the other mean be 'x'.

∴ The numbers 5, 10, x and 80 are in proportion.

∴ product of means = product of extremes

or 10 × x = 5 × 80

or x = $\dfrac{5 \times \cancel{80}^{8}}{\cancel{10}_{1}}$ = $\dfrac{5 \times 8}{1}$ = 40

The required other mean = 40

Exercise - 2

1. State whether the two ratios form a proportion or not.
   (i) 16:12 and 4:3
   (ii) 24:32 and 27:36
   (iii) 3:4 and 4:10
   (iv) 2:7 and 14:50
   (v) 12:48 and 11:44
   (vi) 14:12 and 7:4

2. Find 'x' in the following such that the four numbers form a proportion:
   (i) 10, 4, 5, x
   (ii) 20, 4, 15, x
   (iii) x, 2, 35, 5
   (iv) 20, x, 10, 15

3. Find the third terms of a proportion whose first, second and fourth terms are 80, 32 and 16.

4. If the last three terms of a proportion are 24, 15 and 20, then find the first term.

5. If 35, 24, 24, x are in proportion then find x.

6. If 625, x, x, 4 are in proportion, find x.

7. The extremes of a proportion are 24 and 8. If one of the means be 6 then find the other mean.

8. Find the mean proportion of 18 and 8.

9. Find a number which has the same ratio with 48 as 30 has to 72.

10. If the length and breadth of a rectangle are in the ratio 6:5 and its length is 120 m, then find its breadth.

Answers

1. (i) Yes (ii) Yes (iii) No (iv) No
   (v) Yes (vi) No
2. (i) 2 (ii) 3 (iii) 14 (iv) 30
3. 40   4. 18   5. 16   6. 50
7. 32   8. 12   9. 20   10. 100 m

Unitary Method

"A method in which we find the value one unit and then the required number of units" is called as *Unitary Method*.

For example; let the 15 tons of a metal costs Rs 4,50,000 then the cost of 1 kg of metal would cost

$$Rs\ (4,50,000 \div 15,000)$$
$$= Rs\ 30$$

∴ 1 ton = 1000 kg

Now, it is easy to find the cost the metal for any number of kilograms.

Suppose, we want the cost of 160 kg of the metal then, it is equal to Rs 30 × 160 = Rs 4,800.

The unitary method involves the method of direct variation (or direct proportion).

DIRECT VARIATION

"When two quantities are so related that an increase in one causes a corresponding increase in the other". This principle is called direct variation.

For example: Let us take the case of above example.

If weight of metal increases, the cost is also increased proportionally. i.e.

The more the weight of metal, the more the cost.

Some more examples of direct variation are

(i) The greater the speed, the greater the distance covered.

(ii) The more the invitees at a dinner party, the greater the expenses.

(iii) The more the syllabus for an exam, the greater the time to revise.

(iv) The greater the diameter of a circle, the greater the area.

Let us go back to above example, in which we have to find the cost of 160 kg of the metal. Let it be Rs x. Since, it is a case of direct variation in two quantities, weight and cost.

Since, **WEIGHT**   **COST**

      15,000 kg   Rs 4,50000   ∴ 1 ton = 1000 kg

      160 kg   Rs x   ∴ 15 ton = 15000 kg

Obviously, the ratio of 15000 kg to 160 kg = the ratio of Rs 45000 to Rs x

i.e.   15000:160   = 450000:x

In a proportion, we have product of extremes = product of means

∴   15000 × x   =   450000 × 160

or $\quad x = \dfrac{\cancel{450000}^{30} \times 160}{\cancel{15000}_{1}} = 30 \times 160 = 4800$

Thus, the cost of 160 kg of metal is Rs 4800

EXAMPLE - 16

A man earns Rs 105000 in 5 months. How much will he earn in 8 months?

SOLUTION

Let the required earning be Rs x

EARNING	TIME
Rs 105000	5
Rs x	8

∴ 105000 : x = 5 : 8

∴ [product of means] = [product of extremes]

or $\quad 5 \times x = 105000 \times 8$

or $\quad x = \dfrac{105000 \times 8}{5} = 21000 \times 8 = 168000$

Thus the required earnings = Rs 1,68000.

EXAMPLE

Rajat earns Rs 1,44000 in 6 months. In how many months will he earn Rs, 180000?

SOLUTION

Let the required number of months be 'x'

Now,

EARNINGS	MONTHS
Rs 144000	6
Rs 180000	x

∴ 144000 : 180000 = 6 : x

Since, product of extremes = product of means

∴ 144000 × x = 6 × 180000

or x $= \dfrac{\cancel{6}^{1} \times \cancel{180000}^{15}}{\cancelto{24000}{144000}_{2}} = \dfrac{15}{2} = 7.5$

So, the required number of months = 7.5

EXAMPLE - 18

5 dozen bananas cost Rs 200. How many bananas can be purchased for Rs 320?

SOLUTION

Let the required number of bananas be x dozens.

Now,

	QUANTITY OF BANANAS	COST OF BANANAS
	5	Rs 200
	x	Rs 320

∴ 5:x = 200:320

Since, product of means = product of means

∴ x × 200 = 320 × 5

or x $= \dfrac{\cancel{320}^{8} \times \cancel{5}^{1}}{\cancelto{40}{200}} = 8$

∴ Required quantity of bananas is 8 dozen.

SPEED TIME AND DISTANCE

SPEED

Somesh covers a distance of 5 km on bicycle in 1 hour. We say: his speed is 5 km per hour (or 5 km/hr). An insect covers 3 metre distance in 1 minute. We say that its speed is 3 metre per minute (or 3 m/minute).

In general,

The speed of a body is the distance moved in unit time. We denote it as km/hr, m/sec, etc.

Speed can be calculated using the formula:

$$\text{Speed} = \dfrac{\text{Distance}}{\text{Time}}$$

Ratio and Proportion | 187

Time can be calculated using the formula:
$$\text{Time} = \frac{\text{Distance}}{\text{Speed}}$$
Distance can be calculated using the formula:
$$\text{Distance} = \text{Speed} \times \text{Time}$$

NOTE

1 km = 1000 metre and 1 hour = 3600 sec

$$\therefore 1 \text{ km/hour} = \frac{1000 \text{ m}}{3600 \text{ sec}} = \frac{5}{18} \text{ m/sec}$$

To convert km/hr into m/sec, multiply by $\frac{5}{18}$

To convert m/sec into km/hr, multiply by $\frac{18}{5}$

EXAMPLE

Prabhat cycles at 18 km/hr. How long will he take to cover a distance of 750 metres?

SOLUTION

Here speed = 18 km/h

$$= 18 \times \frac{5}{18} \text{ m/sec} = 5 \text{ m/sec}$$

Distance = 750 m

$$\text{Time} = \frac{750 \text{ m}}{5 \text{ m/sec}} = 150 \text{ sec} \qquad \text{Using, Time} = \frac{\text{Distance}}{\text{Speed}}$$

$$= \frac{150}{60} \text{ minutes} = 2.5 \text{ minutes}$$

EXAMPLE

Sumita covers 1200 metre distance in 4 minutes. Find her speed in km/hr.

SOLUTION

$$\therefore 4 \text{ minutes} = 4 \times 60 \text{ seconds} = 240 \text{ seconds}$$

$$\therefore \text{Using, speed} = \frac{\text{Distance}}{\text{Time}}, \text{ we get}$$

$$\text{Speed} = \frac{1200}{240} \text{ m/sec} = 5 \text{ m/sec}$$

To find speed as km/sec

$$5 \text{ m/sec} = 5 \times \frac{18}{5} \text{ km/hr} = 18 \text{ km/hr}$$

EXAMPLE

How far will an aeroplane flying at 450 km/hr fly in 35 seconds?

SOLUTION

speed = 450 km/h = $450 \times \frac{5}{18}$ m/sec = 125 m/sec

Time = 35 seconds

Using, Distance = Speed × Time, we get

Distance = 450 km/hr × 35 seconds

= 125 m/sec × 35 sec

= 125 × 35 m = 4.375 m

= $\frac{4375}{1000}$ km = 4.375 km

Alternate method

Distance = 450 km/hr × $\frac{35}{3600}$ hr

= $\frac{450 \times 35}{3600}$ km = 4.375 km

EXERCISE - 3

1. If 10 ballpoint pens cost Rs 6900, how much would 6 pens cost?
2. If 15 bicycles cost Rs 84000. How many bicycles can be bought for Rs 100800?
3. If $5\frac{1}{4}$ kg of sweets contain $1\frac{1}{2}$ kg of basin, 6 kg of basin would be consumed by how much sweets?
4. Find the rent for 5 shops in a month, if the annual rent of these 5 shops is Rs 180000.
5. A car can travel 240 km in 12 litres of petrol. How much distance will it travel in 29 litres of petrol?
6. Rohan earns Rs 270000 in 9 months. At the same rate, in how many months does he earn Rs 420000?
7. A wire having weight as 1.3 kg, is 3 metres long. What would be the weight of the same kind of 7.8 m long wire?
8. The quarterly school fee of class VI is Rs 9630. What will be the fee for 8 months?

9. Express the following speeds in metres/second:
   (i) 180 km/hr   (ii) 90 km/hr   (iii) 108 km/hr
   (iv) 18 km/hr   (v) 36 km/hr   (vi) 54 km/hr

10. Express the following speeds in km/hr:
    (i) 10 m/sec   (ii) 15 m/sec   (iii) 20 m/sec
    (iv) 3 m/sec   (v) 50 m/sec   (vi) 30 m/sec

Answers

1. Rs 4140   2. 18 bicycles   3. 21 kg   4. Rs 15000
5. 580 km   6. 14 months   7. 3.38 kg   8. Rs 25680
9. (i) 50 m/sec   (ii) 25 m/sec   (iii) 30 m/sec   (iv) 5 m/sec
   (v) 10 m/sec   (v) 15 m/sec
10. (i) 10 km/hr   (ii) 54 km/hr   (iii) 72 km/hr   (iv) 10.8 km/hr
    (v) 180 km/hr   (v) 108 km/hr

Miscelleneous Exercise

1. Write the ratio of:
   (i) 4 cm to 5 cm   (ii) 5 kg to 3 kg
   (iii) 5 cm to 60 mm   (iv) 3 kg to 2000 gm

2. Express the following fractions as ratios:
   (i) $1\frac{1}{2}$ (ii) $\frac{5}{2}$ (iii) $\frac{3}{5}$ (iv) $\frac{7}{10}$ (v) $\frac{9}{8}$ (vi) $\frac{6}{7}$

3. Write the following ratios as fractions:
   (i) 1:2   (ii) 3:4   (iii) 7:9   (iv) 15:23
   (v) 41:38   (vi) 9:11

4. Express each of the following ratios in its lowest terms:
   (i) 6:3   (ii) 4:12   (iii) 10:15   (iv) 6:15
   (v) 200:500   (vi) 1.5:4.5

5. Fill in the blanks using '>' or '<':
   (i) 3:5 _____ 7:4   (ii) 6:8 _____ 9:13
   (iii) 1:7 _____ 7:10   (iv) 9:5 _____ 6:7
   (v) 3:20 _____ 18:13   (ii) 13:14 _____ 11:18

6. (a) Rewrite the following ratios in ascending order:

   5:7,   6:1,   3:5

   (b) Rewrite the following ratios in descending order:

   7:8,   2:3,   5:4

7. Divide 160 chocolates between two groups of children in the ratio 3:5. How many chocolates does each group receive?

8. Find 'x' in the following:

   (i) 2:3 :: 4:x   (ii) 3:4 :: x:6   (iii) x:3 :: 4:6

   (iv) 4:5 :: x:25   (v) 5:30 :: 8:x   (vi) 5:10 :: x:80

9. In a school, the number of boys is 2250, and the ratio of number of boys to the number of girls is 3:4. Find the number of girls in the school.

10. The distance between Jaipur and Ajmer is 130 km. The distance between Delhi and Bhopal is 780 km. A train takes 12 hours to travel from Delhi to Bhopal. How much time will it take to travel from Ajmer to Jaipur?

11. Samantha cycles at 36km/hr. How long will she take to cover a distance of 2400metres?

Answers

1. (i) 4:5   (ii) 5:3   (iii) 5:6   (iv) 3:2

2. (i) 3:2   (ii) 5:2   (iii) 3:5   (iv) 7:10

   (v) 9:8   (vi) 6:7

3. (i) $\dfrac{1}{2}$ (ii) $\dfrac{3}{4}$ (iii) $\dfrac{7}{9}$ (iv) $\dfrac{15}{23}$ (v) $\dfrac{41}{38}$ (vi) $\dfrac{9}{11}$

4. (i) 2:1   (ii) 1:3   (iii) 2:3   (iv) 2:5

   (v) 2:5   (vi) 1:3

5. (i) <   (ii) >   (iii) <   (iv) >

   (v) <   (vi) >

6. (a) (3:5),   (6:1),   (5:7)

   (b) (5:4),   (7:8),   (2:3)

7. 60 and 100

8. (i) x = 6   (ii) x = 4.5   (iii) x = 2   (iv) x = 20
   (v) x = 48   (vi) x = 40

9. 3000   10. 2 hrs   11. 4 minutes

Hots

1. An office at 9.00 a.m. and closes at 5:40 p.m. with a lunch break of 40 minutes. What is the ratio of lunch break to the total period in the office?

2. In the following figure, each division represents 1cm. Express numerically the ratios of the following:

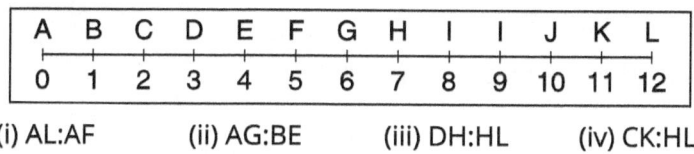

   (i) AL:AF   (ii) AG:BE   (iii) DH:HL   (iv) CK:HL

Answers

1. 1:13
2. (i) 12:5   (ii) 2:1   (iii) 4:5   (iv) 9:5

Mental Maths

1. What do we call the first term of a ratio?
2. What is the fractional form of the ratio 10:22 in the lowest terms?
3. What is the equivalent ratio of 2:3 having the consequent as 39?
4. What is the equivalent ratio of 4:5 having its antecedent as 20?
5. Write the lowest terms of the ratio 100:175
6. Are the ratios 7:9 and 119:153?
7. If $\frac{3}{5} > \frac{4}{7}$, then put the appropriate symbol > or < in the following box:?

$$\frac{4}{7} \quad \Box \quad \frac{3}{5}$$

8. Is it true to say:

   The ratio of (3 boys) to (5 girls) is as the ratio of (18 chairs) to (30 chairs)?

9. Is it true to say that in a:c :: c:d, $c^2 = a \times d$?

10. In a proportion, the product of extremes is equal to what?

11. By which number do we multiply to convert m/sec speed into km/hr?

ANSWERS

1. Antecedent 2. $\frac{5}{11}$ 3. 26:39 4. 20:25
5. 4:7 6. Yes 7. < 8. Yes
9. Yes 10. Product of means 11. $\frac{18}{5}$

MULTICHOICE QUESTIONS

1. The ratio of 8 birds to 24 birds is:
   (i) 1:5   (ii) 1:3   (iii) 5:1   (iv) 3:1

2. The second term of a ratio is called as:
   (i) Antecedent   (ii) Consequent
   (iii) Fractional term   (iv) Linear term

3. The ratio number of sides of a square to the number of its diagonals is:
   (i) 1:2   (ii) 2:1   (iii) 4:1   (iv) 1:4

4. In an almirah of a school library the ratio of Maths books to English books is 4:7. Which of the following is the total number of the books in the almirah?
   (i) 21   (ii) 22   (iii) 18   (iv) 19

5. The greatest ratio among 3:2, 8:5, 121:75 and 25:40 is
   (i) 121:75   (ii) 8:5   (iii) 25:40   (iv) 3:2

6. The smallest ratio among 21:8; 15:4; 11:16 and 9:24 is
   (i) 11:16   (ii) 9:24   (iii) 21:8   (iv) 15:4

7. A 100 cm of wire is cut in the ratio 7:3. How much longer is the longer piece than the shorter piece.
   (i) 70 cm   (ii) 50 cm   (iii) 40 cm   (iv) 30 cm

8. A recipe needs sugar and flour in the ratio 1:3. If 1.5 kg of flour is used in the recipe, which of the following is the amount of sugar needed?
   (i) 1 kg   (ii) $\frac{1}{2}$ kg   (iii) 0.7 kg   (iv) 0.2 kg

9. Which of the following is same as 4:5?

   (i) 28:25 (ii) 6:30 (iii) 28:25 (iv) 16:20

10. A train travel 384 km in 6 hours and a bus travels 160 km in 4 hours at uniform speed. Which of the following represent the ratio of the [distance travelled in 1 hr by the bus] to the [distance travelled in one hour by the train]?

    (i) 5:8 (ii) 8:5 (iii) 4:5 (iv) 1:2

ANSWERS

1. (ii)   2. (ii)   3. (ii)   4. (ii)
5. (iv)   6. (ii)   7. (iii)   8. (ii)
9. (iv)   10. (i)

WORKSHEET

1. Fill in the blanks:

   (i) A ratio is a comparison of two quantities of the same kind by _____.

   (ii) An equality relation between two ratios is called the _____.

   (iii) When quantities are so related that an increase in one causes a corresponding increase in the other, then this relation is called the _____.

   (iv) To find the ratio of two quantities, first they must be expressed in _____ units.

   (v) The ratio of 5 gm to 25 gm is the same as 200 gm to _____ kg.

2. Match the column A to the column B:

COLUMN-A	COLUMN-B
(a) If a, b, c and d are in proportion then	(i) vary directly
(b) If a, b, c and d are in proportion then, extremes are:	(ii) consequent
(c) An equality of two ratios is called as:	(iii) ad = bc
(d) If increase in one quantity causes the increase in the other quantity then the two quantities are said to:	(iv) proportion
(e) The second term of a ratio is called as _____.	(v) 'a' and 'd'

3. Write 'true' or 'false' for each of the following statements.

(i) The ratio of 1 hour to 60 minutes is 1:1

(ii) The ratio (3.5):(7.0) is in its lowest form

(iii) The ratio 4:7 is different from the ratio 7:4

(iv) A ratio is always smaller than 1

(v) If b:a = c:d then a, b, c, d are in proportion.

(vi) If a:c = b:d then a, c, b, d are in proportion.

(vi) The four quantities of a proportion must be of the same kind.

(vii) The first two terms of a proportion should be of the same kind and the last two terms together can be of other kind.

(viii) Two quantities can be compared only if they are of the same kind and same units.

(ix) Two numbers in the ratio 3:5 have sum equal to 80. The larger part is 50.

(x) The ratio $1\frac{1}{6} : 2\frac{1}{6}$ in simplest form is 6:13

4. In a school, the ratio of the large class rooms to small class rooms is 4:3. If the number of large class room is 15, then find the number of small classrooms.

5. For preparing a cake for 6 persons, the recipe calls for 1 cup of milk for every $2\frac{1}{2}$ cups of flour. For how many persons the cake would be sufficient if $4\frac{2}{3}$ cups of both flour and milk is taken?

6. Fill in the blanks:

DISTANCE	TIME	SPEED
(i). _____	4 hrs	25 km/hr
(ii). 225 km	_____	45 km/hr
(iii). 200 km	4 hrs	_____
(iv). 180 km	_____	40 km/hr

ANSWERS

1. (i) division   (ii) proportion
   (iii) direct proportion   (iv) same   (v) 1

2. (a) (iii)   (b) (v)   (c) (iv)
   (d) (i)   (e) (ii)

3. (i) True (ii) False (iii) True
   (iv) False (v) False (vi) True
   (vii) True (viii) True (ix) True
   (x) False
4. 20 classrooms 5. 8 persons
6. (i) 100 km (ii) 5 hours (iii) 60 km/hr
   (iv) 4.5 hours

Chapter 8
Fundamental Concept of Algebra

Introduction

In arithmetic, we use symbols 0, 1, 2, 3, 4, 5, 6, 7, 8, and 9. These symbols have specific (individual) value.

The value of 1 is always 'one' and not 0, 2, 3, 4, 5, ... etc.

The value of 2 is always 'two' and not 0, 1, 3, 4, 5, ... etc.

The value of 3 is always 'three' and net 0, 1, 2, 4, 5, ... etc.

----------   ----------------------   ----------   ----------------------

----------   ----------------------   ----------   ----------------------

The value of 9 is always 'nine' and not 0, 1, 2, 3, 4, 5, ... etc.

The value of 0 is always 'zero' and not 1, 2, 3, 4, 5, ... etc.

It is like that for large numbers also.

There is another branch of mathematic, called 'ALGEBRA' in which numbers are used as in 'arithmetic', but letter (a, b, c, ... x, y, z, etc) are also used. In algebra, these letters do not have specific (individual) value. A letter can have any value. The value of

'x' can be 1, 2, 3, 4, ... 10, 11, 12, ...

'y' can be 1, 2, 3, 4, ... 10, 11, 12, ...

'z' can be 1, 2, 3, 4, ... 10, 11, 12, ..., etc.

These letters, in algebra, behave like 'numbers', so they are called *"literal numbers"* or simply *"literals"* since, the literals behave like common numbers, so, they obey all the basic - operations (+, -, ×, ÷), which are obeyed by arithmetic numerals.

Difference Between 'Numerals' and 'Literals'

A *literal* can have a lot of different values. The value of 'x' can be 5 in one situation and 20 in other situation; i.e. a 'literal' can be associated with an endless variety of values.

A *numeral* can have a specific value. The numeral '5' is always a 5. It is never a 6 whichever may be the situation.

NOTE

Literals can stand for numbers and not for 'number of things.' There is no more sense in saying that "an almirah contains x books" than there is in sayings that "it has 18 books."

Apart from operations, literals, also obey other symbols such as:

$=$	means,	**"is equal to"**
$\neq$	means,	**"is not equal to"**
$>$	means,	**"is greater than"**
$<$	means,	**"is less than"**
$\not>$	means,	**"is not greater than"**
$\not<$	means,	**"is not less than"**
$\therefore$	means,	**"therefore"**
$\because$	means,	**"because or since"**
$\Rightarrow$	means,	**"implies that"**

DID YOU KNOW?

Where does the word *"ALGEBRA"* come from ?

Al Khwakizmi wrote a book —

"Hisab al – jabr a *almukabla*"

But those who didn't speak Arabic, abbreviated the title as *"Algebra"*

Thus, 'algebra' is an extension of arithmetic and is defined as:

A branch of mathematics which deals with numbers and letters and obey all basic rules and operations such as **addition, subtraction, multiplication** *and* **division.**

OR **Algebra is generalized arithmetic**

CONSTANTS AND VARIABLES

Every numeral has a specific (individual) value. Such values are *same* everywhere in all situations i.e. remain unchanged or constant. Therefore such numerals are called *constants*. On the other hand 'literals' have variable values so they are called *variables*.

For example, 5, (−7), $\frac{3}{4}$ etc, are CONSTANTS, whereas a, b, c, d, m etc, are VARIABLES.

EXAMPLE – 1

Identify the variables and constants (numerals) in the following:

(i) 9   (ii) −7   (iii) x   (iv) xy   (v) −7x
(vi) $\frac{1}{2}$ (vii) 0   (viii) $\frac{3}{4}$xy   (ix) $\frac{-lm}{2a}$ (x) $\frac{3}{4mn}$

SOLUTION

(i) ∵ 9 is a numeral, ∴ 9 is a CONSTANT

(ii) ∵ −7 is a numeral, ∴ −7 is a CONSTANT

(iii) x is a literal, ∴ x is a VARIABLE

(iv) ∵ x and y both are literals

∴ xy is a VARIABLE

(v) ∵ x is a literal, ∴ it is a VARIABLE

⇒ −7x is a VARIABLE

A combination of a 'variable' and a constant is a variable

(vi) ∵ $\frac{1}{2}$ is a numeral, ∴ It is a CONSTANT

(vii) ∵ '0' is a numeral, ∴ it is a CONSTANT

(viii) ∵ $\frac{3}{4}$xy is a combination numeral $\left(\frac{3}{4}\right)$ and variables (x and y)

∴ It is a VARIABLE

(ix) ∵ $\frac{-lm}{2a}$ is combination of numeral $\left(-\frac{1}{2}\right)$ and literals (l, m, and a),

∴ it is a variable

NOTE

(i) **Literal numbers obey all the rules of addition (+), subtraction (−), multiplication (×) and division (÷). i.e.**

x + y means 'y' is added to 'x'; x − y means 'y' is subtracted from 'x'; x × y means 'x' is multiplied by 'y' and x ÷ y means 'x' is divided by 'y.'

(ii) **Sometimes, we use the sign of multiplication '·' instead of '×.' But incase of literal numbers, usually we omit the sign of multiplication between two literal numbers or between more than two literals or between a digit and a literal numbers. For example:**

5 × l can be written as 5 × l or 5l

3 × a × b can be written as 3.a.b or 3ab

x × y × z can be written as x.y.z or xyz

a × b × c × d × 3 can be written as a.b.c.d.3 or 3abcd

(iii) 3·5 means '3 decimal 5' and

3.5 mean '3 multiplied by 5' sometimes 3.5 (Three decimal five) is mistaken as 3·5 (3 multiplied by 5). To avoid this confusion, we do not use "(⌴)" between two digits to show multiplication.

(iv) xy means product of 'x' and 'y.' This is true only in literal numbers but not in numerals; since 45 does not mean the product of 4 and 5.

MATHEMATICAL STATEMENTS

We express our ideas by using sentences. We know that "a *sentence* is a set of words which are grammatically arranged to convey a meaningful sense. For example:

(i) Give me a piece of paper.
(ii) Let us attend the morning assembly.
(iii) There are 18 school-buses in our school.
(iv) Five students are absent today.
(v) Two and seven together make nine.
(vi) Eighteen is greater than eleven.
(vii) Three more than eight is eleven.
(viii) There are eight planets in our solar system.

These sentence make some *statement*. Statements involving one or more numbers (numeral or literal) are called mathematical statement. For example; the statements (iii), (iv), (v), (vi), (vii) and (viii) are mathematical statements (because numerals are involved in these statements).

TO WRITE MATHEMATICAL STATEMENTS IN ALGEBRAIC FORM

The statements involving literals, or literals and numerals, alone or along with symbols (+, −, ×, ÷, ∵, ∴, =, > or < etc.) are called algebraic statements.

Example – 2

Write the following mathematical statements in algebraic form:

    (i)   x boys are more than 13 girls in a group of students.

    (ii) x and 7 are together equal to 15.

    (iii) Twice of y is equal to 42.

    (iv) Half of n and twice of m is 32.

    (v) 78 less than 6 times p is 48.

    (vi) 62 and $\frac{3}{2}$ times x is greater than 100.

    (vii) $\frac{1}{2}$ x less than y is 10 more than z

    (viii) 3x and 3 less than y is 15.

Solution

Above statements converted to algebraic form as:

    (i) Students in the group = 13 girls + x boys

    (ii) $x + 7 = 15$ (iii) $2y = 42$

    (iv) $\frac{1}{2}n + 2m = 32$ (v) $6p - 78 = 48$

    (vi) $\left(68 + \frac{3}{2}x\right) > 100$ (vii) $y - \frac{1}{2} x = z + 10$

    (viii) $3x + y - 3 = 15$

Example – 3

Write the following algebraic form into mathematical statements:

    (i) $x - 3 > 15$ (ii) $3x = 72$ (iii) $x + 15 < 10$

    (iv) $\frac{x}{y} - 3 = 20$ (v) $y - 7 = 5$

Solution

    (i)   3 less than x is greater than 15

    (ii) 3 times x is 72

    (iii) 15 more than x is less than 10

    (iv) Quotient of x and y reduced by 3 is 20

    (v) 7 reduced from y is 5

POWERS ON VARIABLES

The repeated product of a literal with itself is written in the exponential form. We know that $5 \times 5 = 5^2$, $3 \times 3 \times 3 \times 3 = 3^4$

Similarly, we can write

$x \times x \times x \times x = x^3$ [∵ **x is a literal and literals between like numbers**]

$a \times a \times a \times a = a^4$

$7 \times p \times p \times p = 7p^3$

$5 \times a \times a \times b \times b \times b \times c \times c \times c \times c = 5a^2b^3c^4$

In x^3, the base is 'x' and exponent (or index) is 3. Thus, the exponent in power of a literal indicates the number of times the literal has been multiplied by itself.

∴ $x^9 = x \times x \times x \times x...$ [repeatedly multiplied by 9]

In $x^9 = x \times x \times x \times x \times x...$[9 times]

x^9 is exponential form and $(x \times x \times x \times x \times x ....$ 9 times) is the expanded form.

x^1 means x and p^1 means p

EXAMPLE – 4

Express the following in exponential form:

(i) $x \times x \times x \times x \times x \times x$

(ii) $a \times a \times a \times a \times 15 \times c \times c$

(iii) $b \times b \times b \times c \times c \times c \times c$

(iv) $9 \times y \times y \times y \times y \times y \times y \times z$

(v) $m \times m \times m \times$ .... 15 times

(vi) $10 \times p \times p \times p \times$ .... 20 times $\times q \times q \times q \times$ .... 15 times

SOLUTION:

(i) $x \times x \times x \times x \times x \times x = x^6$

(ii) $a \times a \times a \times a \times 15 \times c \times c$ $= a^4 \times 15c^2 = 15a^4c^2$

(iii) $b \times b \times b \times c \times c \times c \times c$ $= b^3 \times c^4 = b^3c^4$

(iv) $9 \times y \times y \times y \times y \times y \times y \times z = 9 \times y^6 \times z = 9y^6z$

(v) $m \times m \times m \times$ .... 15 times $= m^{15}$

(vi) $10 \times p \times p \times p \times \ldots 20$ times $\times q \times q \times q \times \ldots 15$ times
$$= 10 \times p^{20} \times q^{15} = 10p^{20}q^{15}$$

Example – 5

Express the following in expanded (product) form:

(i) $p^5 q^3$ (ii) $a^4 b^3 c^5$ (iii) $7m^2 n q^3$ (iv) $18 l^4 m^2 n^3$

(v) $20 a^6 b c^4$ (vi) $12 x^4 y^5 z$

Solution

(i) $p^5 q^3 = p^5 \times q^3$
$$= p \times p \times p \times p \times p \times q \times q \times q$$

(ii) $a^4 b^3 c^5 = a^4 \times b^3 \times c^5$
$$= a \times a \times a \times a \times b \times b \times b \times c \times c \times c \times c \times c$$

(iii) $7 m^2 n q^3 = 7 \times m^2 \times n \times q^3$
$$= 7 \times m \times m \times n \times q \times q \times q$$

(iv) $18 l^4 m^2 n^3 = 18 \times l^4 \times m^2 \times n^3$
$$= 18 \times l \times l \times l \times l \times m \times m \times n \times n \times n$$

(v) $20 a^6 b c^4 = 20 \times a^6 \times b \times c^4$
$$= 20 \times a \times a \times a \times a \times a \times a \times b \times c \times c \times c \times c$$

(vi) $12 x^4 y^5 z = 12 \times x^4 \times y^5 \times z$
$$= 12 \times x \times x \times x \times x \times y \times y \times y \times y \times y \times z$$

Example – 6

'x' students travel in school-bus No. 1. 'y' students travel in school-bus No. 2, whereas 3 times z students travel in the school-bus No.3. How many students in these three-school-buses altogether.

Solution

Number of students travelling in school-bus No.1 = x

Number of students travelling in school-bus No.2 = y

Number of students travelling in school-bus No.3 = 3z

∴ Total number of students travelling in the three school-bus = x + y + 3z.

EXAMPLE – 7

If the cost of a pen is Rs x and the cost of a pencil is Rs y, then find the total cost of 5 pens and 7 pencils.

SOLUTION

	Cost of 1 pen	= Rs x	
∴	Cost of 5 pens	= 5 × (Rs x)	= Rs 5x
	Cost of 1 pencil	= Rs y	
∴	Cost of 7 pencils	= 7 × (Rs y)	= Rs 7y

∴ Total of 5 pen and 7 pencils = Rs (5x + 7y)

EXERCISE – 1

1. Identify the variables and constants (numerals) in the following:

    (i) 2xy   (ii) 36   (iii) 9·45

    (iv) $\frac{1}{3}a$ (v) 3.2   (vi) $-\frac{7}{5}z$

    (vii) $\frac{115}{2}xy$ (viii) $\frac{69}{p}$ (ix) $-6xyz$

    (x) $6 - 7x$

2. Write the following mathematical statements in the algebraic form:

    (i) The perimeter of a square of side 's' is 4 times the side

    (ii) Four is less than seven times y

    (iii) Three and thrice of x make sixteen

    (iv) Eight is greater than five times x

    (v) Each side of an equilateral triangle is x. Its perimeter 'P' is three times the side

    (vi) Seven reduced by x is greater than y

    (vii) Sum of x and twice of y is greater than z

    (viii) $\frac{4}{5}$th of a number x greater than twice the number y

    (ix) The quotient of x and 5 is 10

    (x) The product of two numbers x and y is less than 100

3. Write the following algebraic form into mathematical statements:

   (i) $x + 8 > 11$
   (ii) $2x - y < 10$
   (iii) $x + (y + 5) = 20$
   (iv) $5y - 7 > 12$
   (v) $x - 2y < 7$
   (vi) $\dfrac{4}{y} - 8 = 25$
   (vii) $a + 3 = b$
   (viii) $3x - y > z$
   (ix) $\dfrac{x}{6} > 7$
   (x) $\dfrac{5}{c} - \dfrac{1}{2} = 42$

4. Look at the following shapes and answer the given questions:

   (a) Write the length of the knife.

   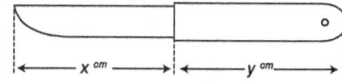

   (b) Write the length of the pencil.

   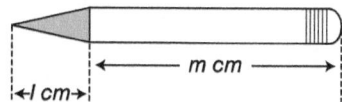

   (c) What is the length of the handle of the screwdriver.

   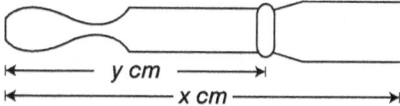

   (d) What is the total cost of these items?

   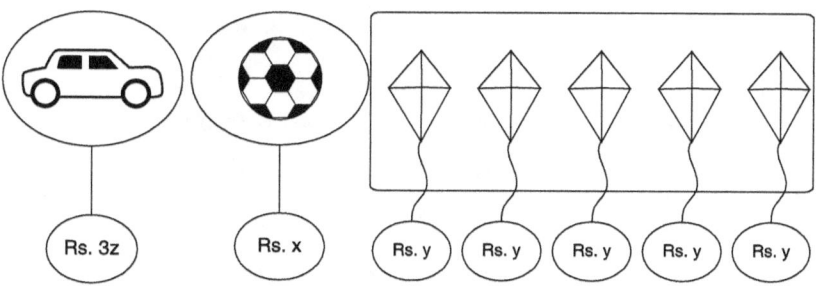

(e) Write the total cost of these fruits.

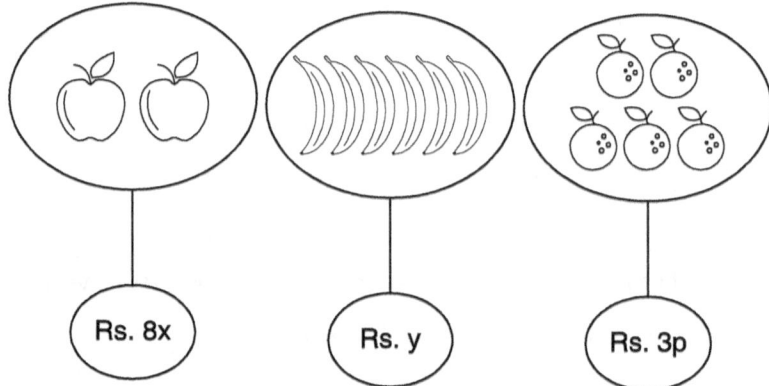

Rs. 8x   Rs. y   Rs. 3p

5. Express the following in exponential form:
   (i) $18 \times a \times a \times a \times a \times a \times b \times b \times b$
   (ii) $6 \times x \times x \times x \times y \times y \times y \times y$
   (iii) $3 \times p \times p \times p \times p \times 4 \times q \times q \times q \times q$
   (iv) $2 \times l \times l \times l \times l \times m \times m \times m$
   (v) $16 \times b \times b \times b \times c \times c \times c \times d \times d \times d$
   (vi) $8 \times c \times c \times d \times d \times d \times d \times d \times d \times d$
   (vii) $y \times y \times y \times \frac{1}{2} \times c \times c \times c \times d \times d \times d$
   (viii) $n \times n \times 7 \times m \times m \times m \times m \times p \times p$
   (ix) $12 \times p \times p \times q \times q \times q \times q \times r \times r \times r \times r$
   (x) $9 \times x \times x \times x \times y^3 \times z^3$

6. Express the following in expanded form:
   (i) $5x^2y$ (ii) $9yz^4$ (iii) $4x^3y^4z^2$
   (iv) $10a^2b^2$ (v) $2b^2c$ (vi) $12d^3eg^2$
   (vii) $3p^5q^3r$ (viii) $8l^3m^2n^3$ (ix) $7p^2qr^4$
   (x) $6a^2bc^4$

7. Praneeta scored x marks in English and 2y marks in maths. If she scored 48 marks in computer then what the total score in these three subjects?

8. Three times a number x is 15 more than twice the number y. Express this state using literals, numerals and basic symbols of operations.

9. The perimeter 'P' of a rectangle is given by 2 (length + breadth). If its l and breadth is b then express 'P' using numerals and symbols.

10. If the cost of an apple is Rs x and that of a banana is Rs y then find the cost of 10 apples and 12 bananas.

Answers

1. (i) variable (ii) constant (iii) constant (iv) variable
   (v) constant (vi) variable (vii) variable (viii) variable
   (ix) variable (x) variable

2. (i) Perimeter = 4s (ii) 4 < 7y (iii) 3 + 3x = 16
   (iv) 8 > 5x (v) P = 3x (vi) 7 − x > y
   (vii) x + 2y > z (viii) $\frac{4}{5}$x > 2y (ix) $\frac{x}{5}$ = 10
   (x) xy < 100

3. (i) The sum of x and 8 is greater than 11
   (ii) Twice a number x reduced by y is less than 10
   (iii) Sum of y and 5 added to x is 20
   (iv) Product of 5 and reduced by 7 is greater than 12
   (v) x reduced by twice y is less than 7
   (vi) 8 subtracted from the quotient of 4 and y is 25
   (vii) 3 added to a is b
   (viii) y subtracted from the product of 3 and x is greater than z
   (ix) Quotient of x and 6 is greater than 7
   (x) $\frac{1}{2}$ subtracted from the quotient of 5 and c is 42

4. (a) (x + y) cm (b) (l + m) cm (c) (x − y) cm
   (d) Rs (x + 5y + 3z) (e) Rs (8x + y + 3p)

5. (i) $18a^5b^3$ (ii) $6x^2y^4$ (iii) $12p^4q^4$ (iv) $2l^4m^3$
   (v) $16b^3c^3d^3$ (vi) $8c^2d^7$ (vii) $\frac{1}{2}c^3d^3y^3$ (viii) $7n^2m^4p^2$
   (ix) $12p^2q^4r^4$ (x) $9x^2y^3z^3$

6. (i) 5 × x × x × x × y (ii) 9 × y × z × z × z × z
   (iii) 4 × x × x × x × x × y × y × y × y × z × z
   (iv) 10 × a × a × b × b (v) 2 × b × b × c

(vi) $12 \times d \times d \times d \times e \times g \times g$

(vii) $3 \times p \times p \times p \times p \times p \times q \times q \times q \times r$

(viii) $8 \times l \times l \times l \times m \times m \times n \times n \times n$

(ix) $7 \times p \times p \times q \times r \times r \times r \times r$

(x) $6 \times a \times a \times b \times c \times c \times c \times c$

7. $x + 2y + 48$
8. $3x = 2y + 15$
9. $p = 2(l + b)$
10. $10x + 12y$

ALGEBRAIC EXPRESSION

Any numeral, variable or combination of numerals and variables connected by one or more of the symbols $(+, -, \times, \div)$ is called a ALGEBRAIC – EXPRESSION. For example:

$13x$; $2 - x$; $x + 3y$; $5x - \dfrac{2}{y}$; $3x - 5 + 2y$; $5x^2 - 2xy - \dfrac{9}{y}$ are algebraic expressions.

TERMS

The various parts of an algebraic expression separated by + or − are called the terms of the algebraic expression. For example:

ALGEBRAIC EXPRESSION	TERMS
(i) $7x$	$7x$
(ii) $9 - 7x + y$	$9, -7x$ and y
(iii) $2x + \dfrac{7}{y}$	$2x$ and $\dfrac{7}{y}$
(iv) $2x - \dfrac{7}{y}$	$2x$ and $-\dfrac{7}{y}$
(v) $4x^2 - 9xy + 8x^5 - \dfrac{7}{y}$	$4x^2, -9xy, 8z^5$ and $-\dfrac{7}{y}$

NOTE

(i) Only + and − signs separate the terms of a polynomial. '×' and '÷' do not separate them. That is why $7x$ is a single term, similarly $\dfrac{7}{x}$ is also a single term.

(ii) The sign + or − is an essential part of a term. For example, in the expression $2x - \dfrac{7}{y}$, the terms are $(+2x)$ and $\left(\dfrac{-7}{y}\right)$. The '+' sign is understood to be there with $2x$, '+' is not attached when the term is either 'alone' or the expression begin with it.

TYPES OF ALGEBRAIC EXPRESSIONS

MONOMIAL

An algebraic expression containing only one term is called a monomial.

For example; $5x$, -9, $\frac{5}{9}a^2b$, etc are all monomials.

BINOMIAL

An algebraic expression containing exactly two terms is called a binomial.

For example; $a + y$, $7 - x$, $ab - 7xyz$ are all binomials.

TRINOMIAL

An algebraic expression containing three terms is called a trinomial.

For example; $2p + q - 3r$, $\frac{x}{2} + \frac{3}{4}y - 5z$, $p + \frac{1}{2}q - 7$ etc. are trinomials.

QUADRINOMIAL

An algebraic expression containing four terms is called a quadrinomial.

For example; $3x - y + 2z - 7$, $a + b - 7c - 2d$, $21 + p - q - \frac{1}{2}r$ are all quadrinomial.

NOTE

(i) Any algebraic expression having two or more than two terms is also called a *'multinomial.'* All binomials, trinomials, quadrinomials are multinomials. For example, $2x + 3$, $7 - 3x + 2x$, $10x - y + z - 6$, $2x + 3y - 5z - \frac{6}{7} + x^2 - y^2$ are all multinomials.

(ii) A monomial is called also a 'simple expression.'

(iii) '$2x + 5x$' is not a binomial as $2x + 5x = 7x$

'$7 + 2x - \frac{1}{2}$' is not a trinomial as $7 - \frac{1}{2} = \frac{13}{2}$ so, $7 + 2x - \frac{1}{2}$ is written as $2x + \frac{13}{2}$, which is a binomial.

EXAMPLE – 8

Write the number of terms in each of the following and also identify *monomials, binomials, trinomials* and *quadrinomials* from the following expressions:

(i) $5x + 6 - 2x^2$
(ii) $3x + 4y - 5z - 10$
(iii) $6p$
(iv) $2x^2 - 9yz + 9z^2 - \frac{10}{x}$
(v) $5p^2 - q$
(vi) $7l$

(vii) $\dfrac{2x^2}{z} + \dfrac{3y}{x} - 5y^2z^2$ (viii) $\dfrac{1}{2}x^2 - \dfrac{3}{y^2}$

SOLUTION

ALGEBRAIC EXPRESSION	NUMBER OF TERMS	KIND OF THE EXPRESSION
(i) $5x + 6 - 2x^2$	3	Trinomial
(ii) $3x + 4y - 5z - 10$	4	Quadrinomial
(iii) $6p$	1	monomial
(iv) $2x^2 - 9yz + 9z^2 - \dfrac{10}{x}$	4	Quadrinomial
(v) $5p^2 - q$	2	Binomial
(vi) $7l$	1	monomial
(vii) $\dfrac{2x^2}{z} + \dfrac{3y}{x} - 5y^2z^2$	3	Trinomial
(viii) $\dfrac{1}{2}x^2 - \dfrac{3}{y^2}$	2	Binomial

DID YOU KNOW?

The word "QUAD" refers to "FOUR", and "POLY" means "MANY."

FACTORS OF A TERM

Each of the quantity (literal or numeral) multiplied together to get a product is called a *FACTOR* of the product.

For example; 9pq is a product, which obtained by multiplying q, p and 9 together.

Therefore, p is a factor of 9pq

        q is a factor of 9pq

        9 is a factor of 9pq

NOTE

In $-8p^2q$; -8 is the numerical factor and p, q, p^2, pq and p^2q are literal factors.

COEFFICIENT

When two factors form a product, either factor is the *co-efficient* of the other, For example;

In the product 9p, $\begin{cases} 9 \text{ is the co-efficient of p.} \\ \text{p is the co-efficient of 9.} \end{cases}$

In the product $-4xy$, $\begin{cases} -4x \text{ is the co-efficient of y.} \\ -4y \text{ is the co-efficient of x.} \end{cases}$

REMEMBER

The sign associated with a term is part of the co-efficient, such as the co-efficient of x^3 in $-6x^3$ is -6

Numerical Co-efficient

The 'number' factor of a product is called its **numerical co-efficient** of the remaining part of the factor. For example;

In $-18q^2$, the numerical co-efficient of q^2 is '-18'

Literal Co-efficient

The letter factor (or the variable factor) of a product is called the **literal co-efficient**.

For example; The literal co-efficient in 4 l is l.

NOTE

(i) When a term seems to have no co-efficient then the co-efficient is actually 1. Thus in p^2, the co-efficient is 1, and in $-l$, the co-efficient is -1.

(ii) When a term is constant, then that constant itself (along with its associated sign) is the co-efficient. Therefore the co-efficient of -18 is '-18.'

EXAMPLE – 9

Write the numerical co-efficient and the literal co-efficient in each of the following terms:

(i) $5x^2y$ (ii) $\dfrac{2}{11}a^2bc$ (iii) $-\dfrac{3}{4}xyz$ (iv) $-lm$ (v) $\dfrac{-2xy}{7z}$ (vi) $-7p^2q^2r^2$

Term	Numerical Co-efficient	Literal Co-efficient
(i) $-5x^2y$	-5	x^2y
(ii) $\frac{2}{11}a^2bc$	$\frac{2}{11}$	a^2bc
(iii) $-\frac{3}{4}xyz$	$-\frac{3}{4}$	xyz
(iv) $-lm$	-1	lm
(v) $\frac{-2xy}{7z}$	$-\frac{2}{7}$	$\frac{xy}{z}$
(vi) $-7p^2q^2r^2$	-7	$p^2q^2r^2$

LIKE AND UNLIKE TERMS

'9 apples and 3 apples' can easily be put together to say that there are 12 apples. But 9 apples and 3 buns cannot be put together simply, because they are unlike objects. Similarly '**9a and 3a**' are like terms but '**9a and 3b**' are unlike terms. Thus

Any two or more than two algebraic terms that have the same variable part (i.e., both the letters and their exponents) are called like terms, otherwise the terms are called unlike terms. For example;

(i) $4x = y$, $\frac{7}{2}xy$, $-3xy$, are like terms.

[∵ **xy, the variable part is same**]

(ii) $-6p^2q^2r$, $\frac{4}{5}p^2q^2r$, $\frac{-7}{8}p^2q^2r$, are like terms.

[∵ **p^2q^2r, the variable part is same**]

(iii) $2x^2y$, $-\frac{3}{2}xy^2$, $6xy$, are unlike terms.

[∵ **x^2y, xy^2, and xy are not same**]

(iv) $\frac{lm}{p}$, $\frac{mp}{l}$, $\frac{lp}{m}$ are unlike terms.

[∵ $\frac{lm}{p}$, $\frac{mp}{l}$, and $\frac{lp}{m}$ **are not same**]

CONSTANT TERM

The term of an algebraic expression having no literal (or literals) is called its *constant term*.

For example,

In algebraic expression $3p^2 - 7q + 10$, the constant term is 10.

In algebraic expression $-6x^2 + \frac{x}{z} - 7$, the constant term is -7.

In algebraic expression $9pq - \frac{7}{r} + \frac{5}{9}$, the constant term is $\frac{5}{9}$.

NOTE

All algebraic expressions do not have a constant term, such as '$7x^2y - \frac{3}{5}z$' has no constant term.

EXAMPLE – 10

Group the like terms together in the following:

(i) $7a^2b$, $2a^2b$, $-5b^2a$, $-6a^2b$, $\frac{1}{2}b^2a$

(ii) $-3x^2y$, $-xy^2$, $\frac{1}{2}x^2y$, $-6xy^2$, $-x^2y$

(iii) pq^2, $-2pq^2$, $-7p^2q$, $-8p^2q$

(iv) $3a^2b^2c$, $-2ab^2c^2$, $2a^2b^2c$, $7ab^2c^2$

(v) $\frac{1}{2}l^2mn^2$, $-2l^2m^2n$, $-3l^2m^2n$, $5l^2mn^2$

SOLUTION

(i) $(7a^2b, 2a^2b, -6a^2b)$ and $(-5b^2a, \frac{1}{2}b^2a)$

(ii) $(-3x^2y, \frac{1}{2}x^2y, -x^2y)$ and $(-xy^2, -6xy^2)$

(iii) $(pq^2, -2pq^2)$ and $(-7p^2q, -8p^2q)$

(iv) $(3a^2b^2c, 2a^2b^2c)$ and $(-2ab^2c^2, 7ab^2c^2)$

(v) $(\frac{1}{2}l^2mn^2, 5l^2mn^2)$ and $(-2l^2m^2n, -3l^2m^2n)$

POLYNOMIALS

*All algebraic expressions (**monomials or multinomials**) are polynomials provided none of its terms has:*

a variable with negative exponent,

a variable in the denominator,

a variable inside a square root.

For example:

$4x^2 - 3x + 9$ and $2x^5y^2 + 6x^3y - 9x + 8$ are polynomials. But

$4x + 3y^{-2}x + 6$ is not a polynomial, because the term $3y^{-2}x$ has a negative exponent.

$3x - \frac{2}{x+2} + 6$ is not a polynomial, because the term $\frac{-2}{x+2}$ has a variable in the denominator.

$4x^2 - x + 6\sqrt{x}$ is not a polynomial, because the term $6\sqrt{x}$ has a variable inside a square root.

POLYNOMIAL IN ONE VARIABLE

A polynomial is said to be in one variable if its terms have numerals and the same variable. For example: $5x^2 - \frac{1}{2}x + 6$

Degree of polynomial in one variable:

The greatest index (power) of the variable present in the polynomial is called the *degree* of the polynomial. For example;

The degree of the polynomial '3x' is 1

[∵ **3x can be written as 3x¹**]

The degree of the polynomial $3x^2 - 2x - 4$ is 2

The degree of the polynomial $7x^6 + 2x^4 - 3x^2 + 8$ is 6

POLYNOMIAL IN TWO VARIABLE

A polynomial is said to be in two variables if its terms involve in two variables. For example;

$$3xy^2 + 5x^2y - x^3 + 4$$

is a polynomial in two variables.

Degree of polynomial in two variable:

If the polynomial is in two (or more) variables then the sum of the powers of the variables in each term is taken and the greatest sum is the *degree* of the polynomial.

For example;

(i) $4x^3 - x^2y + 6xy^2$ is polynomial in x and y.

The sum of powers of variables in various terms are: 3, 2+1, 1+2 or 3, 3, 3.

∴ The degree of the polynomial is 3.

(ii) $7x^2 + 2x^2y^2 - xy + 4$ is a polynomial is x and y.

The sum of the powers of variables in various terms are: 2, (2 + 2), (1 + 1), i.e. 2, 4, 2

Thus, the degree of the polynomial = 4

Note

A constant term alone is also a polynomial. For example:

8 is a monomial

The degree of the polynomial 8 is '0'

[$\because$ **8 = 8 × 1 = 8 × x^0 = $8x^0$ and any literal number raised to power zero in always 1.**]

Example – 11

Group the like terms in the following polynomials:

(i) $3x^2y - 2xy^2 + 5x^2y - 2xy^2$

(ii) $\frac{1}{2}p^2q^2 + 3pqr - 4p^2q^2 - pqr$

(iii) $10lm^2n - 8lmn^2 + lm^2n - 3lmn^2$

(iv) $\frac{5}{6}ab^2 + \frac{1}{2}a^2b - a^2b + ab^2 + 3a^2b$

Solution

(i) $(3x^2y + 5x^2y)$ and $(-2xy^2 - 2xy^2)$

(ii) $\left(\frac{1}{2}p^2q^2 - 4p^2q^2\right)$ and $\left(3pqr - \frac{1}{2}pqr\right)$

(iii) $(10lm^2n + lm^2n)$ and $(-8lmn^2 - 3lmn^2)$

(iv) $\left(\frac{5}{6}ab^2 + ab^2\right)$ and $\left(\frac{1}{2}a^2b - a^2b + 3a^2b\right)$

Exercise – 2

1. Write all the terms in each of the following algebraic expressions:

   (i) $x + 8$ (ii) $6m + 2n - 7$ (iii) $\frac{3x}{y}$

   (iv) $4xy + 7yz + 3yz - 8$ (v) $2xyz - 1$ (vi) $\frac{7}{3} + 4x^2 - 3x$

2. Indentify monomials, binomials, trinomial and quadrinomials from the following algebraic expressions:

   (i) $3m^2n^2 - \frac{c}{3}$ (ii) $\frac{2ax}{3y}$ (iii) $5x^2y$

   (iv) $x - y + z$ (v) $x^4 + 2x^3 - \frac{1}{x} + 5$ (vi) k^2

   (vii) $5abc - 8ab + 10bc$ (viii) $5x^2 - 8y^2$

3. Write the algebraic expression whose terms are:

   (i) $6, -3x$ (ii) $-4, 6x^2, -8x^5$

(iii) $2a^2, -3b^2, 4c^2, 8$ (iv) $3x^2, \dfrac{4}{x}, \dfrac{6}{x^2}, -1$

(v) $2x, -\dfrac{6}{7}, \dfrac{7x}{y^2}, -2$ (vi) $2xy, -\dfrac{2}{x^2}, -7x+10$

4. Group the like terms in the following:

(i) $2x^2y^3$, $-3x^2y^3$, $4xy^2$, $7x^2y^3$

(ii) $2a^2b$, $4ba^2$, $12a^2b$, $3ab^2$

(iii) $-2x^3y^2$, $7x^2y^3$, $4x^2y^3$, $2x^3y^2$

(iv) $8x^2y^2z$, $2xy^2z^2$, $-5x^2y^2z$, $7zx^2y^2$, $-11xy^2z^2$

(v) $-5a^3b^2c$, $4cb^3a^2$, $12a^3b^2c$, $17a^3b^2c$

5. Write the numerical co-efficient and the literal co-efficient in each of the following terms:

(i) $54m^2n^2$ (ii) $-9\dfrac{x}{y}$ (iii) $\dfrac{3}{7}xy^3$

(iv) $\dfrac{-11}{c}$ (v) $-7ab^2c^2$ (vi) $\dfrac{3}{4}xyz^2$

6. Write the co-efficient of x in the following:

(i) $-x$ (ii) $-2x$ (iii) x

(iv) $-7px$ (v) $\dfrac{1}{7}xy$ (vi) $\dfrac{-bx}{y}$

7. Write the co-efficient xy^2 in the following:

(i) $-xy^2$ (ii) $8xy^2$ (iii) $-7xy^2$

(iv) $\dfrac{2}{9}xy^2$ (v) $3xy^2z$ (vi) $\left(\dfrac{-3p}{q}\right)xy^2$

8. Write the co-efficient of -1 in the following:

(i) $-px$ (ii) $-7xy$ (iii) $-\dfrac{x}{y}$

(iv) $-6x^2y$ (v) $\dfrac{-10}{x^2}$ (vi) $7x$

9. Write the numerical co-efficient of each term of the following algebraic expressions:

$6x^5 - 4x^3y^2 + 9x^2 - \dfrac{3}{4}x + \dfrac{3}{4}$

10. Indentify, which of the following algebraic expressions are polynomials:

(i) $5 + 6x^2 + \sqrt[3]{4}\, x^3$ (ii) $5xy^2 - \dfrac{4}{5}x^2y - \dfrac{2}{x} + \dfrac{3}{y}$

(iii) $1 + 3p + 4p^2 + 5p^4 - 6p^6$ (iv) $m + mn - 7mn^2 + 9m^2n - 11$

(v) $2\sqrt{x} - 3x + 7x^2 - \dfrac{1}{10}x^3$ (vi) $3y^4 - 2y^3 - 7y^2 + 5y + \dfrac{3}{8}$

11. Write the degree of each of following polynomials:
    (i) $3x^2 - 3x^5 + 7x^6$
    (ii) $18 + 8p - 12p^2 + 20p^3 - p^4$
    (iii) $\frac{\sqrt{7}}{8}x - 11$
    (iv) $13x^5 - 11x^3 + 9x$
    (v) $1 + m + m^2 - m^3$
    (vi) $-3x$
    (vii) 9
    (viii) $-\frac{6}{5}$

Answers

1. (i) x, 8   (ii) 6m, 2n, −7   (iii) $\frac{3x}{y}$
   (iv) 4xy, 7yz, 3yz, −8   (v) 2xyz, −1   (vi) $\frac{7}{3}$, $4x^2$, −3x

2. (i) Binomial   (ii) Monomial   (iii) Monomial
   (iv) Trinomial   (v) Quadrinomial   (vi) Monomial
   (vii) Trinomial   (viii) Binomial

3. (i) $6 - 3x$ (ii) $-4 + 6x^2 - 8x^5$ (iii) $2a^2 - 3b^2 + 4c^2 + 8$
   (iv) $3x^2 + \frac{4}{x} + \frac{6}{x^2} - 1$ (v) $2x - \frac{6}{7} + \frac{7x}{y^2} - 2$ (vi) $2xy - \frac{2}{x^2} - 7x + 10$

4. (i) $(2x^2y^3, -3x^2y^3, 7x^2y^3)$ and $4xy^2$
   (ii) $(2a^2b, 12a^2b, 4ba^2)$ and $3ab^2$
   (iii) $(-2x^3y^2, 2x^3y^2)$ and $(7x^2y^3, 4x^2y^3)$
   (iv) $(8x^2y^2z, -5x^2y^2z, 7zx^2y^2)$ and $(2xy^2z^2, -11xy^2z^2)$
   (v) $(-5a^3b^2c, 12a^3b^2c, 17a^3b^2c)$ and $(4cb^3a^2)$

5.

	(i)	(ii)	(iii)	(iv)	(v)	(vi)
Numerical co-efficient	54	−9	$\frac{3}{7}$	−11	−7	$\frac{3}{4}$
Literal co-efficient	m^2n^2	$\frac{x}{y}$	xy^3	$\frac{1}{c}$	ab^2c^2	xyz^2

6. (i) −1   (ii) −2   (iii) 1
   (iv) −7p   (v) $\frac{1}{7}y$ (vi) $-\frac{b}{y}$

7. (i) −1   (ii) 8   (iii) −7
   (iv) $\frac{2}{9}$ (v) 3z   (vi) $\frac{-3p}{q}$

8. (i) px   (ii) 7xy   (iii) $\frac{x}{y}$
   (iv) $6x^2y$ (v) $\frac{10}{x^2}$ (vi) −7x

9.

Term	$6x^5$	$-4x^3y^2$	$9x^2$	$-\dfrac{3}{4}x$	$\dfrac{3}{4}$
Numerical co-efficient	6	-4	9	$-\dfrac{3}{4}$	$\dfrac{3}{4}$

10. (i), (iii), (iv) and (vi) are polynomials

11. (i) 6   (ii) 4   (iii) 1   (iv) 5

    (v) 3   (vi) 1   (vii) 0   (viii) 0

SUBSTITUTION

An algebraic expression has numerals and literals. A literal can stand for any numerical value. Since numerals have fixed values and literals (variable) have different values, therefore the value of an expression depends upon the values of variables (literals). For example;

$2x + 3$ is an algebraic expression.

Value of 2 is always 2

Value of 3 is always 3

Value of x can be any numerical value.

$\therefore$ If $x = 1$,   then $2x + 3 = 2(1) + 3$ $= 5$

    If $x = 2$,   then $2x + 3 = 2(2) + 3$ $= 7$

    If $x = 3$,   then $2x + 3 = 2(3) + 3$ $= 9$

    If $x = -1$,   then $2x + 3 = 2(-1) + 3$ $= 1$

    If $x = -2$,   then $2x + 3 = 2(-2) + 3$ $= -1$

    If $x = \dfrac{1}{2}$,   then $2x + 3 = 2(\dfrac{1}{2}) + 3$ $= 4$

and so on.

Thus, *the process of finding the value of an algebraic expression by replacing the variables (literals) by their numerical values is called substitution.*

EXAMPLE – 12

If $x = 9$ and $y = -3$ then find the value of the following expression:

$$\dfrac{x}{3} - 5y + 10$$

SOLUTION

The given expression is $\dfrac{x}{3} - 5y + 10$ and $x = 9$ and $y = -3$

∴ Substituting the values of x and y i.e. x = 9 and y = −3 in the given expression we get:

$\dfrac{x}{3} - 5y + 10 = \dfrac{9}{3} - 5(-3) + 10 = \dfrac{\cancel{9}^{3}}{\cancel{3}_{1}} - (-15) + 10$

$= 3 + 15 + 10 = 28$

Example – 13

Find the value of $\dfrac{3p}{q} + 2r - 10$ when $p = \dfrac{2}{3}$, q = 4, and r = (−2)

Solution

∵ $p = \dfrac{2}{3}$, q = 4 and r = −2

∴ Substituting the values of p, q and r in $\dfrac{3p}{q} + 2r - 10$, we get

$\dfrac{3p}{q} + 2r - 10 = \dfrac{3\left(\dfrac{2}{3}\right)}{4} + 2(-2) - 10$

$= \dfrac{\cancel{3}^{1} \times \cancel{2}^{1}}{\cancel{3}_{1} \times \cancel{4}_{2}} + (-4) - 10 = \dfrac{1}{2} - 4 - 10$

$= \dfrac{1 - 8 - 20}{2} = \dfrac{-27}{2} = 13\dfrac{1}{2}$

Example – 14

If x = 3, y = −1 and z = 0 then find the value: $\dfrac{2x^2 - 5y}{7y - 2z}$

Solution

Substituting x = 3, y = −1 and z = 0 in $\dfrac{x^2 - 5y}{7y - 2z}$; we have

$\dfrac{x^2 - 5y}{7y - 2z} = \dfrac{(3)^2 - 5(-1)}{7(-1) - 2(0)}$

$= \dfrac{9 + 5}{-7 - 0} = \dfrac{14}{-7} = -\dfrac{\cancel{14}^{2}}{\cancel{7}_{1}}$

$= -\dfrac{2}{1} = -2$

Exercise – 3

1. If x = 5 and y = 2 then find the value of 2x + 3y
2. If x = −2 and y = 1 then find the value of $3x^2 − 4y + 1$
3. If x = 3, y = 2 and z = −4, find the value of 2x + 3y − 5c − 2
4. If p = 3, q = 2 and r = −1, find the value of $3p^2 − 2p + qr + 1$
5. If x = −2 and y = 3 then find the value of $xy^3 + x^2y + x + 44$
6. If x = 1, y = 2, find the value of $x^2 + 2xy + y^2$
7. Find the value of the expression $(x + y + z)^2 − 3xyz$. When x = 1, y = 2 and z = 3
8. If x = −1, y = 2 and z = 3, then find the value of $(x + y + z)^2 + 2xyz − 4$
9. If x = −1, y = 2 and z = 0 then find the difference between the values of $\frac{x+y+z}{-2}+xyz$ and $\frac{x+y+z}{xy} - 2z$
10. If x = 10, y = −10 and z = 0 then find the sum of the values of 2x + 3y − 4z and 2x − 3y + 4z

Answers

1. 16	2. 9	3. 30	4. 20	5. 0
6. 9	7. 18	8. 0	9. 0	10. 40

Miscellaneous Exercise

1. Identify the variables and constants in the following

    (i) − pq   (ii) 18   (iii) $\frac{1}{18}$ (iv) $2x^2$

    (v) $\frac{3}{x}$

2. Write the following statements in algebraic statements:

    (i) $\frac{3}{4}x$ less than y is 10 more than 2z

    (ii) 9x and 9 less than twice y is less than 15

3. Express the following in exponential form:

    (i) p × p × p × q × q × r × r × r × r

    (ii) 10 × m × m × m × n × n × n

4. Express the following in expanded form:

    (i) $9p^2q^3$ (ii) $-18lm^2n^3$

5. "In group of less than 15 students, there are x girls and y boys." Express this statement as an algebraic expression.
6. Count the number of terms in each of the following and state which of them are monomial, Binomial, trinomial or quadrinomial
   (i) $p^2 - \dfrac{6}{q} + r - 10$
   (ii) $xyz - x^2 + y^2$
   (iii) $pq - 7p$
   (iv) $\dfrac{7xyz}{3}$
   (v) $6 - \dfrac{abc}{d}$
7. Write the numerical co-efficient and literal co-efficient of the following terms:
   (i) $\dfrac{-3pq}{4r}$
   (ii) $-7xy^2z^3$
8. Group the like terms together in the following:
   (i) $p^2q, \ -p^2q, \ 8p^2q, \ 10pq^2, \ -5p^2q$
   (ii) $-xy^3, \quad 3x^2y^2, \quad -8x^2y^2, \ -7xy^3, \quad xy^3$
9. Write the degree of the following polynomials:
   (i) $-3x + 5x^2 - 7x^3 - 7x^4 + 10$
   (ii) $\dfrac{x}{2} - \dfrac{x^2}{3} + \dfrac{x^3}{4} - \dfrac{x^4}{5} + \dfrac{x^5}{7} + 15$
10. Which of the following are polynomials:
    (i) $7x^2 + \dfrac{3}{2}xy + 7y^2 - 10$
    (ii) $10p^2 + \dfrac{q^2}{r} - 7rp + 2pr + 15$
    (iii) $1 + \dfrac{x^2}{y} + 2x$
    (iv) $\sqrt{x} + 2x + 3x^2 - 6x^3 + 12$
    (v) $1 - 2x - 3x^2 - 4x^3 + \sqrt{5}x^4$
11. Find the degree of the following polynomial:
    $3xy - x^2y + \dfrac{1}{2}x^3y^2 - \dfrac{4}{5}x^5y^4 + 10$
12. If $x = -1, y = 4$ then find the value of:
    $xy + \dfrac{1}{3}(x^2y - 1) + 4xy$
13. Which of the following expressions a not polynomial?
    (i) $3 - 2x$
    (ii) $x^2 + y^3 - x^3 + 4y^6 - 7xy + 7$
    (iii) $\dfrac{x^3 + x^2 + 12}{x + 2}$
    (iv) $5 - \dfrac{3}{1-x}$
14. If $x = 1, y = -2$ and $z = -5$, then find the value of $x^3 - 3xy + z + y$
15. If $x = 0, y = -1$ and $z = 2$ then find the value of $\dfrac{x+y-z}{x-y+z}$

ANSWER

1. Constants are: (ii), (iii); variables are: (i), (iv), (v).
2. (i) $\dfrac{3}{4}x - y = 2z + 10$
   (ii) $9x + (2y - 9) < 15$
3. (i) $p^3q^2r^4$
   (ii) $10m^3n^3$

4. (i) $9 \times p \times p \times q \times q \times q$ (ii) $-18 \times l \times m \times m \times n \times n \times n$

5. (x girls + y boys) < 15 students

6.
Expression	(i)	(ii)	(iii)	(iv)	(v)
No. of Terms	4	3	2	1	2
Type of Expression	qudrinomial	Trinomial	Binomial	Monomial	Binomial

7.
Term	Literal co-efficient	Numerical co-efficient
(i)	$\dfrac{pq}{r}$	$-\dfrac{3}{4}$
(ii)	$x^2y^2z^3$	-7

8. (i) $(p^2q, 8p^2q, -5p^2q)$ and $(-pq^2, 10\,pq^2)$

   (ii) $(-xy^3, -7xy^3, xy^3)$ and $(3x^2y^2, -8x^2y^2)$

9. (i) 4   (ii) 5   10. (i), and (v) are polynomials

11. Degree of the polynomial = 9   12. 19

13. (iii) and (iv) are not polynomials 14. 0   15. -1

Hots

Mrs. Geetanjali was having 'x' bananas. She distributed them to her three children such that

   Amita gets 'p' bananas

   Sahil gets 2 more than Amita and

   Parul gets 3 bananas less than Sahil.

Now, answer the following questions:

(i) How many bananas did Parul get?

(ii) How many bananas did Sahil get?

(iii) How many bananas did Parul get less than that of Amita?

(iv) How many bananas did Parul get less than that of Sahil?

(v) If x is expressed in terms of 'p' then what the co-efficient of 'p'?

Answers

(i) p − 1   (ii) p + 2   (iii) 1   (iv) 3   (v) 3

Mental Maths

1. What is the exponential form of $p \times p \times p \times q \times q \times r \times r \times r$?
2. Whose expanded form is: $2 \times x \times x \times x \times x \times y \times y \times y \times y \times z \times z$?
3. What is the numerical co-efficient of $-p$?
4. What is the literal co-efficient of $-100\,m$
5. Is '2' is a monomial?
6. Can a monomial be a multinomial?
7. Can a multinomial be a monomial?
8. Can a monomial be a polynomial?
9. What is the degree of a (non zero) constant term?
10. What is the degree of the polynomial $5 - 3m^2n^3p$?
11. Can $\frac{1}{2}\left(\frac{x}{y}\right) - 3z$ be a polynomial?
12. Are ab^2 and a^2b like terms?
13. Does the algebraic expression $7x^2y - \frac{p}{y} + \frac{5}{p}$ has a constant term?
14. What is the co-efficient of xy in $-9x^2y$
15. If $x = 1$ and $y = -1$ then, are the values of $y + x$ and $x - y$ equal?

Answers

1. $p^3q^2r^3$	2. $2x^2y^4z^2$	3. -1	4. m
5. yes	6. No	7. No	8. Yes
9. 0	10. 6	11. No	12. No
13. No	14. $-9x$	15. No	

Multi Choice Questions

1. $2ax + 5ax - 3ax + \frac{1}{2}ax$ is a
    (i) Quadrinomial (ii) Binomial
    (iii) monomial (iv) Trinomial
2. $3ab^2 + 2a^2b - ab^2 + 3a^2b$ is a
    (i) Quadrinomial (ii) Trinomial
    (iii) monomial (iv) Binomial

3. $1 - 7pq^2 + 5p^2q + 9pq^2$ is a
   (i) Trinomial
   (ii) monomial
   (iii) Quadrinomial
   (iv) Binomial

4. $\dfrac{3x^2y^2}{z}$ is an algebraic expression having
   (i) one - term
   (ii) 2 - terms
   (iii) 3 - terms
   (iv) 4 - terms

5. In $-6x^2y$, the numerical factor is:
   (i) 6
   (ii) -6
   (iii) $-6y$
   (iv) $6y$

6. The degree of the polynomial '3' is:
   (i) 0
   (ii) 1
   (iii) 3
   (iv) none of these

7. The degree of the term $-6x^2y^3z$ is
   (i) 0
   (ii) 5
   (iii) 3
   (iv) 6

8. Which of the following is not a polynomial?
   (i) $x + 2x^2 + 3x^3 + 4x^4 + 5x^5 + 6$
   (ii) $\dfrac{1}{2}x + 3x^2 + 4x^3 + 5x^4 + 6x^5 + 10$
   (iii) $\dfrac{1}{x} + x + 2x^2 + 3x^3 + 4x^4 + 5x^5$
   (iv) $\dfrac{7}{\sqrt{3}}x + 4x^2 + \dfrac{1}{2}x^3 + 5x^4 + \dfrac{1}{3}x^4 - 9$

9. Which of the following terms has degree as '6'?
   (i) $-6a^2b^4c$
   (ii) $9a^3b^2c$
   (iii) $\dfrac{7}{8}a^2b^2c^3$
   (iv) $-10a^2b^3c^3$

10. If $x = 1$, $y = -2$, and $z = 0$, then which of the following has its value as '9'
    (i) $4\dfrac{yz}{x}$
    (ii) $4(x + y + z)$
    (iii) $\dfrac{2x}{y} - 3z + 10$
    (iv) $7x - 5y + z - 7$

11. Which of the following expressions is not a multinomial?
    (i) xyz
    (ii) $7x + y + z$
    (iii) $\dfrac{2x}{3} + \dfrac{3y}{4} + \dfrac{4z}{7}$
    (iv) $2 + x + y + xy + \dfrac{x}{y}$

12. Which of the following is a multinomial?
    (i) $6x + 2x + 4y$
    (ii) $-2p - 3p + 5p + 4p$
    (iii) $7x + 2x - 3y + 5x$
    (iv) $ab^2 + b^2a + 2ab^2 - 2b^2a$

13. Which of the following is a polynomial?
    (i) $a\sqrt{x} + b\sqrt{y} + c\sqrt{y} - 8$
    (ii) $ax^2 + bx^3 + cx^4 - \frac{3}{4}$
    (iii) $\frac{1}{x} + \frac{1}{y} + \frac{1}{z} - 3 + xyz$
    (iv) $\frac{x^2}{y} + \frac{y^2}{z} + \frac{z^2}{x} - 3x^2y^2z^2 + 20$

14. Which of the following is the value of:
    $\frac{1}{2}x^2 - \frac{2}{3}y^2 + 3xy + 15$ When $x = -1$ and $y = 3$
    (i) $\frac{1}{2}$
    (ii) $15\frac{1}{2}$
    (iii) $-14\frac{1}{2}$
    (iv) $30\frac{1}{2}$

15. If $a = 2$, $b = -2$ and $c = 1$ then which of the following have same value?
    (A) $a^2 + a^2 + 1$
    (B) $2(a) + 2(b) + 1$
    (C) $4(a) - 4(a) + 1$
    (D) $4(a) + 4(a) + 1$
    (i) A and B
    (ii) B and D
    (iii) C and D
    (iv) A and C

ANSWERS

1. (iii)   2. (iv)   3. (i)   4. (i)
5. (ii)   6. (i)   7. (iv) 8. (iii)
9. (ii)   10. (iii) 11. (i) 12. (iii)
13. (ii)   14. (i)   15. (ii)

WORKSHEET

1. Combine the like terms and match *column – A* with *column – B*:

Column A	Column B
(i) $3x^2y, 4y^2x, 4x^2y, 3yx^2$	(a) $(-xy^2, 2xy^2, xy^2)$ and $(-x^2y, x^2y)$
(ii) $2y^2, 3xy - 6y^2, -xy$	(b) $(3xy^2, 2xy^2)$ and $(-x^2y, -3x^2y)$
(iii) $x^2y, 2xy^2, -3x^2y, -4xy^2$	(c) $(3x^2y, 4x^2y, 3yx^2)$ and $(4x^2x)$
(iv) $-xy^2, -x^2y, 2xy^2, xy^2, x^2y$	(d) $(2y^2, -6y^2)$ and $(3xy, -xy)$
(v) $3xy^2, -x^2y, 2xy^2, -3x^2y$	(e) $(x^2y, -3x^2y)$ and $(2xy^2, -4xy^2)$

2. Write True or False for the following.
   (i) Letters used in algebra are called 'literals.'
   (ii) $3abc$ means $3 + a + b + c$
   (iii) In $4a^2b^2c^2 = 4 \times a \times a \times b \times b \times c \times c \times c$; $4a^2b^2c^2$ is the exponential form of $4 \times a \times a \times b \times b \times c \times c \times c$
   (iv) $6ba$ and $-ab$ are unlike terms
   (v) $7ba^2$ and $-a^2b$ are like terms
   (vi) $x + y + z \div 15 + 5$ has 5 terms
   (vii) The degree of the polynomial $11xy^2 + 13x^5y^2 - 7$ is 7
   (viii) $\dfrac{6ax}{5y}$ is a monomial in three variables a, x and y.
   (ix) $8 + 5x^2 - 6x^2y + 4xy^2$ is a polynomial of three terms
   (x) The value $\dfrac{x^2 - y^2 - z^2}{12} + 10$ at $x = 1$, $y = 2$ and $z = 3$ is 11

3. Fill in the blanks:
   (i) An algebraic expression having two or more than two terms is called a _____.
   (ii) Each part of an algebraic expression which are separated by (+) or (−) is called the _____ of that algebraic expression.
   (iii) If a polynomial has only (non-zero) constant term, then its degree is _____.
   (iv) In the algebraic expression $11xy - \dfrac{2}{x^3} - \dfrac{2}{3}$, the constant term is _____.
   (v) In $-9y^2x$, the numerical factor is _____.
   (vi) The algebraic expression $2 + 2x + \dfrac{1}{x} - 4x$ has _____ terms.
   (vii) The degree of the polynomial $9 + 9x$ is _____.
   (viii) '18' is a polynomial of degree _____.
   (ix) The value of the expression $2x^2 - 2x + y$ at $x = 1$ and $y = 0$ is _____.
   (x) No term in a _____ can be of the form $\dfrac{1}{x}$, $\dfrac{1}{x^2}$, $\sqrt{x}$ etc.

4. Find the value of $y - x^2 + \dfrac{x}{2} + 6$, when
   (i) $x = 1$ and $y = 0$ (ii) $x = -1$ and $y = -2$

(iii) $x = 2$ and $y = 2$ (iv) $x = -2$ and $y = -1$
(v) $x = 0$ and $y = 1$

5. Write the algebraic expression whose terms are:
   (i) $1, -2x, \dfrac{3}{y}, 5y^2$ (ii) $p, -2p^2, \dfrac{p^3}{4}, -10$ (iii) $-x\ 20y, -z^2, -xyz$

Answers

1. (i) → (c);   (ii) → (d);   (iii) → (e);
   (iv) → (a); (v) → (b);

2. (i) True   (ii) False   (iii) True
   (iv) False (v) True   (vi) False
   (vii) True (viii) True   (ix) False
   (x) True

3. (i) Multinomial   (ii) Term   (iii) Zero
   (iv) $\dfrac{-2}{3}$ (v) -9 (vi) 3
   (vii) 1   (viii) 0   (x) 0
   (ix) Polynomial

4. (i) $5\dfrac{1}{2}$ (ii) $2\dfrac{1}{2}$ (iii) 6
   (iv) 0   (v) 7

5. (i) $1 - 2x + \dfrac{3}{y} + 5y^2$ (ii) $p - 2p^2 + \dfrac{p^3}{4} - 10$ (iii) $-x + 20y - z^2 - xyz$

Chapter 9
Basic Geometrical Ideas

Introduction

'Geometry' is a branch of mathematics in which the properties of points, lines, surfaces and solids are studied. Looking around us, we notice that most of the man-made objects have geometrical forms. Our book, our pencil, wheels of our bicycle, our playground, our lunch-boxes, sandwich etc. have some geometrical form. Geometry helps us to analyze, measure and create objects.

Concept

Concept means an idea, imagination

Fundamental Concepts of Geometry

There are three fundamental concepts: *point, line* and *plane.*

These three terms do not have specific definitions but we can try to visualize them as ideas suggested by certain practical situations.

Did You Know?

The word "geometry" is derived from the Greek words: *'Geo'* meaning 'earth' and *'metron'* meaning 'measuring'

Point

A dot marked by a five tip of a pencil on a paper gives an idea of a point. A point has no dimension, so it cannot be seen or felt or moved. It has a position only it is denoted by capital letters. In the figure points P, Q, R and S are represented.

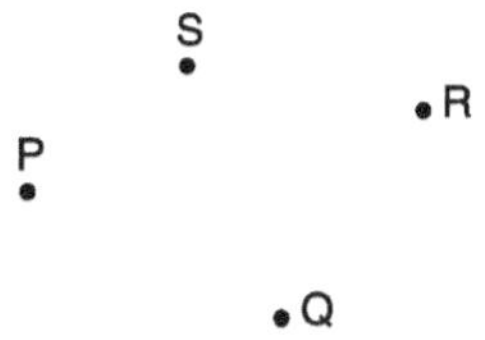

```
              S
              •
                      • R
         P
         •

              • Q
```

PLANE

A flat and smooth surface gives an idea of a plane. The size of a plane is unlimited because it is thought to extend endlessly in all the directions. So, a plane has no length, breadth or thickness. A part of a plane can be represented by a piece of paper, a large surface, a floor, a surface of a wall etc. A model of a plane is given below:

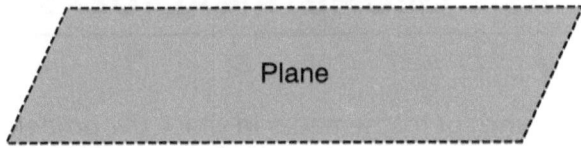

We can draw certain plane shapes on a plane, such as circle, triangle, square, rectangle etc.

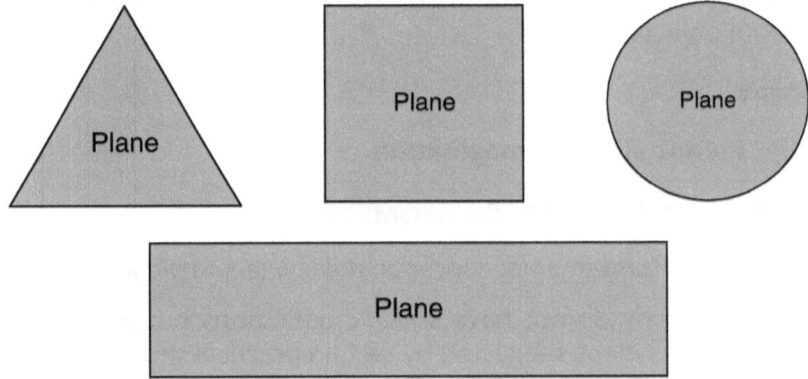

DID YOU KNOW

The surface of a paper in your note book is a part of plane. A complete plane cannot be drawn.

SURFACE

The external part of an object is known as surface of that object. A surface has length and breadth, but no thickness. A surface may be flat or curved. Floor, wall, mirror, blackboard etc. are flat surfaces and the surface of a ball is a curved surface.

LINE

A line is a set of points in a plane which extends endlessly in both directions.

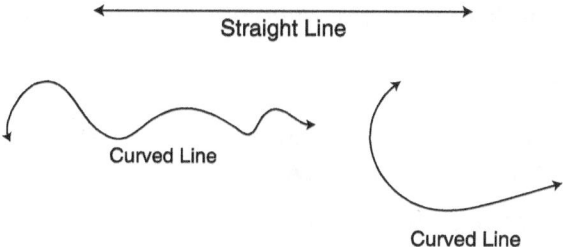

If it is a straight path then the line represented by it is 'straight-line' or simply a 'line'. Lines can be curved also. A line has no end points, therefore, it cannot be drawn wholly on a paper. Only a portion of it is drawn and arrowheads on both directions are marked to indicate that it extends indefinitely in both directions.

To name a line, we mark any two points on it, for example: if 'A' and 'B' two points are marked it is denoted by $\overleftrightarrow{AB}$ or $\overleftrightarrow{BA}$

Sometimes, a line is also denoted by small letters, for example, 'line-p', 'line-q', 'line l' etc.

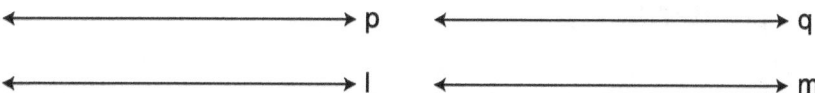

Since a line has no end points, so it can never be measured, i.e. a line has an unlimited length.

LINE SEGMENT

Take a line 'l'. Mark any two points P and Q on it. Then the portion of the 'line l' between the points P and Q is called the "line segment PQ".

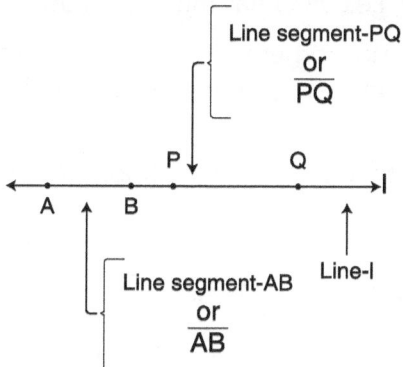

Similarly, if we mark another pair of points, say A and B, then the portion AB is also a 'line-segment AB'. There can be more than one line segments on the same line.

The symbolic name the 'line segment PQ' is $\overline{PQ}$ or $\overline{QP}$. Similarly the 'line segment AB' is denoted by $\overline{AB}$ or $\overline{BA}$.

A line segment has two end-points.

Since, the end-points of line segment are known, it has a definite length.

[The length of the line segment AB] = [The distance between the points A and B]

REMEMBER

The end-points of a line segment are always the parts of the line-segment. Therefore, they are included in the length of a line segment.

We can draw only one segment joining two given points.

RAY

If we mark a point 'A' a 'line l' the part of the line-l that extends indefinitely in one direction from 'A' is called a ray. The point 'A' is called 'initial point' (or end-point) of the ray.

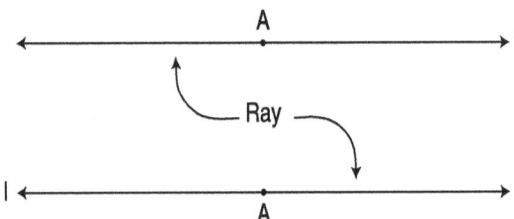

Since, a ray extends in one direction indefinitely, therefore, its length cannot be measured, i.e., a ray has indefinite length.

To name a ray, we mark another point (i.e. a point other than initial point) on it. Thus a ray is completely known, if its initial point and one more point on it are known.

Let mark a point B other than the initial point A, then the ray is named as 'ray AB'

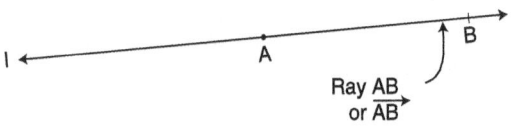

The symbolic name of 'ray AB' is $\overrightarrow{AB}$. The ray $\overrightarrow{AB}$ means that the ray starts from 'A' as its initial point and it passes through the point B and extending indefinitely in the direction from A to B.

NOTE

I. **The initial point of a ray is always the part of the ray.**
II. **Rays- $\overrightarrow{AB}$ and $\overrightarrow{BA}$ are different. The $\overrightarrow{AB}$ has 'A' as its initial point and its direction is from A to B, while the ray BA i.e. $\overrightarrow{BA}$ has 'B' as its initial point and its direction is from 'B to A'**
III. **Many rays can be drawn from a point.**

OPPOSITE RAYS

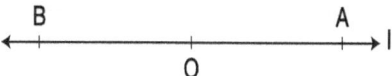

The rays with same initial point and extending indefinitely in opposite directions of the same line, are called opposite rays. In the above figure $\overrightarrow{OA}$ and $\overrightarrow{OB}$ lie on the same line l, so they are opposite rays.

NOTE

An unlimited number of rays can be drawn with a given initial point, as shown in the figure, OA, OB, OC, OD, OE, OF, OG, OH, etc. are rays from the common initial point O.

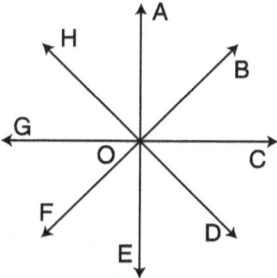

INCIDENCE PROPERTIES IN A PLANE

The universal truths with respect to points and lines are called incidence properties.

PROPERTY - I

In a plane, an infinite number of lines can be drawn passing through a given point.

Mark a point O on a sheet of paper. Draw many lines passing through O, as shown in the adjoining figure.

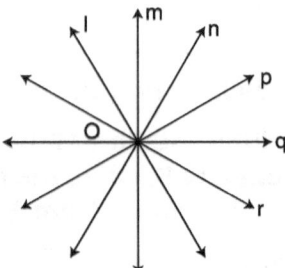

An unlimited number of lines can be drawn in a plane passing through a given point.

PROPERTY - II

Exactly one line can be drawn passing through two distinct given points in a plane.

Mark Two distinct points A and B on the sheet of paper.

Let us try to draw many lines passing through A and B.

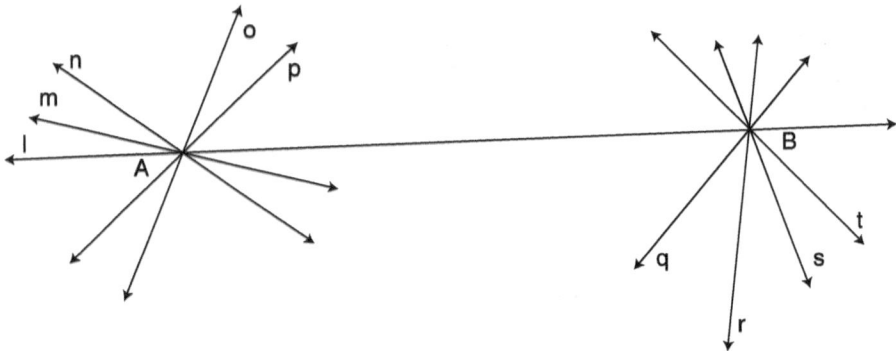

We observe that only one line (here 'l') passes through both points A and B. Thus, we can say that two distinct points in a plane determine a unique line.

PROPERTY - III

An infinite number of points lie on a line in a plane. In the figure there are many points A, B, C, D, E,... are marked on the line l.

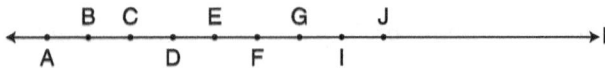

Distinction Between *A Line*, *A Line-Segment* and *A Ray*

LINE	LINE SEGMENT	RAY
• A line has no end point $\underset{A B}{\longleftrightarrow}$	A line segment has two end points $\underset{A B}{\rule{3cm}{0.4pt}}$	A ray has one end point $\underset{A B}{\longrightarrow}$
• Symbolic representation is $\overleftrightarrow{AB}$	Symbolic representation is $\overline{AB}$	Symbolic representation is $\overrightarrow{AB}$
• A line has no end points	A line segment has two end points	A ray has only one end point.
• A line has no definite length	A line segment has a definite length.	A ray has no definite length.
• $\overleftrightarrow{AB}$ and $\overleftrightarrow{BA}$ represent the same line	$\overline{AB}$ and $\overline{BA}$ represent the same line segment.	$\overrightarrow{AB}$ and $\overrightarrow{BA}$ represent two different rays
• In the figure, $\overleftrightarrow{AB}$ is a line.	In the figure, $\overline{AB}$ is a line segment.	In the figure, $\overrightarrow{AB}$ is a ray.

Open and Closed Curves

The curves, in which initial and end points coincide with each other are called closed figures. In the adjoining figure some closed curves are shown.

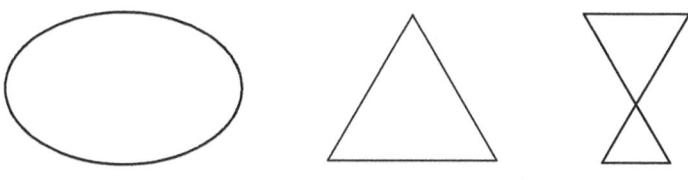

Closed curves

On the other hand the curves which have initial and end points are called open figures, as show in the adjoining figure.

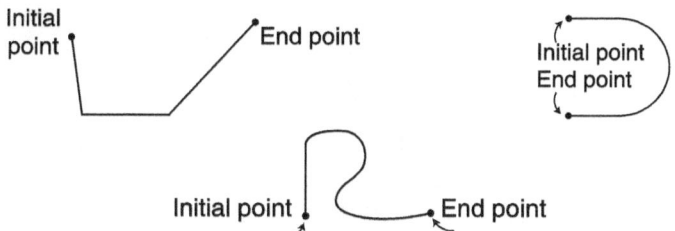

INTERIOR AND EXTERIOR OF CLOSED FIGURES

There are three parts of the plane into which a plane figure (drawn on a plane) divides the plane:

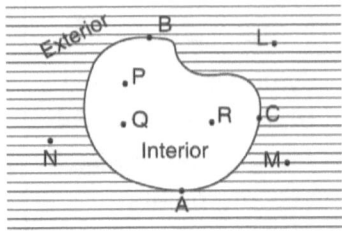

(I) **INTERIOR** of the plane closed figure: It is the part of the plane included by the boundary of the figure. The unshaded portion is the *interior* of the figure. The points P, Q and R are in the interior the figure.

(II) **EXTERIOR** of the plane, closed figure: The portion of the plane, which is outside the figure, is called the *exterior* of the figure. The shaded region is the exterior of the given figure.

Points L, M, and N are in the exterior of the figure.

CURVILINEAR AND LINEAR BOUNDARIES

All closed plane shapes bounded by boundaries. These boundaries can be curved or straight lines.

A closed figure bounded by curved lines is said to have *curvilinear boundaries*.

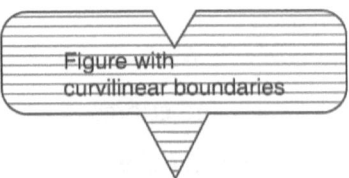

A closed figure bounded by straight lines is said to have *linear boundaries*

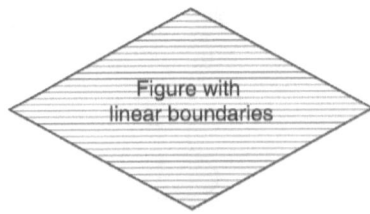

Angle

An *angle* is formed by two different rays starting from the (common) same point. This common starting point (initial point) of the rays is called the *vertex* of the angle and the two rays are known as its *arms*.

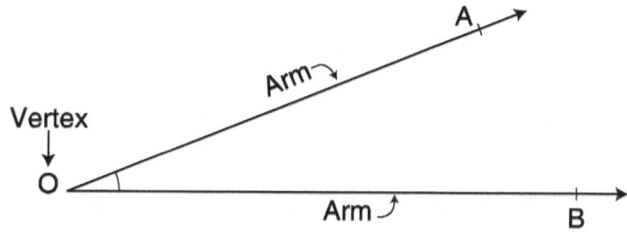

In the adjoining figure, the rays $\overrightarrow{OA}$ and $\overrightarrow{OB}$ from an angle whose vertex is O and arms are $\overrightarrow{OA}$ and $\overrightarrow{OB}$.

Notation and Naming an Angle

We use the symbol ∠ for an angle. In naming an angle, the vertex is always written in the middle of the two points on its arms one on each arm.

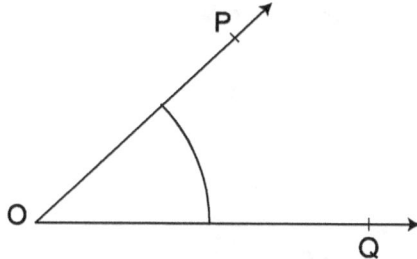

In the figure the angle is named on ∠POQ or ∠QOP.

$\overrightarrow{OA}$ is fixed whereas $\overrightarrow{OB}$ can rotate about the vertex in the plane.

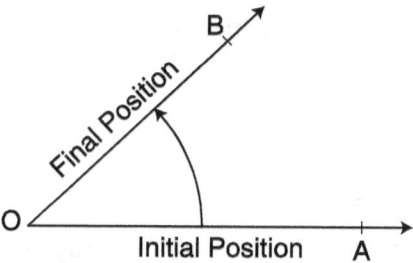

Let the two rays coincide at the starting position (position of $\overrightarrow{OA}$) or the initial position.

Let $\overrightarrow{OB}$ is allowed to rotate about O in the anticlockwise direction. Let it reach at the position $\overrightarrow{OB}$. In this position, $\angle AOB$ is formed, between the initial-position and the 'final position'. A curved arrow starting from the initial position to $\overrightarrow{OB}$ indicates the formation of an angle.

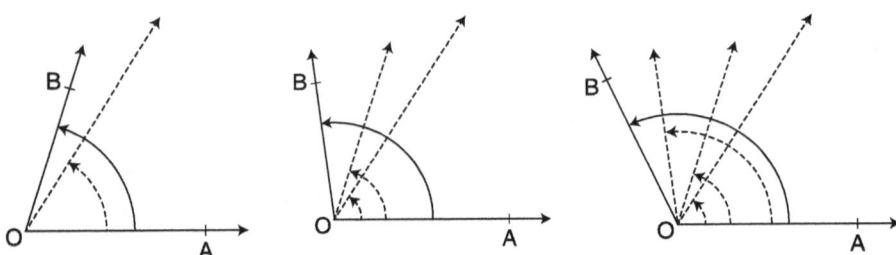

Different positions of $\overrightarrow{OB}$ indicate the formation of different angles as shown in the adjoining figure.

Note

An angle such as $\angle AOB$ or $\angle POQ$ can also be denoted by their vertex only, such as $\angle O$. But this is not possible, in case, there are more than one angles at the same vertex. In the adjoining figure four angles are sharing the same vertex.

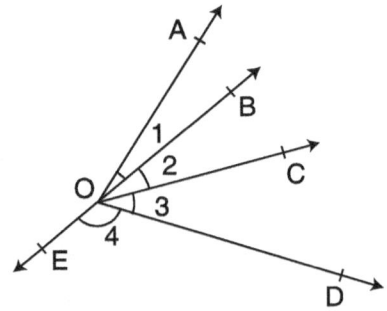

In such cases we can use numerals to denote them. Thus $\angle AOB$, $\angle BOC$, $\angle COD$ and $\angle DOE$ can conveniently be denoted as $\angle 1$, $\angle 2$, $\angle 3$ and $\angle 4$ respectively

Example

Interior and Exterior of an Angle

Let us consider the angle AOB in a plane. The rays $\overrightarrow{OA}$ and $\overrightarrow{OB}$ divide the plane in three different parts as shown in the following figure:

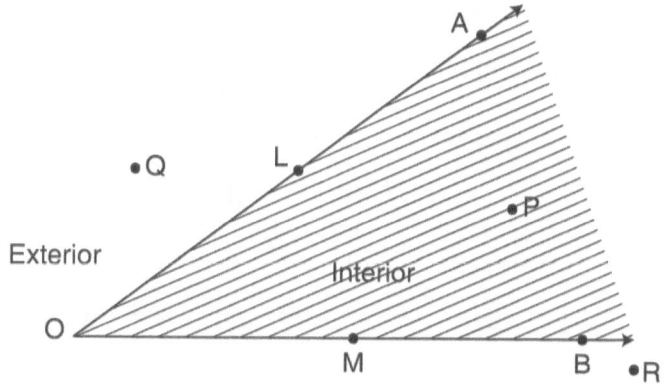

These parts are:

(i) **Interior of the angle**

The part of the plane within the two arms of ∠AOB is called the interior of ∠AOB, as shown by the shaded region of the plane. The point P is in the interior.

(ii) **Exterior of the angle**

The part of the plane which is outside the two arms of ∠AOB is called its exterior. In the figure, the unshaded part is its exterior. The points Q and R are lying in the exterior of ∠AOB.

(iii) **Boundary of the angle**

The portion of the plane covered by the arms (here $\overrightarrow{OA}$ and $\overrightarrow{OB}$) of the angle is called the boundary of the angle. The points A, L, O, M and B are lying on the boundary of ∠AOB.

EXAMPLE

Name each of the following angles in three different ways:

SOLUTION

(i)   (ii)

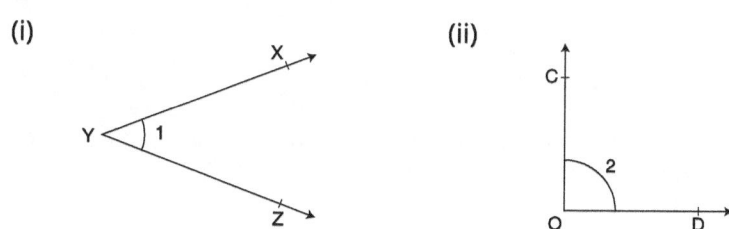

(b) List the numbers which are in the

    (i) interior of the angle

(ii) exterior of the angle

(iii) on the boundary of the angle.

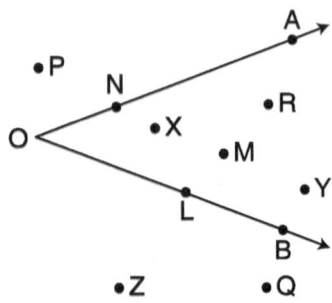

(a) (i) The given angle can be written as

∠XYZ, ∠Y and ∠1

(ii) The given angle can be written

as ∠COD, ∠O and ∠2

(b) (i) The points in the interior are:

M, R, X and Y.

(ii) The points in the exterior are:

P, Q, and Z

(iii) The points on the boundary are:

A, B, L, N and O

Angular Region

The interior of an angle together with its boundary form the angular region.

Triangle

A triangle is a *geometrical plane figure formed by three line segments having common end points when taken in pairs*

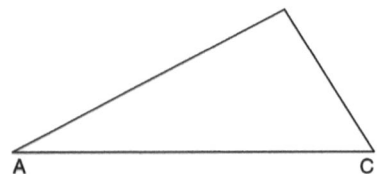

In the figure, the three line segments $\overline{AB}$, $\overline{BC}$ and $\overline{CA}$ meet at B, C and A respectively when they meet in pairs, form a triangle.

We use the symbol '△' to represent a triangle. A triangle is named by using the points of intersection of the line segments. Above triangle is 'triangle ABC'.

Symbolic representation of 'triangle ABC' is △ ABC.

The △ ABC can be named as:

△ BCA or △ CAB or △ ACB or △ CBA or △ BAC

PARTS OF A TRIANGLE

A triangle has six parts:

Three sides and *Three angles*.

In the adjoining triangle:

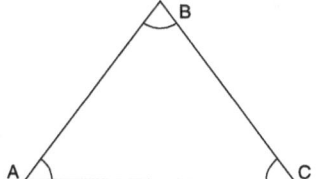

(i) Three sides of it are AB, BC and CA.
(ii) Three angles are ∠BAC, ∠ABC and ∠BCA. We can also denote these angle as ∠A, ∠B and ∠C respectively.

VERTICES OF A △

The point of intersection of two adjacent sides of a triangle is called a vertex. A triangle namely △ ABC has three vertices A, B and C.

ELEMENTS OF A △

The six components of a triangle, namely, the three sides and three angles are known the elements of a triangle

The vertex A is opposite to the side BC,

The vertex B is opposite to the side AC,

The vertex C is opposite to the side AB.

It is important to note that in a △ABC the sum of the lengths of any two sides is always greater than the length of the third side.

i.e.

(i) (AB + BC) > AC

(ii) (BC + AC) > AB

(iii) (AC + AB) > BC

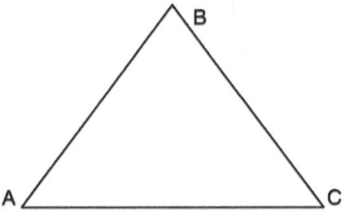

This is known as *Triangle Inequality Property*.

EXTERIOR AND INTERIOR OF A TRIANGLE

A triangle divides the plane into three distinct parts as shown in the adjoining figure:

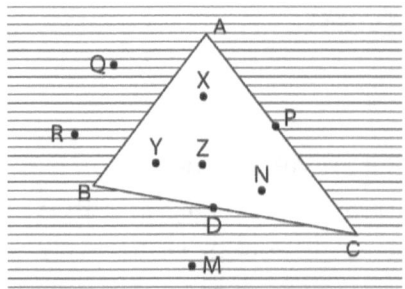

(i) Points A, B, C, D, P etc. lie on the BOUNDARY forming the triangle
(ii) The region outside the boundary (as shown by the shaded region) is called the EXTERIOR of the triangle. In the figure point Q, M, R etc. lie in the exterior of the triangle.

ALTITUDE OF A TRIANGLE

An altitude of a triangle is a perpendicular segment from a vertex to its opposite side. In the adjoining figure AX is an altitude of triangle ABC.

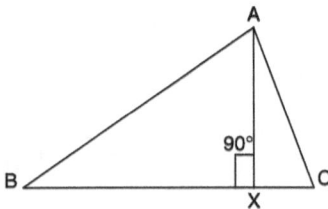

In a triangle there can be exactly three altitudes. In the figure, AX, BY and CZ are the altitudes of triangle ABC.

The three altitudes intersect at a point. This point is called the *'Orthocentre'* of the triangle. O is the orthocentre of triangle ABC.

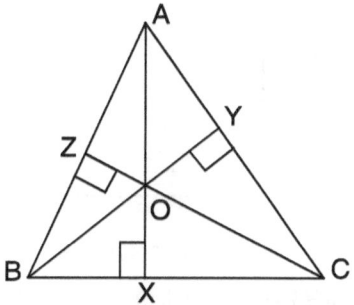

MEDIAN

A median of a triangle is a line segment joining a vertex to the midpoint of the opposite side. In the adjoining figure AP is a median.

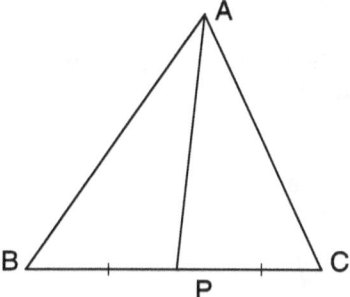

Every triangle has exactly three medians, one from each vertex. These medians intersect at a point. This point is called the *'centroid'* of the triangle. In the adjoining figure, AP, BQ and CR are medians and they meet at I. I is the centroid of the triangle ABC.

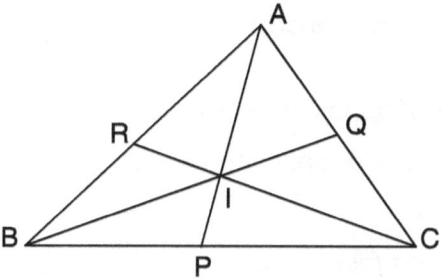

QUADRILATERAL

A closed figure formed by 4 line segments is called a quadrilateral.

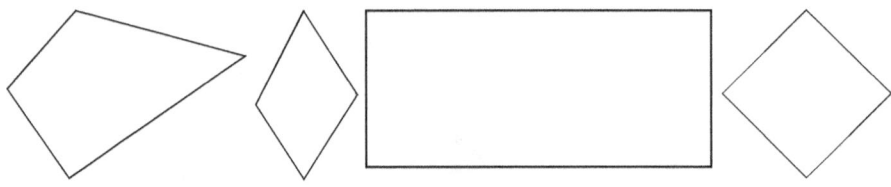

Quadrilaterals

All these plane shapes are quadrilaterals. Each of them is bound by four line segments. A quadrilateral has four sides, four vertices and four angles.

ELEMENTS OF A QUADRILATERAL

SIDES

A quadrilateral has *four* sides. In the figure, $\overline{AB}$, $\overline{BC}$, $\overline{CD}$ and $\overline{DA}$ are the sides of quadrilateral ABCD.

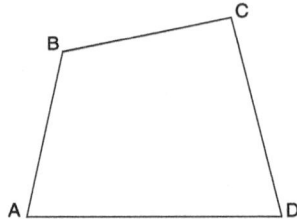

$\overline{AB}$ and $\overline{CD}$ are opposite sides $\overline{BC}$ and $\overline{DA}$ are also opposite sides.

VERTICES

A point, at which two sides of a quadrilateral meet, is called a vertex. There are *four* vertices of a quadrilateral. In the figure, points A, B, C and D are its vertices.

A and C form a pair of opposite vertices.

B and D also form a pair of opposite vertices

ANGLES

Each quadrilateral has *four* angles. In the figure, $\angle A$, $\angle B$, $\angle C$ and $\angle D$ are its angles.

∠A and ∠C form a pair of opposite angles.

∠B and ∠D form another pair of opposite angles.

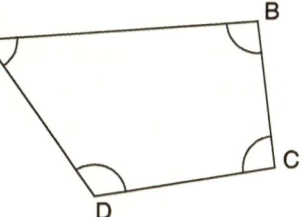

Diagonals

The line segments joining the opposite vertices are called diagonals of the quadrilateral. In the adjoining figure AC and BD are two diagonals of the quadrilateral ABCD

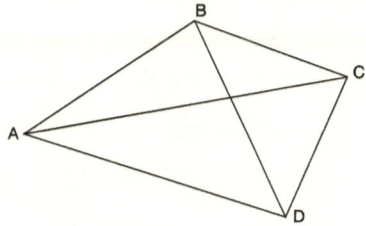

Adjacent Sides

Two sides of a quadrilateral having a common end point (vertex) are called adjacent sides.

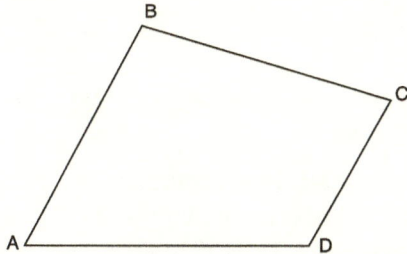

In the figure, $\overline{AD}$ and $\overline{AB}$ is a pair of adjacent sides. Other pairs of adjacent sides are: $\overline{AD}$ and $\overline{DC}$; $\overline{DC}$ and $\overline{CB}$; $\overline{CB}$ and $\overline{BA}$.

Opposite Sides

In the above quadrilateral the sides $\overline{AD}$ and $\overline{BC}$ are opposite sides. A pair of opposite sides in a quadrilateral does not have any common vertex. $\overline{AB}$ and $\overline{CD}$ is another pair of opposite sides.

Note

Like adjacent sides and opposite sides, a quadrilateral has adjacent-angles and opposite angles. In quadrilateral ABCD, one pair opposite angles is ∠A and ∠C.

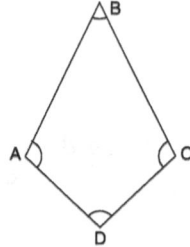

The other pair of opposite angles is ∠B and ∠D. Also, ∠A and ∠B; ∠B and ∠C; ∠C and ∠D; ∠D and ∠A are adjacent angles.

Interior and Exterior of a Quadrilateral

A quadrilateral divides the surface of the paper into three parts:

(i) The interior

(ii) The exterior

(iii) The quadrilateral itself.

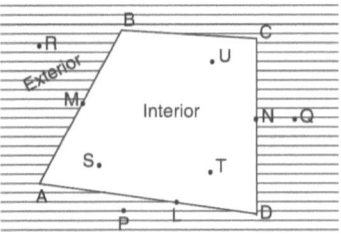

In the figure, the shaded region is the *exterior* of the quadrilateral ABCD. Points P, Q and R are in its exterior.

The unshaded portion bounded the boundary is the *interior* of the quadrilateral ABCD points S. T and U are in its interior. Points P, Q and R are in its exterior. Whereas the points A, M, B, C, N, D and L are on its boundary.

Circle

A circle is a geometrical plane curved shape consisting of all the points (in a plane) which are at fixed distance from a fixed point on the plane.

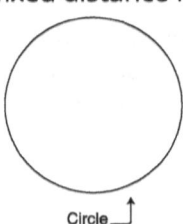

Circle

Elements of a Circle

Centre

Since, circle is a set of points which are equidistant from a fixed point. This fixed point on the plane is called the *centre* of the circle. In the adjoining figure, the point O is the centre of the circle.

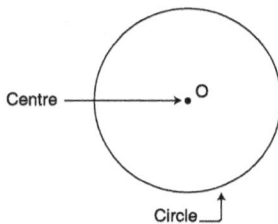

Radius

The fixed distance of any point on the circle to the fixed point is called *radius* of the circle. It is equal to the length of a line segment whose one end point is on the circle and other at the centre of the circle. In the fig $\overline{OA}$ is the radius of the circle.

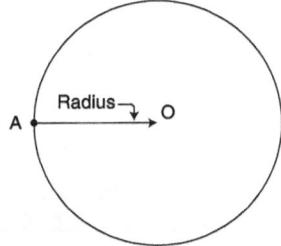

Circumference

The total length of the curved boundary of the circle is called its circumference. It is also called the perimeter of the circle.

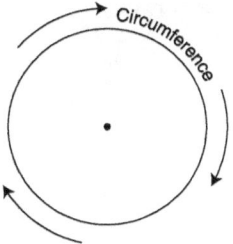

Interior and Exterior of a Circle

A circle divides the plane (in which it lies) into three distinct parts:

(i) The interior of the circle

(ii) The exterior of the circle

(iii) The circle itself.

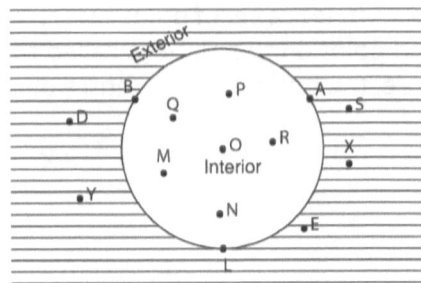

All those points which lie inside the boundary of the *circle* form the interior of the circle. In the figure, points such as P, Q, R, M, N etc. lie in the interior of the circle.

Remember

The distance of every point (in interior of a circle) from the centre of the circle is always less than the radius of the circle. If 'r' be the radius and 'o' be centre then OP < r, OQ < r, OM < r etc.

The shaded region shown in the figure is *exterior* of the circle. All those points which are outside the boundary of the circle form the exterior of the circle. Points E, S, D, X, Y etc. are in the exterior of the circle.

Remember

The distance of any point (in the exterior of the circle) from the centre is always greater than the radius, i.e. OE > r, OS > r, OD > r etc.

The part of the plane between the interior of the circle and the exterior of the circle form the boundary of the circle, in the figure, points such as A, B, L. lie on the circle.

The distance of each such point from the centre is always equal to the radius of the circle, i.e. OA = r, OB = r, OL = r, etc.

DID YOU KNOW?

The centre of a circle never lie on the circle. It is always in the interior of the circle.

CIRCULAR REGION

The part of the plane consisting of the circle and its interior region is called the *circular region*. The circular region consists of all the points of the plane whose distance from the centre are either less than or equal to the radius of the circle.

In the above figure, points such as O, P, Q, R, M, N, L, A and B lie in the circular region.

SOME IMPORTANT TERMS ASSOCIATED TO CIRCLES

CHORD

The line segment with its end-points lying on a circle is called the chord of the circle.

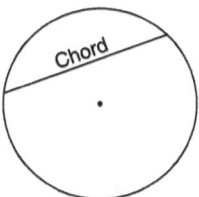

NOTE

The longest chord is the line segment passing through the centre. It is named as *diameter* of the circle.

DIAMETER

Any line segment passing through the centre of a circle, and having its end-points on the circle, is called a diameter of the circle. In the figure, AB is a diameter.

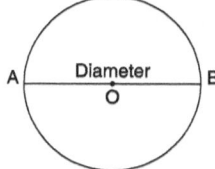

Obviously, OA is a radius is i.e. OA = r

Similarly, OB is a radius is i.e. OB = r

$\therefore$ OA + OB = r + r = 2r

$\Rightarrow$ AB = 2r

or Diameter = 2(radius)

REMEMBER

There can be an infinite number of diameters of a circle. Since, all the diameters pass through the centre. So all the diameters are concurrent the centre is their point of concurrence.

ARC

Any part of the boundary of a circle is known as an arc of the circle. In the figure the curved portion AB is an arc of the circle.

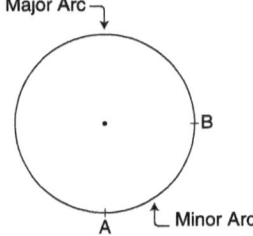

The symbolic representation of arc AB is $\overset{\frown}{AB}$.

The portion of the circle between the two points A and B can be looked at in two ways as shown in the following figures. The smaller arc is called a *minor arc* and the larger arc is called a *major arc*.

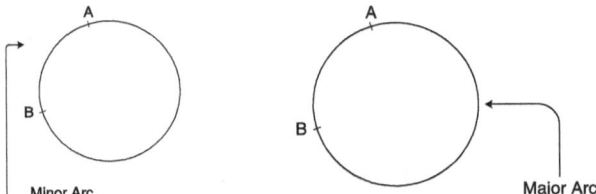

SEMI-CIRCLE

The end point of a diameter of a circle divide the circle into two equal parts. Each of these parts is called a semi-circle.

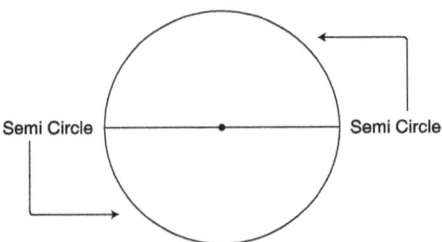

Segment of a Circle

The region enclosed by an arc and the corresponding chord is called a segment.

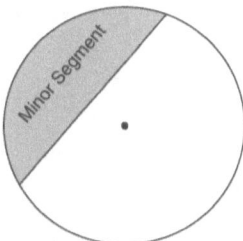

Every chord divides a circular region into two segments. The smaller segment is called minor-segment and the larger segment is called the major segment.

Remember

The centre of this circle always lies in the major-segment.

Sector of a Circle

The portion of the circular-region bounded by two radii and an arc of a circle is called a *sector* of the circle.

If the sector is formed by a major arc, then it is called a *major sector*

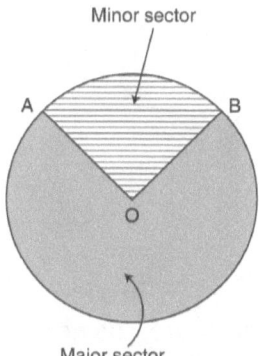

If the sector is formed by a minor arc, it is called a minor sector, as shown in the above figure.

Quadrant of a Circle

The sector of a circle formed by two mutually perpendicular radii, as show in the figure, is called a quadrant. In the figure BO ⊥ AO, So the region enclosed by the two radii (AO and BO) and the minor arc AB is a quadrant.

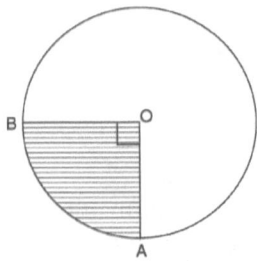

Note

In a circle, there are 4 quadrants.

Concentric Circles

Two or more circles with the same centre are called *concentric circles*. In the adjoining figure the center O is common to all the circle shown in it.

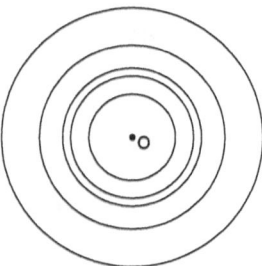

Exercise - 1

1. Take five points A, B, C, D, and E. Join them in order.
2. If 4 points P, Q, R and S are collinear. How many lines can be drawn passing through all of them?
3. 'A' is a point on your paper. How many lines can be drawn through A?
4. Two points P and R are given on a plane paper. How many lines be drawn passing through both P and R?

•R

•P

5. How many line segments can be drawn to join three non-collinear point?
6. Name all the line segments in the following figure:

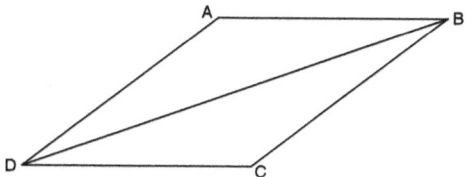

7. Name all the line segments in the following figure:

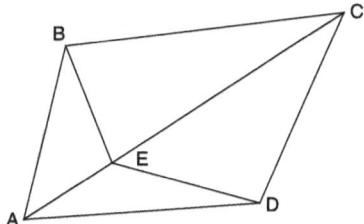

8. From the adjoining figure write all the lines which intersect at:

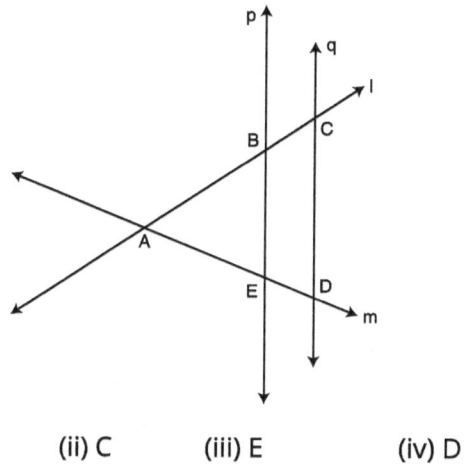

(i) A (ii) C (iii) E (iv) D

9. How many lines can be drawn four distinct points?

10. Name the collinear points in the figure given below.

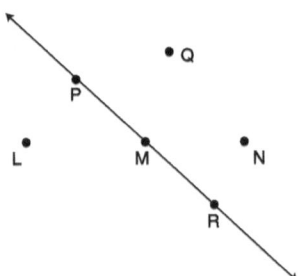

11. Name all the rays in the following figure whose initial point is B

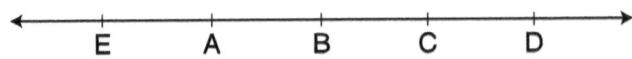

12. In the adjoining figure which points are non-collinear?

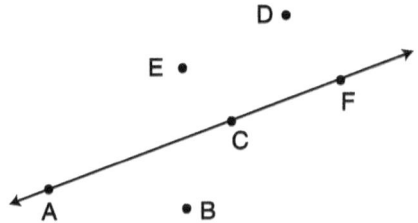

13. Name the points which are:

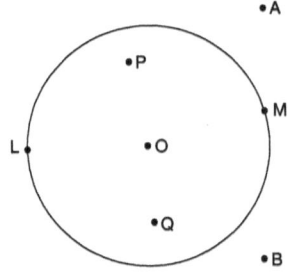

(i) In the interior of the circle.

(ii) In the exterior of the circle.

14. In the following figure, write

(i) A pair of parallel lines

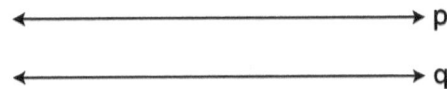

(ii) A pair of intersecting lines.

15. In the adjoining figure can you identify a point of concurrence? If yes, name it.

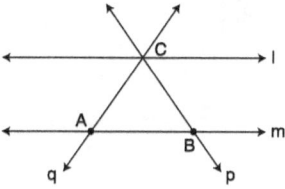

16. In the adjoining figure, name centre, and radius of the circle.

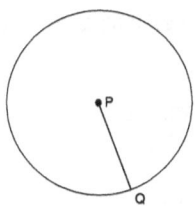

17. In the following figure, name the following

    (i) Diameter of the circle
    (ii) Chords of the circle
    (iii) All radii of the circle

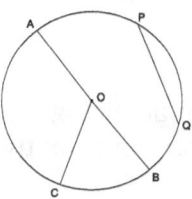

18. Look at the adjoining. What are the shaded regions called?

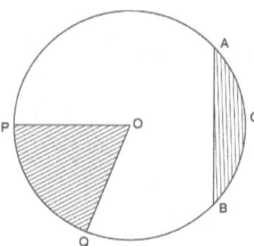

19. In the adjoining, what do we call the shaded

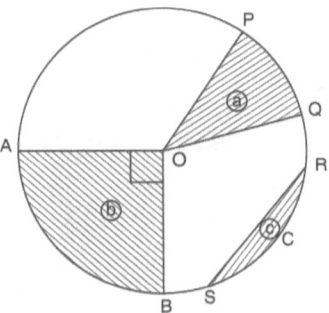

   a. _____

   b. _____

   c. _____

20. In the adjoining figure which points are:

    (i) In the interior of the circle

    (ii) In the exterior region of the circle

    (iii) On the circle

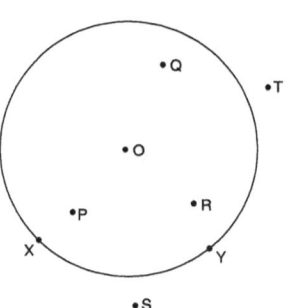

21. In the adjoining figure name all those points which are in the circular region of the given circle.

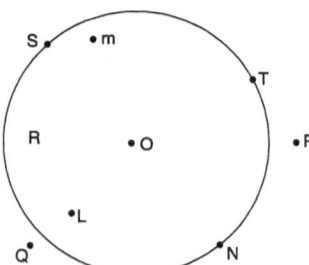

22. In the adjoining figure, name the following:

(i) vertices of the △ PQR
(ii) Sides of the △ PQR
(iii) Altitude from the vertex P
(iv) Side opposite to vertex R

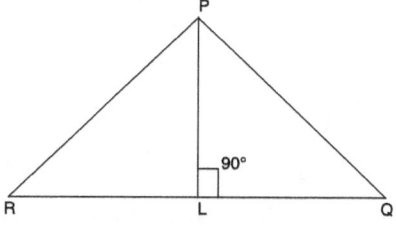

23. In the figure, medians AP, BQ and CR meet at I. What is the point I called.

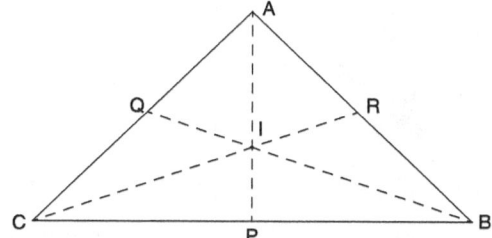

24. PQRS is a quadrilateral Name:

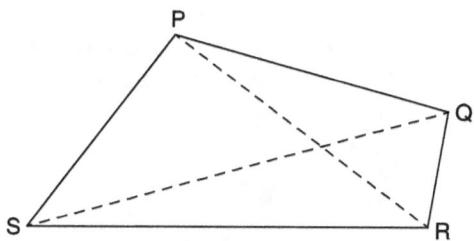

(i) Vertices of the quadrilateral PQRS

(ii) Sides of the quadrilateral PQRS

(iii) Diagonals of the quadrilateral PQRS

(iv) Side opposite to $\overline{QR}$

(v) Side adjacent to $\overline{PQ}$

(vi) Angle opposite to angle Q

Answers

2. only one   3. Indefinite   4. Only 1   5. 3
6. $\overline{AB}, \overline{BC}, \overline{CD}, \overline{DA}, \overline{BD}$ 7. {$\overline{AB}, \overline{BC}, \overline{CD}, \overline{DE}$, AE, BE, CE and AD.

8. (i). l and m (ii). l and q (iii). m and p (iv). m and q
9. 4   10. P, M, and R
11. $\overline{BC}$, $\overline{BD}$, $\overline{BA}$ and $\overline{BE}$ 12. B, E and D
13. (i). O, P and Q (ii). A and B   14. (i). p, q (ii). r and s
15. yes, 'C'   16. centre: P; radius: $\overline{PQ}$
17. (i). $\overline{AB}$ (ii). $\overline{AB}$ and $\overline{PQ}$ (iii). $\overline{OA}$, $\overline{OB}$ and $\overline{OC}$
18. POQ is a sector; APB is a segment
19. a : sector; b : quadrant; c segment
20. (i). O, P, Q and R   (ii). S and T   (iii). X and Y
21. O, L, M, N, R, S and T
22. (i). P, Q and R (ii). $\overline{PQ}$, $\overline{QR}$ and $\overline{RP}$ (iii). $\overline{PL}$ (iv). $\overline{PQ}$
23. centroid
24.(i). P, Q, R and S (ii). $\overline{PQ}$, $\overline{QR}$, $\overline{RS}$ and $\overline{SP}$ (iii). $\overline{PR}$ and $\overline{SQ}$
    (iv). $\overline{SP}$ (v). $\overline{SR}$ (vi). $\underline{S}$

HOTS

1. What is common in the following two figures. Are both of them closed shapes? If not then which one is not a closed figure?

   (i)   (ii)

   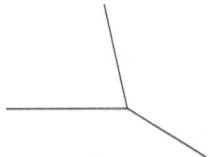   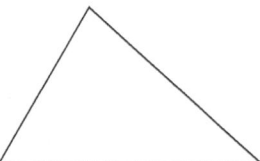

2. (i) Count and name all the line segments in the following figure:

   A———B———C———D———E

   (ii) If one more point F is added (marked) (as shown in the figure given below), then how many more line segments add to the list. Name all the increased line segments.

Answers

1. Three line segments. (i). is not closed.

2. (i) Ten, $\overline{AB}$, $\overline{AC}$, $\overline{AD}$, $\overline{AE}$, $\overline{BC}$, $\overline{BD}$, $\overline{BE}$, $\overline{CD}$, $\overline{CE}$ and $\overline{DE}$.
   (ii) Five: $\overline{AF}$, $\overline{BF}$, $\overline{CF}$, $\overline{DF}$ and $\overline{EF}$

Mental Maths

1. Does a line segment contain indefinite number of points?
2. Does a ray has definite length?
3. How many line segments are there in the following figure?

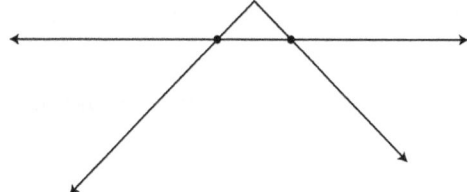

4. How many line segments are there in the following figure?

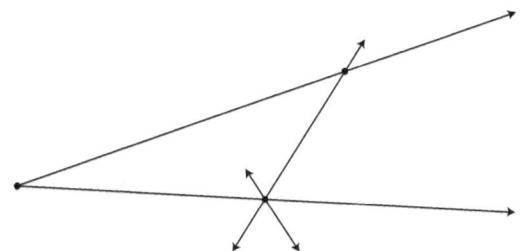

5. How many line segments are there in the following figure:

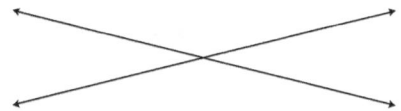

6. Name the point not lying on l in the following figure:

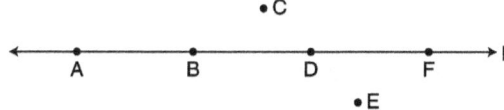

7. Name the lines, concurrent at the point B?

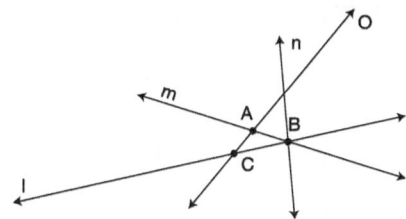

8. Does the centre of a circle lie in the interior of the circle?

9. Do both end points of a radius lie in the interior of the circle?
10. What is the longest chord of circle is called?
11. All medians of a triangle meet at a point what is this called?
12. All the altitudes of a triangle. What is this point called?
13. How many diameters can be drawn in a circle?
14. Are all radii of a circle concurrent?
15. Are all diameters of a circle concurrent?
16. In the adjoining figure, identify the, set of equal line segments:

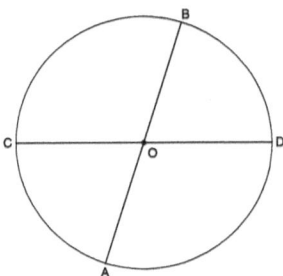

17. What do we call the set of circles drawn with same centre and different radii?

ANSWERS

1. Yes	2. No	3. Three
4. Three	5. Zero	6. Point C
7. l, m and n	8. Yes	9. No
10. diameter	11. centroid	12. Orthocentre
13. Infinite	14. Yes	15. Yes

16. $\overline{OA}$, $\overline{OB}$, $\overline{OC}$, $\overline{OD}$, $\overline{AB}$ and $\overline{CD}$

17. Concentric circles

MULTI CHOICE QUESTIONS

1. Which of the following is the number of lines passing through five non-collinear points?
    (i) 5   (ii) 8   (iii) 10   (iv) 15

2. Which of the following is the number of line segments in the figure, given below:

   P   Q   R   S   T

   (i) 5   (ii) 10   (iii) 15   (iv) 20

3. Which of the following shapes have curved boundary?

   (i) 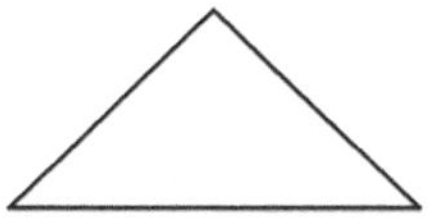 [triangle]   (ii) 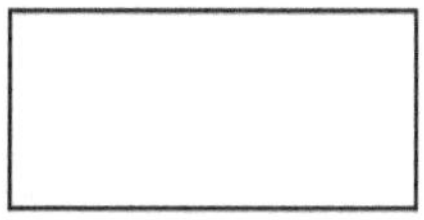 [rectangle]

   (iii) [square]   (iv) 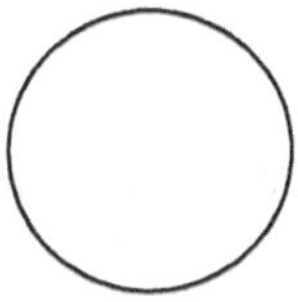 [circle]

4. Three lines l, m and p intersect at a point O. Which of the following is the number of more lines (other than l, m and p) that can be drawn through O?

   (i) None   (ii) Two   (iii) Three   (iv) Infinite

5. Which of the following is not represented in the following figure.

   C   B   A

   (i) Point   (ii) Line segment   (iii) Ray   (iv) Straight line

6. Surface of which of the following does not contain a line?

   (i) Black board   (ii) Book   (iii) Football   (iv) Floor

7. A ray has:

   (i) two end points   (ii) one end point

   (iii) no end point   (iv) none of these

8. Which of the following is the number of lines which can be drawn through three non collinear points, joining two points at a time?

   (i) 1   (ii) 2   (iii) 3   (iv) 4

9. The centre of a circle lies:

   (i) on the circle   (ii) out side the circle

   (iii) inside the circle   (iv) none of them.

10. Which of the points in the adjoining figure is a point of concurrence?

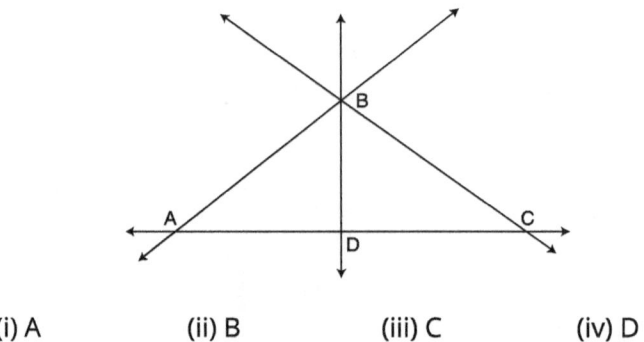

(i) A (ii) B (iii) C (iv) D

11. In the adjoining figure which of the following is true?

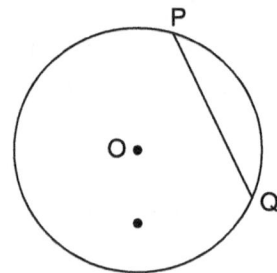

(i) $\overline{PQ}$ is a radius (ii) $\overline{PQ}$ is a diameter
(iii) $\overline{PQ}$ is a chord (iv) $\overline{PQ}$ is a segment

12. In the adjoining figure, which line segment is a radius?

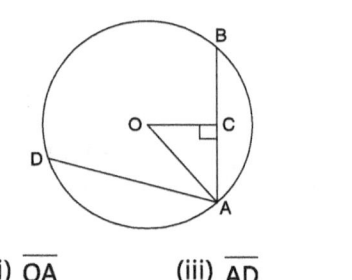

(i) $\overline{OC}$ (ii) $\overline{OA}$ (iii) $\overline{AD}$ (iv) $\overline{AB}$

13. In the adjoining figure which line segment is the diameter of the circle?

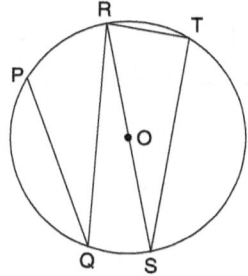

(i) $\overline{RQ}$ (ii) $\overline{PQ}$ (iii) $\overline{ST}$ (iv) $\overline{RS}$

14. The end points of a diameter lie:

   (i) On the circle   (ii) in the interior of the circle

   (iii) $\begin{cases}\text{in the exterior}\\ \text{of the circle}\end{cases}$ (iv) None of these

15. Which of the following is the longest chord of a circle, shown in diagram?

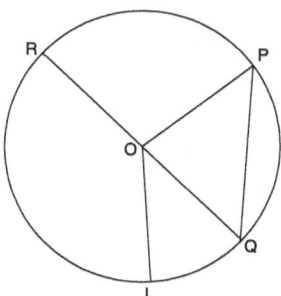

(i) Its radius $\overline{OP}$ (ii) $\overline{PQ}$ (iii) $\overline{QR}$ (iv) Is radius $\overline{OL}$

16. Which the following is represented by the points O, A, B, P and S?

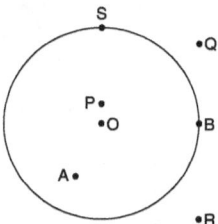

(i) Interior of the circle   (ii) Exterior of the circle

(iii) Circular region of the circle   (iv) The circle.

17. Which of the following can form a quadrant of a circle?

    (i) two mutually perpendicular radii

    (ii) two mutually perpendicular diameters

    (iii) two mutually perpendicular chords

    (iv) any two mutually perpendicular line segments.

18. How many quadrants can make a circle?

    (i) 2   (ii) 4   (iii) infinite   (iv) 3

19. All diameters of a circle are concurrent at:

    (i) the centre of the circle

    (ii) one of the end of a diameter

    (iii) any point in the interior of the circle

    (iv) any point on the circle

20. A four sided closed figure is called:

    (i) a triangle   (ii) a circle   (iii) an angle   (iv) a quadrilateral

21. A triangle has _____ medians.

    (i) 1   (ii) 2   (iii) 3   (iv) 4

22. A quadrilateral has _____ diagonal(s).

    (i) 1   (ii) 2   (iii) 3   (iv) 4

ANSWERS

1. (iii)	2. (ii)	3. (iv)	4. (i)	5. (iii)
6. (iii)	7. (ii)	8. (iii)	9. (iii)	10. (ii)
11. (iii)	12. (ii)	13. (iv)	14. (iv)	15. (iv)
16. (iii)	17. (i)	18. (ii)	19. (i)	20. (iv)
21. (iii)	22. (ii)			

WORK SHEET

1. How many:

    (i) altitudes can there be in a triangle?

    (ii) pairs of opposite angles does a quadrilateral have?

(iii) pairs of adjacent sides does a quadrilateral have?

(iv) pairs of opposite sides does a quadrilateral have?

2. Do $\overline{AB}$ and $\overline{BA}$ represent the same straight line?
3. Do $\overrightarrow{PQ}$ and $\overrightarrow{OP}$ represent the same ray?
4. Do $\overline{XY}$ and $\overline{YX}$ represent the same line-segment?
5. Name the concurrent lines in the following figure. Also name the point of concurrency.

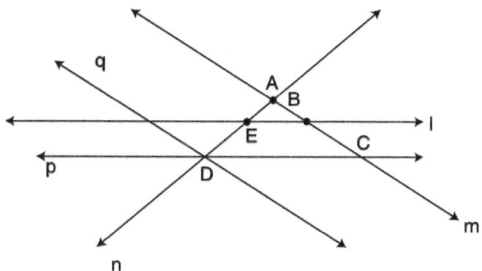

6. Name the line segments in the figure shown in the adjoining diagram.

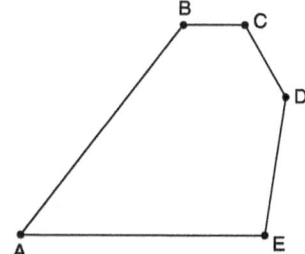

7. In the adjoining figure write name of all points

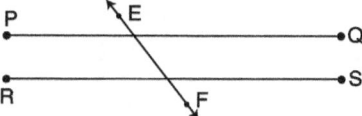

8. How many line segments are there in the following figure? Name all of them.

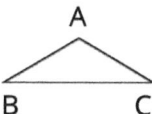

9. What is the difference between concurrency and colinearity?
10. For each of the following pairs of lines write 'parallel' or 'intersecting'.

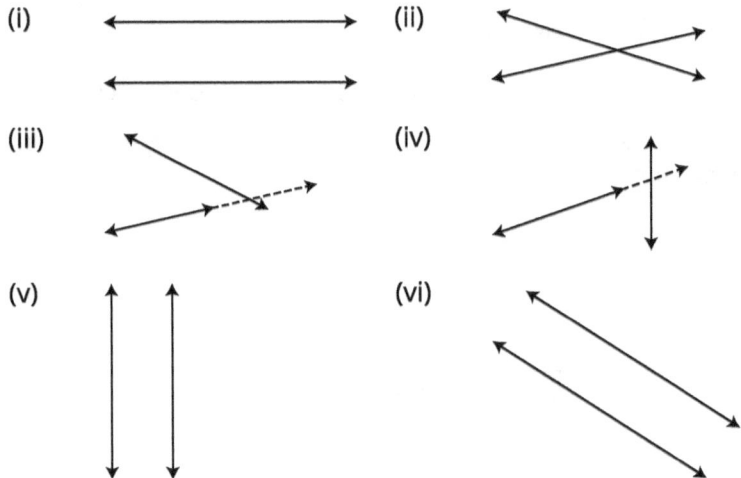

11. In the adjoining figure write names of all lines passing through D?

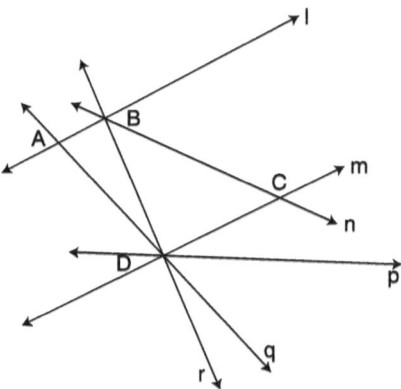

12. Look at the figure given along side and then match *column-A* with *column-B*

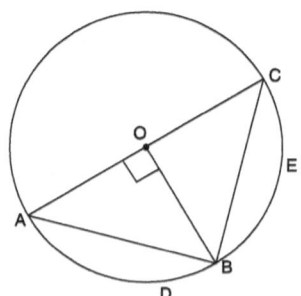

Basic Geometrical Ideas | 267

Column-A	Column-B
(a) Three chords	(i) $\overline{AC}$
(b) Three equal line segments	(ii) ADB and BEC
(c) A diameter	(iii) $\overline{AB}$, $\overline{AC}$ and $\overline{BC}$
(d) Two segments	(iv) AOB and BOC
(e) Two quadrants	(v) $\overline{OA}$, $\overline{OB}$, $\overline{OC}$

13. **Match the 'column-A' with 'column-B'**

Column A	Column B
(a) A part of line having one end point called its initial point. Extends to infinity in one direction only. Its length is unlimited.	(i) The concurrent lines l, m, n and p are passing through point O.
(b) In a plane, three or more lines passing through the same point.	(ii) A line segment AB
(c) In a plane, three or more points lying on the same line.	(iii) A ray PQ its initial point is P
(d) A part of line having two end points. It has a definite length.	(iv) A line AB
(e) Having no breadth, or thickness but only length and extends up to infinity in both directions. It has unlimited length and has no end points	(v) Points A, B, C and D are collinear points

14. **Write 'True' and 'False'**

   (i) A point has no dimensions but has thickness.

   (ii) A line has no end points. Therefore it has not a definite length.

   (iii) Only one line segment can be drawn passing through a given point in a plane.

   (iv) A line, line segment and ray all are called one dimensional (1-D) figures as they have only length (i.e. only one dimension)

   (v) The area (amount of region) enclosed by a closed shape is called its perimeter.

   (vi) The area enclosed inside the circle incuding its boundary is called the circular-region

   (vii) The distance between two parallel lines not necessarily the same everywhere.

   (viii) The symbol '⊥' stands for '**is perpendicular to**' such as l ⊥ m.

   (ix) Circle is a simple closed curve which made by curved line only.

   (x) Two different lines in a plane either intersect at exactly one point or are parallel.

15. **Fill in the blanks:**

   (i) A _____ determines a location. It does not have length, breadth, width or thickness.

   (ii) A _____ has length only, It does not have breadth or thickness.

   (iii) A plane cannot be drawn on a piece of paper, because a paper is a _____ of a plane.

   (iv) The fundamental geometrical concepts depend on the three basic concepts point, line and _____.

   (v) Three or more points in a plane are called collinear if and only if all of them lie on the same line, and this line is called the line of _____

   (vi) Three or more lines (in a plane) are called concurrent if and only if all of them pass through the same point, and this point is called the _____ points

(vii) Two rays having the same initial point but directed in the opposite directions are called _____ rays.

(viii) There is a unique _____ joining two given points A and B.

(ix) An unlimited number of _____ can be drawn passing through a given point.

(x) There is exactly one line passing through _____ distinct points in a plane.

16. Classify all points shown in the adjoining figure lying (i) in the interior of the circle (ii) in the exterior of the circle and (iii) on the circle (iv) in the circular-region of the circle.

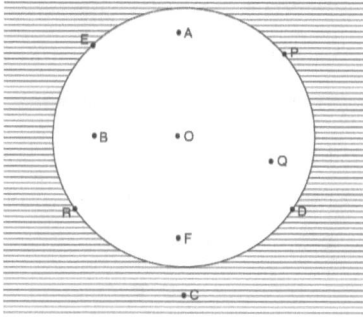

17. Define (i) Circular region (ii) Semi-circular region (iii) Segment (iv) Quadrant (v) Concentric circles.

18. Look at the adjoining figure and fill in the blanks using <, > or =; such that O is the centre of the circle and 'r' is its radius.

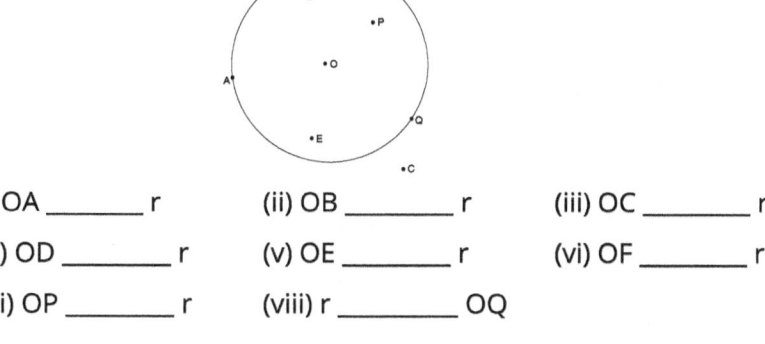

(i) OA _____ r   (ii) OB _____ r   (iii) OC _____ r

(iv) OD _____ r   (v) OE _____ r   (vi) OF _____ r

(vii) OP _____ r   (viii) r _____ OQ

ANSWERS

1. (i) 3   (ii) 2   (iii) 4   (iv) 2

2. yes   3. No   4. Yes

5. p, q and n; point D
7. P, Q, R, S, E, and F
8. Three
10. (i) Parallel lines (ii) Intersecting lines
    (iii) Intersecting lines (iv) Intersecting lines
    (v) Parallel lines (vi) parallel lines.
11. m, p, q and r
12. (a) → (iii); (b) → (v); (c) → (i); (d) → (ii); (e) → (iv)
13. (a) → (iii) (b) → (i) (c) → (v) (d) → (ii) (e) → (iv)
14. (i) False (ii) True (iii) False
    (iv) True (v) False (vi) True
    (vii) False (viii) True (ix) True (x) True
15. (i) point (ii) line (iii) part
    (iv) plane (v) collinearity (vi) concurrence
    (vii) opposite (viii) Line segment (ix) lines (x) two
16. (i) A, B, F, O, Q (ii) C, S
    (iii) D, E, P and R (iv) A, B, F, O, Q, D, E, P and R
17. (i) = (ii) < (iii) > (iv) >
    (v) < (vi) = (vii) < (viii) =

Chapter 10
Understanding Elementary Shapes

Measurement of a Line Segment

Since a line segment in a plane has definite length. Therefore, it can be measured. We use a ruler (scale).

In order to measure the length of a given line-segment, the perpendicular edge having scale marks, is placed along the line segment such that the zero mark of the scale coincide with the left end point of the line segment.

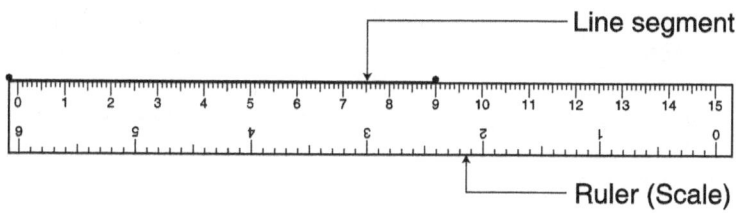

The reading corresponding to the other end point of the line-segment. Tells us the length.

Drawing a Line Segment of Given Measurement

Let us draw a line segment of 6.4 cm with the help of a ruler.

We mark a point 'A' on the sheet of paper and then place the ruler flat on the sheet such that the zero coincide with the point 'A'.

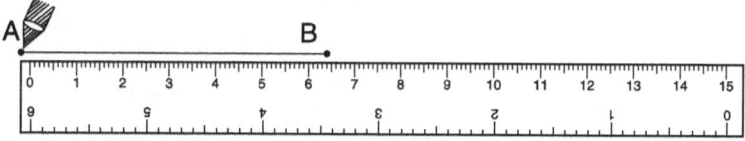

Mark another point B which reads 6.4 cm. Now join the two points A and B by moving the pencil tip along the edge of the scale. Thus, $\overline{AB}$ = 6.4 cm is the required line segment.

PAIR OF LINES IN A PLANE [Parallel Lines and Intersection Lines.]

For two lines in a plane, there can be two possibilities, that they are

(i) Either intersect each other, or

(ii) Do not intersect each other

Let there be two lines p and q in a plane,

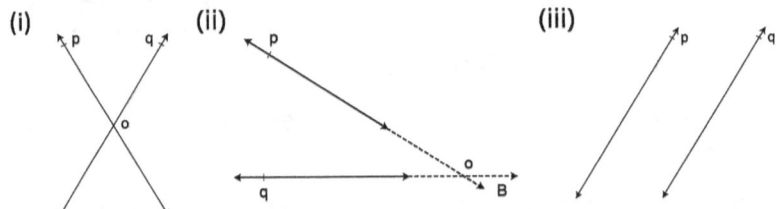

as shown in the fig (i), (ii) and (iii). We observe that in (i) and (ii) lines intersect (or seem to intersect) at o. The point o, which common to both l and m, is called the point of intersection. Such type of lines are called *intersecting lines*. In fig (iii), the lines l and m never intersect each other, however for they are extended in either direction, i.e., they (l and m) do not have any common point. Such lines are called *parallel lines*.

CONCURRENT LINES

If three or more lines, in a plane, pass through the same point, then the lines are called concurrent lines. In the figure, the lines l, m, n and p are *concurrent*. This property of lines is called the *concurrency of lines* and the common point (here, o) is called the *point of concurrence*

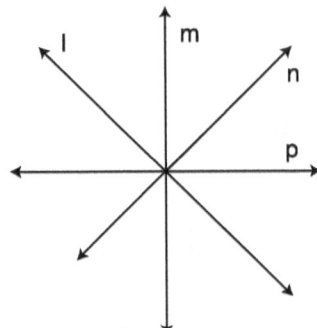

Three or more lines in a plane which are not concurrent are called *non-concurrent* lines. In the adjoining figure l, m and n are non-concurrent lines.

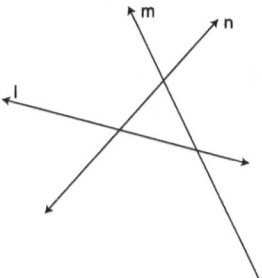

COLLINEAR POINTS

If three or more point, lie on the same line, then the points are called *collinear points*. In the figure the points A, B, C, and D are collinear.

This property is called the *collinearity of points*.

If three or more points are not on the same line, they are called non-collinear points.

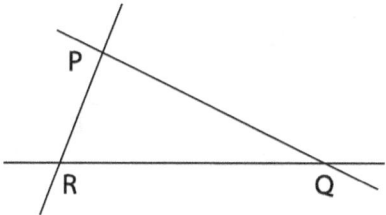

In the figure the points P, Q and R are non-collinear points.

PERPENDICULAR LINES

If two intersecting lines are such that, the angle between them is 90°, then the lines are called perpendicular lines. In the adjoining figure p and q are perpendicular lines.

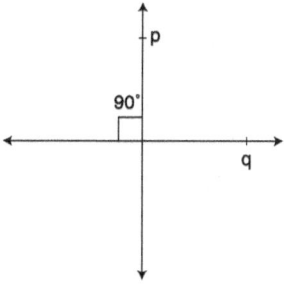

PERPENDICULAR BISECTOR OF A LINE SEGMENT

If a line AB is perpendicular to a line segment PQ at the point o.

Also, if length of OP = Length of OQ, then the line AB is called the *perpendicular bisector* of the line-segment. PQ

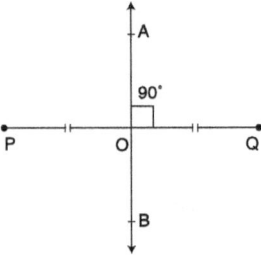

FORMING AN ANGLE

An angle is a formed by rotating a ray about a point (vertex).

Let two ray $\overrightarrow{OA}$ and $\overrightarrow{OB}$ have their common vertex at O and $\overrightarrow{OB}$ rotating about the vertex in clockwise direction from its initial position on.

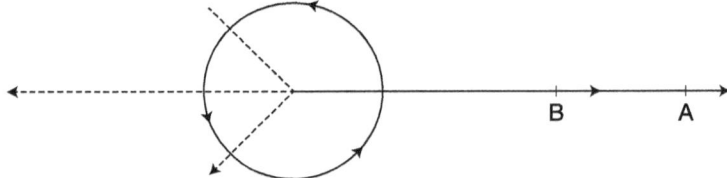

It makes angles of increasing order. Rotating in this way when the $\overrightarrow{OB}$ reaches the position of $\overrightarrow{OA}$ (i.e. from where it started), it would then have made one complete "turn" like a spoke of a wheel or like a hand of a clock.

MEASURING ANGLES IN TURNS

The clock hands rotate like arms of an angle. The minute hand makes 1 complete 'turn' in 1 hour (or 60 minutes).

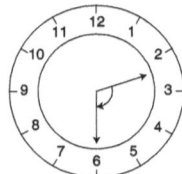

∴ In 1 minute it rotates $\frac{1}{60}$ turn

In 5 minutes it rotates $\frac{1}{60} \times 5 = \frac{1}{12}$ turns

Similarly, in 25 minutes the minute hand rotates $\frac{1}{12} \times 5 = \frac{5}{12}$ turns

MEASURING ANGLES IN DEGREES

It is more convenient to use *'degrees'* instead of 'turns'. A complete turn is divided into 360 equal parts and each part is called a "**degree**"

The symbolic notation of a degree is a small circle '°', written as superscript such as 1 degree = 1°, 2 degrees = 2°, 3 degree = 3° and so on.

One complete turn =180°

DID YOU KNOW

The word 'degree' has come from the latin "gradus" meaning steps

MAGNITUDE OF AN ANGLE

The magnitude or (size) of an angle is the amount of rotation of the rotating arm about its vertex. It is also called the amount of opening between its two arms. The magnitude of an angle does not depend on the length of its arms.

PROTRACTOR

A protractor is a plastic or metallic circular disc used to measure or construct an angle of a given measure. The adjoining figure is a circular protractor which is marked with degrees from 0° to 360°, from left to right.

But normally, we use a semi-circular protractor which is marked in degrees from 0° to 180°. It has two scales, a clockwise scale and an anticlockwise scale.

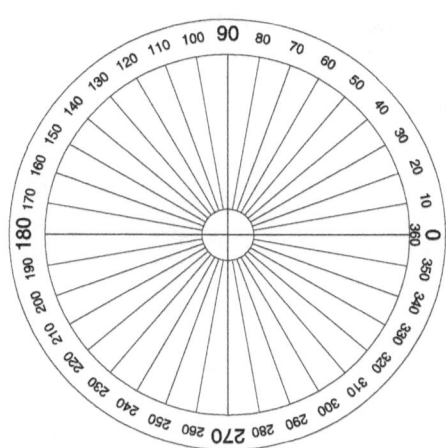

The clockwise scale is on the outer side of the curved edge and the anti-clockwise scale is on the inner side, as shown in the adjoining figure.

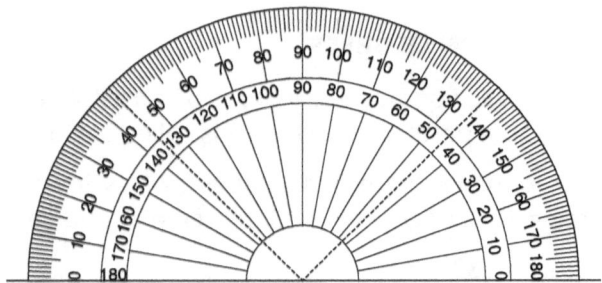

The zeroes of both scales are joined by a straight line which called *zero-line* or the *base-line*. The midpoint of the base line is called the *centre* of the protractor.

MEASURING AN ANGLE USING PROTRACTOR

Let us measure a given $\angle AOB$, using protractor.

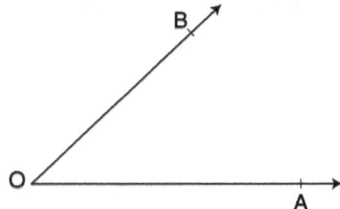

We place the protractor on the angle in such a manner that its centre falls exactly on the vertex O of the given angle and the inner straight edge along one of the arms.

Now read off the mark against the other arm. Here, the arm $\overrightarrow{OB}$ passes through the 60° mark.

Thus, $\angle AOB = 60°$

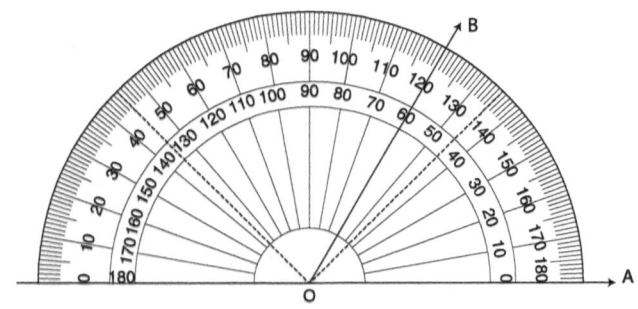

TYPES OF ANGLES

Classification of angles is done using their degree measures: various types of angles are given as:

(i) Zero Angle

If the initial position and the final position of a rotating ray coincide without making any revolution, then the angle formed is a *zero angle*. In the figure ∣AOB = 0°

(ii) Right Angle

When the revolving ray completes $\frac{1}{4}$ of the complete turn, then the angle formed between the two rays (fixed and revolving) is a right angle.

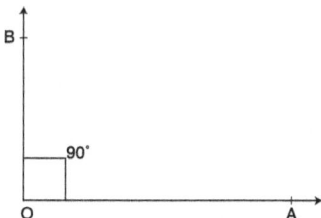

The degree measure of a right angle is 90°. In the figure ∣AOB = 90° = a right angle.

(iii) Straight Angle

An angle formed by two opposite rays is called straight angle.

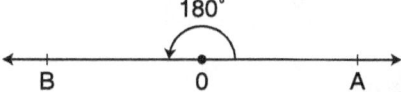

The degree measure of a straight angle in 180°. In the figure ∣AOB = 180° = a straight angle.

(iv) Complete Angle

When the rotating ray, after a complete turn, 360° coincides with the initial position then the angle formed is called a complete angle.

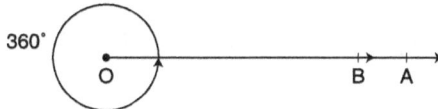

In the figure $\lvert\text{AOB}$ is a complete angle and its measure is 360°.

(v) Acute Angle

Any angle, whose degree measure is greater than 0° and less than 90° is called an acute angle.

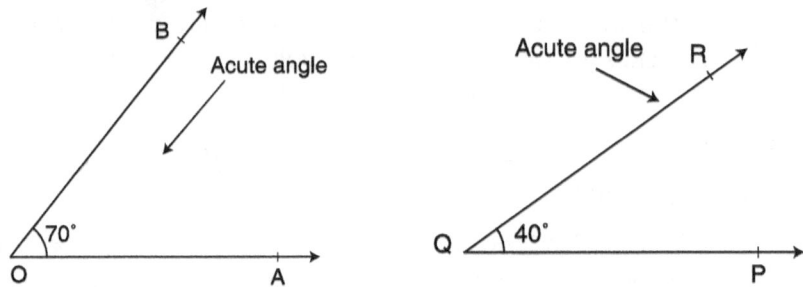

In the adjoining figure, $\lvert\text{AOB} = 70°$ and $\lvert\text{PQR} = 40°$ are acute angles.

(vi) Obtuse Angle

Any angle whose degree measure is greater than 90° but less than 180° is called an obtuse angle.

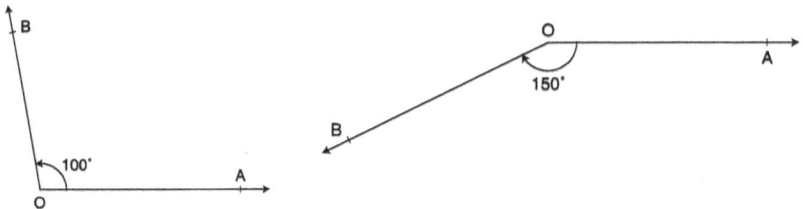

In the adjoining figure $\lvert\text{POQ} = 100°$ and $\lvert\text{AOB} = 150°$ are obtuse angles.

(vii) Reflex Angle

An angle whose measure is more than 180° and less than 360° is called a reflex angle.

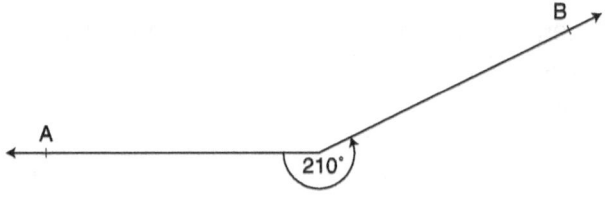

In the figure $\lvert\text{AOB}$ is a reflex angle.

Two Dimensional Shapes (2D Shapes)

A 2-D shape is flat and has no depth or height. It has only length and width and can be drawn on a plane surface. Triangle, rectangle, square, pentagon etc., are two dimension shapes.

Types of Triangles

There are many triangle of different types. We classify them by two different ways:

(I) On the Basis of Sides: Triangles are classified into *three* types on the basis of their sides: **scalene, isosceles** and **equilateral.**

Scalene Triangle

A triangle having all unequal sides is called a *scalene triangle.*

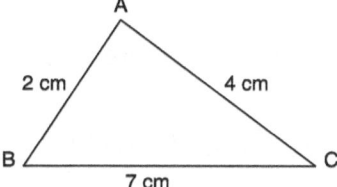

In the figure $\triangle$ ABC is a scalene triangle because AB ≠ BC ≠ CA

Isosceles Triangle

If two sides of a triangle are equal in length, it is called an isosceles triangle.

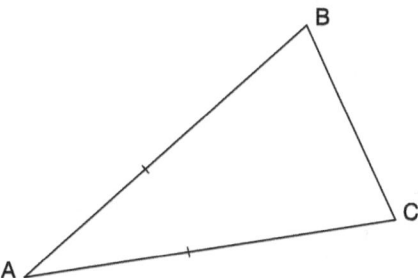

In the figure, AB = AC, therefore, $\triangle$ ABC is an isosceles triangle.

Did You Know?

The word "isosceles" has come from two Greek word *isos* meaning *'equal'* and "*skelos*" meaning *'leg'* so isosceles means *'equal legs.'*

EQUILATERAL TRIANGLE

A triangle is said to be *Equilateral* if all its three sides are equal in length.

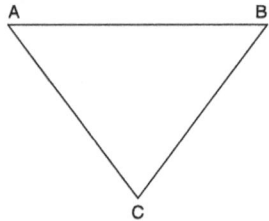

In the figure, △ ABC is equilateral as AB = BC = CA.

REMEMBER

Each angle of an equilateral △ is 60°.

(ii) **ON THE BASIS OF ANGLES**

Triangles are classified into three different types on the basis of their angles:

acute angled; obtuse angled; right angled

ACUTE ANGLED TRIANGLE

A triangle whose three angles are all acute is known as an acute angled triangle.

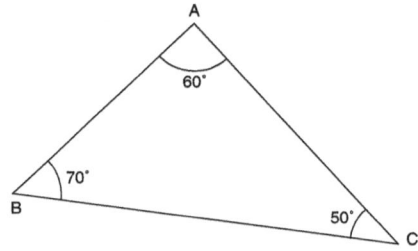

In the figure △ABC is an acute angled triangle [**because each of its angles is less than 90°**]

OBTUSE ANGLED TRIANGLE

A triangle is said to be an obtuse-angled triangle if one of its angles is obtuse.

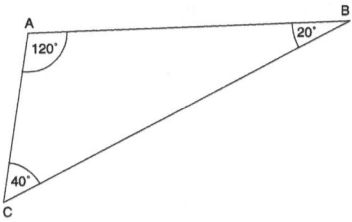

In the figure the degree measure of ⌊BAC is greater than 90° [∵ 120° > 90° and 120° < 180°]

i.e. ⌊BAC is an obtuse angle, and therefore, △ ABC is an obtuse triangle.

RIGHT-ANGLED TRIANGLE

A triangle is said to be right-angled triangle if one of its angle is a right-angle. In the adjoining figure,

△ ABC is a right angled triangle, because

⌊ABC = 90°

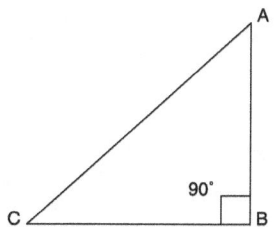

NOTE

A 'right angled triangle' is simply called a 'right-triangle'.

EXERCISE

1. Draw a line segment $\overline{XY}$ = 10.3 cm. Mark two points A and B on this line segment such that $\overline{XA}$ = 3.3 cm and $\overline{YB}$ = 4.4 cm. Measure $\overline{AB}$.

2. Draw any line segment $\overline{AB}$. Take a point c lying between A and B. Measure $\overline{AC}$ and $\overline{BC}$ and AB.

   Is AB = $\overline{AC}$ + $\overline{BC}$?

3. Classify the following angles as:

   (i) Zero angle,   (ii) acute angle   (iii) obtuse angle

   (iv) Reflex angle (v) right angle   (vi) Complete angle

   152°, 195°, 103°, 167°, 0°, 90°, 360°, 100°, 249°, 45°, 200°, 89°, 91°, 10°

4. Measure the following angles, using a protractor. Name these angles.

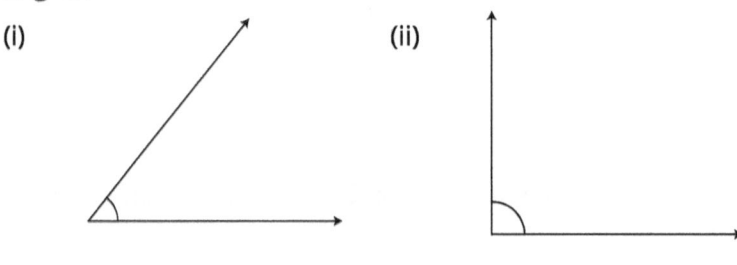

(i)   (ii)

(iii)

(iv)

(v)

5. Name all the elements of the following triangles:

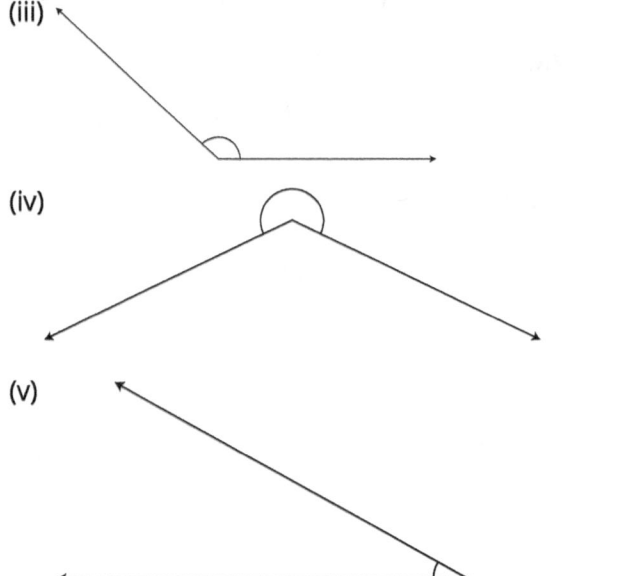

(i)   (ii)

6. List the given points which lie in:

   (i) On the △   (ii) In the interior of the △

   (iii) In the Exterior of the △

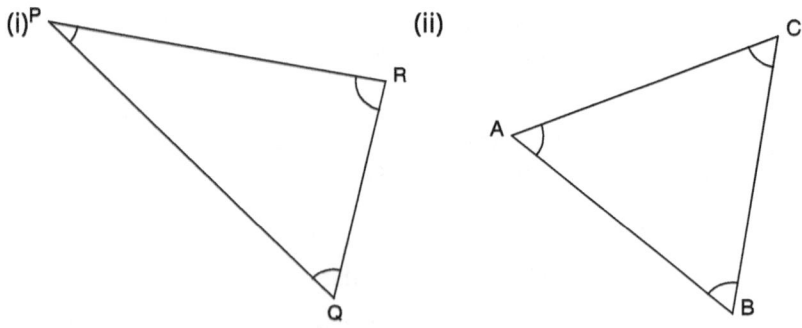

7. Define the triangular-region of a triangle. In the adjoining figure name points which lie in the triangular region.

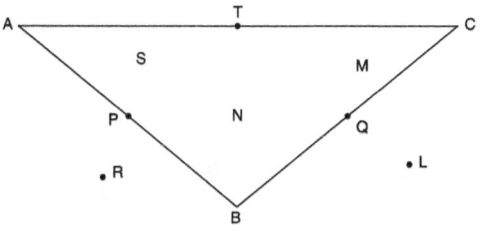

8. Classify the following triangles into

   (i) acute triangle, (ii) obtuse Δ and (iii) Right-triangle:

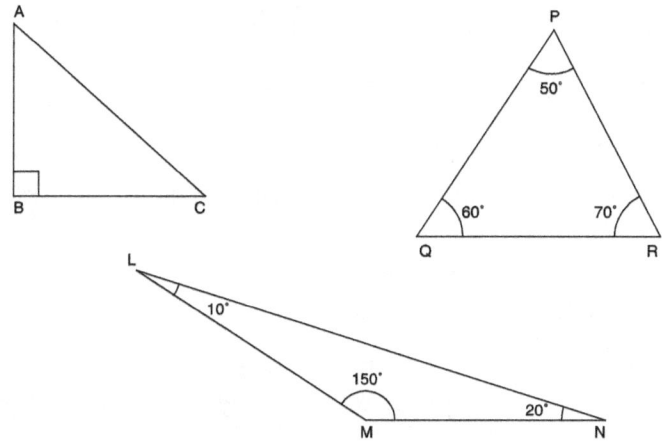

9. Classify the following triangles as (i) Scalene (ii) Isosceles (iii) Equilateral.

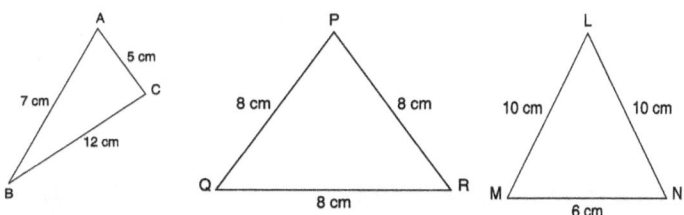

ANSWERS

1. $\overline{AB}$ = 2.6 cm 2. Yes

3. (i) zero angle: 0°   (ii) acute angle: 45°, 10°, 52°, 89°

(iii) Obtuse angle: 103°, 100°, 167°, 91°,

(iv) Reflex angle: 195°, 200°, 249°.

(v) right angle: 90°   (vi) complete angle: 360°

4. (i) 60°, Acute angle   (ii) 90°, right angle

   (iii) 120°, obtuse angle   (iv) 245°, Reflex angle

   (v) 25°, Acute angle.

5. (i) $\overline{PQ}$, $\overline{QR}$, $\overline{RP}$; $\underline{P}$, $\underline{Q}$ and $\underline{R}$

   (ii) $\overline{AB}$, $\overline{BC}$, $\overline{CA}$; $\underline{A}$, $\underline{B}$ and $\underline{C}$

6. (i) A, B, C, P, Q and T   (ii) N, M and S   (iii) L and R

7. A, B, C, P, Q, T, M, N and S

8. (i) right-△   (ii) acute-△   (iii) obtuse-△

9. (i) △ ABC   (ii) △ LMN   (iii) △ PQR

SPECIAL TYPES OF QUADRILATERALS

A quadrilateral is closed figure formed by four sides. It has two pairs of opposite sides. We can divide quadrilaterals broadly into three categories according to the behavior of their sides:

1. **BOTH PAIRS OF OPPOSITE SIDES ARE PARALLEL**

   All quadrilaterals that have both of their pairs of opposite sides as parallel are called *Parallelograms*. In the figure ABCD is a parallelogram. Here, we have AB ∥ CD and AD ∥ BC

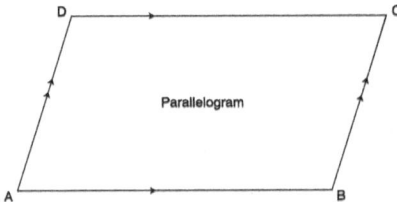

Parallelogram

NOTE

(i) **AB ∥ CD means the line-segment AB is parallel to the line-segment CD.**

(ii) **A parallelogram ABCD is also written as ∥gm ABCD.**

In a Parallelogram,

(a) Opposite *sides* are always equal. In ||^{gm} PQRS,
$\overline{PQ} = \overline{RS}$ and $\overline{PS} = \overline{QR}$

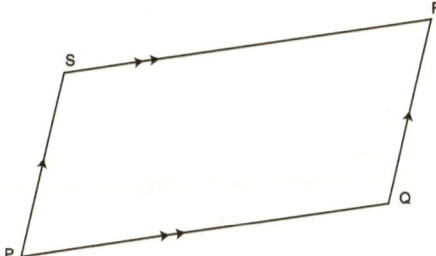

(b) Opposite *angles* are always equal. In ||^{gm} BCDE,
$\angle B = \angle D$ and $\angle C = \angle E$

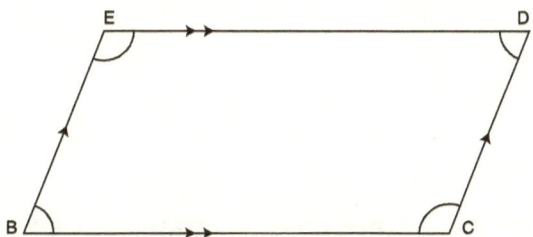

(c) The diagonals bisect each other.

In the ||^{gm} ABCD, the diagonals AC and BD bisect each other at O i.e. $\overline{AO} = \overline{OC}$ and $\overline{BO} = \overline{DO}$

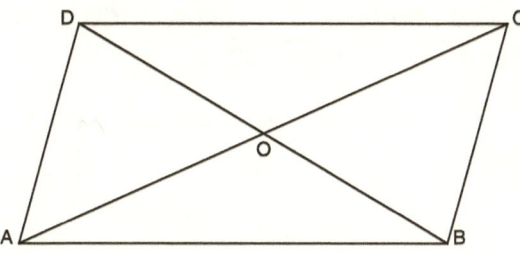

The Parallelograms are further classified as:

(I) **Rectangle:**

If a parallelogram has one angle equal to 90° then it is called a rectangle. In the figure ABCD is a rectangle, such that $\angle A$ = 90°.

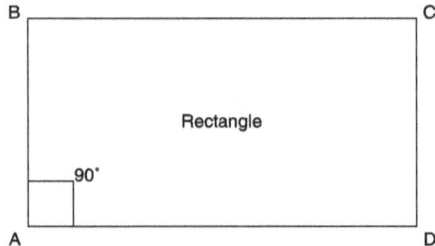

Since, a rectangle is a parallelogram, therefore, it has all the properties of a parallelogram. Also, each angle of a rectangle is always equal to 90°. In the figure, $\angle A = 90°$, $\angle B = 90°$, $\angle C = 90°$ and $\angle D = 90°$.

The diagonals of a rectangle are equal. In the figure $\overline{AC} = \overline{BD}$

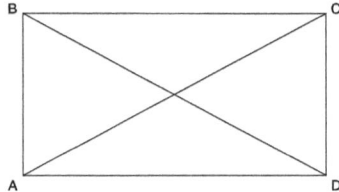

NOTE

(i) All rectangles are parallelograms but all parallelograms are not rectangles.

(ii) A rectangle is also called an equiangular quadrilateral.

(II) SQUARE

If in a parallelogram all sides are equal and one angle is 90°, then it is called a **Square**. In the figure ABCD is a square such that: $\angle A = 90°$ and $\overline{AB} = \overline{BC} = \overline{CD} = \overline{AD}$.

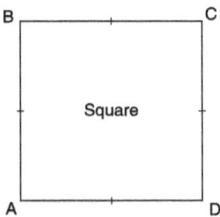

A square is parallelogram, therefore it has all the properties of a parallelogram. It also has the following properties:

Each angles of a square is equal to 90° i.e. $\underline{|A}$ = 90°, $\underline{|B}$ = 90°, $\underline{|C}$ = 90° and $\underline{|D}$ = 90°.

The diagonals of a square are always equal.

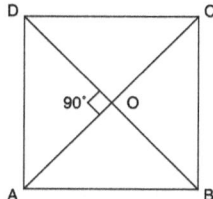

In the figure, $\overline{AC} = \overline{BD}$, the diagonals of a square bisect each other at right angles. In the figure, O is the midpoint of AB as well of BD and OD ⊥ AC and OC ⊥ BD.

NOTE:

(i) Each square is a rectangle but each rectangle is not a square.

(ii) A square is also called a regular quadrilateral.

(III) RHOMBUS

In a parallelogram, if a pair of adjacent sides are equal, then it is called as *rhombus*. A rhombus is a parallelogram, so it has all properties of a parallelogram.

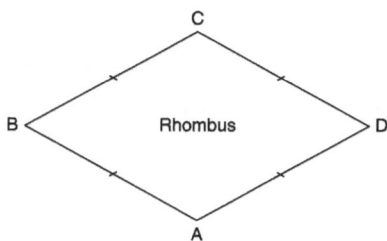

Some additional properties of a rhombus are:

All sides of a rhombus are equal i.e. $\overline{AB} = \overline{BC} = \overline{CD} = \overline{AD}$

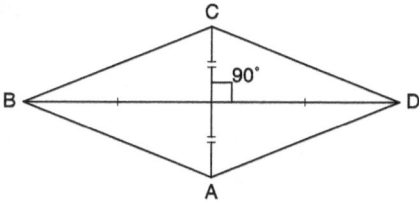

Diagonals of a rhombus are perpendicular to each other i.e. AC ⊥ BD.

2. **ONE PAIR OF OPPOSITE SIDES IS PARALLEL TRAPEZIUM**

   A trapezium is a quadrilateral with only one pair of opposite sides parallel. In the adjoining figure ABCD is a *Trapezium* such that AB || CD.

   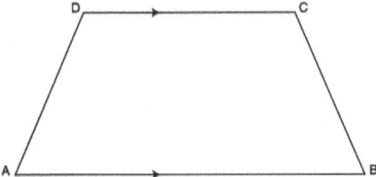

   NOTE

   (i) In a trapezium, the parallel sides are called its bases whereas the non-parallel sides are called its legs.

   (ii) If the legs of a trapezium are equal, then it is called an isosceles trapezium. In the figure $\overline{AD} = \overline{CB}$

   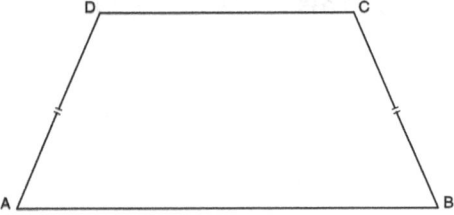

   (iii) The base angles of an isosceles trapezium are equal. In the figure ABCD is an isosceles trapezium, so $\underline{|A} = \underline{|B}$ and $\underline{|C} = \underline{|D}$

   (iv) If a trapezium has two right-angles then it is called a right-trapezium.

3. **NONE-PAIR OF SIDES IS PARALLEL**

   A quadrilateral is called a *kite* if it has two pairs of equal adjacent sides but unequal opposite sides.

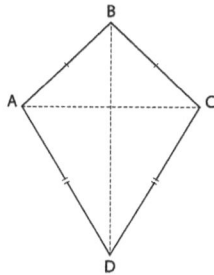

In the adjoining figure, ABCD is a quadrilateral, such that AB = AD, BC = DC but AB ≠ DC and AD ≠ DC

POLYGON

If a simple closed figure is made up of only line-segments, it is called a Polygon. Each of such line-segment is called its side. The point of intersection of any two consecutive sides is called its vertex, and the angle formed by two consecutive sides of a polygon inside the polygon is called an interior angle.

NOTE

In polygon, Number of interior angles = Number of sides of a polygon

EXAMPLES OF POLYGON

(i)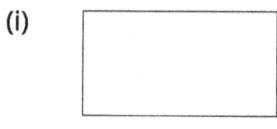

(ii)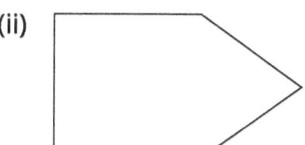

(iii)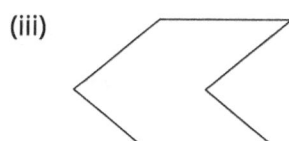

(iv)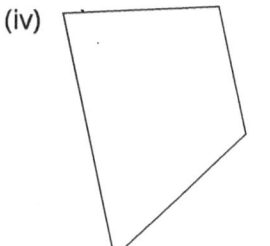

(v)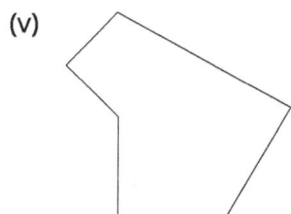

All the above figure are polygons, because:

(i) They are made up of line segments only.

(ii) They are closed curves.

(iii) Their sides (line segments) meet at the end points.

NOTE

These shapes are not polygons:

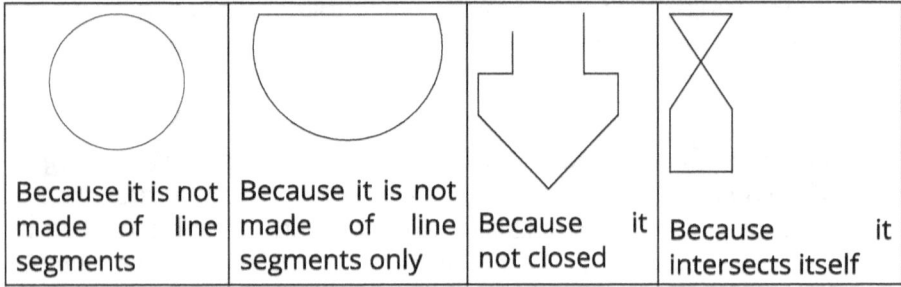

| Because it is not made of line segments | Because it is not made of line segments only | Because it not closed | Because it intersects itself |

NAMING POLYGONS

The simplest polygon is a triangle. Because a polygon cannot be made by less than 3 line segments. A polygon is named according to the number of sides (line-segments) in it.

Numbers of sides	3	4	5	6	7	8	9	10
Name of the Polygon	Triangle or Trigon	Quadrialateral or Tetragon	Pentagon	Hexagon	Heptagon or Septagon	Octagon	Nonagon	Decagon

NOTE

Polygon having more sides are generally written as 13-gon, 14-gon, 15-gon .... n-gon. Though they have their specific names too

CONCAVE AND CONVEX POLYGONS

If each angle of a polygon is less than 180°, then it is said to be *Convex-polygon*. Following polygons are convex-polygons:

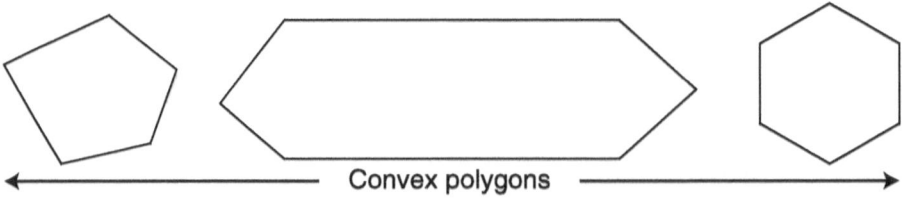

Convex polygons

If at least one angle of a polygon is more than 180°, then it is called a concave polygon. Following polygons are concave polygons:

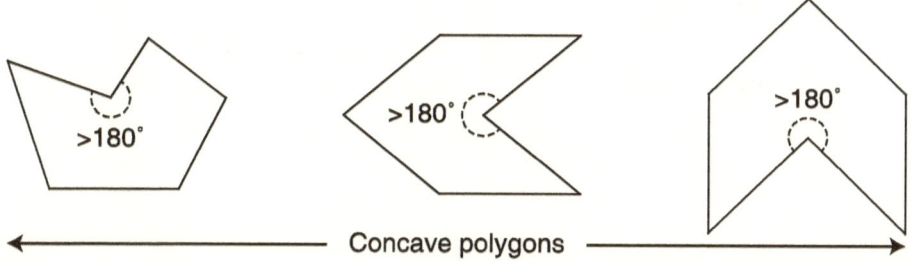

Concave polygons

NOTE

A 'Convex angle' has no pointing inward whereas a concave polygon has at least one "cave" in it.

REGULAR AND IRREGULAR POLYGON

A *regular polygon* polygon is one which is both equilateral and equiangular. For example, a square and an equilateral triangle.

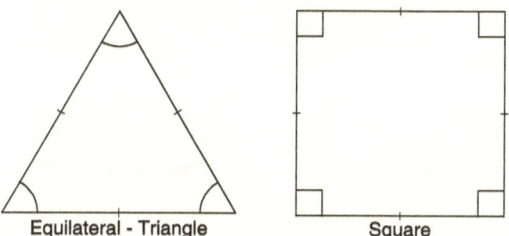

Equilateral - Triangle   Square

When all sides of a polygon are equal, the polygon is said to be an equilateral. When an all angles of a polygon are equal, it is called an equiangular.

Some other examples of regular polygons are give below:

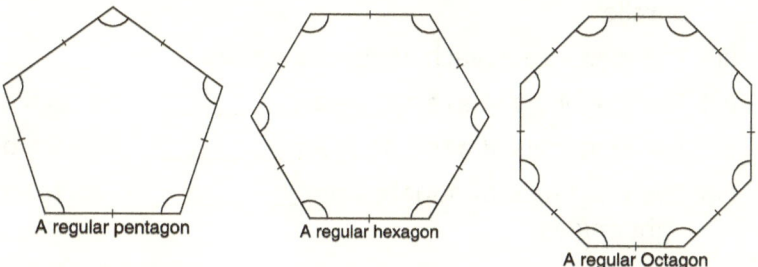

A regular pentagon   A regular hexagon   A regular Octagon

Any polygon having unequal sides or unequal angles or both unequal sides and angles are called *irregular* polygons. For example:

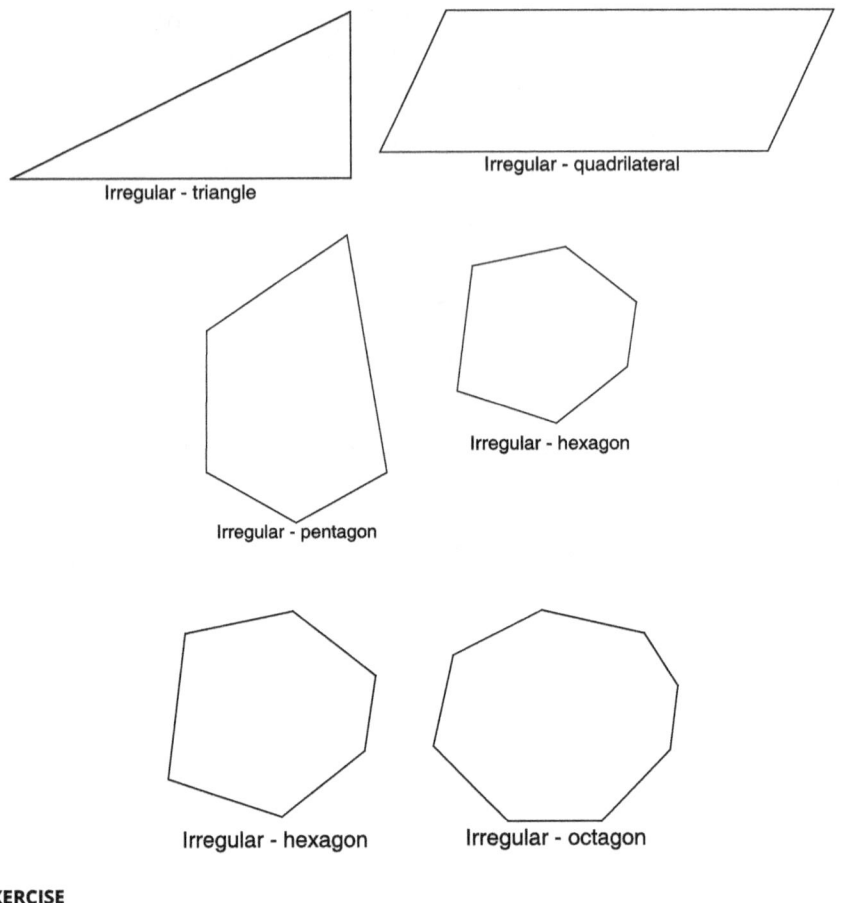

Irregular - triangle
Irregular - quadrilateral
Irregular - pentagon
Irregular - hexagon
Irregular - hexagon
Irregular - octagon

EXERCISE

1. Fill in the blanks:

   (i) A closed figure having _____ sides is called a quadrilateral.

   (ii) The opposite sides of a parallelogram are_____ and parallel.

   (iii) The measure of each angle of a square is _____.

   (iv) The diagonals of a rectangle are _____ to each other.

   (v) The diagonals of a rhombus are _____ to each other.

   (vi) The diagonals of square are _____ to each other at right angles.

   (vii) The diagonals of a kite are _____ to each other.

   (viii) The non-parallel sides of a trapezium are called its_____.

(ix) The non parallel sides of an isosceles trapezium are _____ to each other.

(x) If each of the interior angle of a polygon is less than 180°, then it is called a _____ polynomial.

2. How many pair(s) of opposite sides are parallel in a trapezium?
3. How many pairs of adjacent sides of a kite are equal?
4. Name the regular polygon having four sides.
5. Name the polygon, having number of sides as:

   (i) Five   (ii) Eight   (iii) Three

   (iv) Four   (v) Six   (vi) Seven

6. Which of the following are polygon?

   (i)   (ii)   (iii)

   (iv)   (v)

7. Name the polygons whose diagonals are perpendicular to each other?
8. Name the polygons whose diagonals are equal to each other?
9. Can a square be a rectangle?

ANSWERS

1. (i) Four   (ii) Equal   (iii) 90°

   (iv) Equal   (v) Perpendicular   (vi) Bisect

   (vii) Perpendicular   (viii) Legs   (ix) Equal

   (x) Convex

2. 1   3. 2   4. Square

5. (i) Pentagon   (ii) Octagon   (iii) Triangle

   (iv) Quadrilateral   (v) Hexagon   (vi) Septagon

6. iii and v   7. Square, Rhombus, Kite

8. Parallelogram and Square   9. Yes

THREE-DIMENSIONAL SHAPES (3D SHAPES)

We know that a 2-D figure is drawn on a plane and it is a part of that plane only. It cannot be taken out from the plane. We cannot pick it up and hold it in our hands. For example:

a circle, a triangle, a parallelogram, a square, a parallelogram, a trapezium, etc. are two-dimensional figures.

A 2-D figure has many vertices and edges. But it has only one face, which is the part of the plane enclosed by its edges (sides).

A 3-dimensional object has more than one surface (**except a sphere, which has only one curved surface**) It has 3-dimensions, i.e. length, breadth and height (or depth).

For example: *a cube, a cuboid, a prism, a pyramid, a cone, a cylinder*, etc.

A solid body exists in space. It can be touched and felt. We can pick it up and hold it in our hands.

NOTE

I. '3-D shapes' have length, breadth and height i.e. 3-dimensions, that is why they are called 3-D shapes.

II. Since,'3-D shapes' occupy some space so they have volume.

SOLIDS

The world around us is full of 3D objects. For example, a ludo-dice, a cricket-ball, a gas cylinder, a box, a jug , a book etc. These 3-D objects are called solids. Every solid has a definite shapes and size and thus occupies a fixed amount of space.

Mainly solids are of two types:

(i) Solids having regular shape.

(ii) Solids having irregular shape.

In this chapter, we will study about the solids having regular shape.

PARTS OF A SOLID

Some parts of a solid are:

(I) FACE:

The surface of a solid is called its face. It can be plane, curved or both.

(II) EDGE:

Two faces of a solid generally meet at a line-segment. This line segment is called an edge of the solid.

(III) VERTEX:

Any two or more edges of a solid meet at a point. This point is called a vertex or corner of the solid. A solid may have more than one vertex. The plural of vertex is vertices.

TYPES OF SOLIDS

We divide the regular solids mainly in three categories:

REMEMBER

A simple, closed curve formed by line-segments only is called a *polygon*.

PRISMS	PYRAMIDS	SPHERES
A solid having its base and top identical polygons and side faces as parallelogram is called a prism	A solid having its base as any polygon and side faces as triangle (with a common vertex) is called a pyramid.	A solid, every point of whose surface is equidistant from a fixed point.
EXAMPLES: A cube, a cuboid, a cylinder, etc. are prisms.	**EXAMPLES:** A tetrahedron (triangular pyramid), square pyramid etc.	**EXAMPLE:** A foot ball, a cricket ball, an orange etc.

IDENTIFYING FACES, VERTICES AND EDGES

I – PRISMS

The prisms are named according to the number of sides of the base, such as:

DID YOU KNOW?

If the faces of a solid are, polygonal-regions, then it is called a polyhedron. Two adjoining faces of a polyhedron meet at an edge. Prisms and pyramids are polyhedrons but sphere, cone, cylinder etc. are not polyhedron, because their faces are not polygons.

(I) **SQUARE PRISM**

In a square prism the base is a square. Top is also a square. Lateral faces are rectangular

A square prisms has 12 edges, 8 vertices and 6 faces.

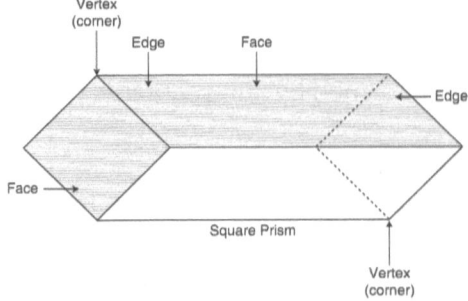

Square Prism

(II) **CUBE**

If the all faces of a square-prism are squares then it is called a cube.

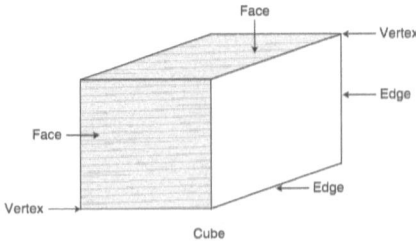

Cube

(III) **RECTANGULAR PRISM**

If a prism has a uniform rectangular cross section, then it is called rectangular prism (or a cuboid)

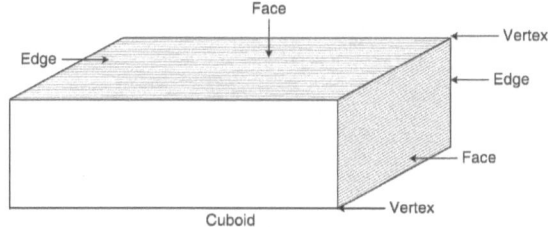

A cuboid has 8 vertices, 12 edges and 6 faces

(IV) **PENTAGONAL PRISM**

If the base and top of a prism are pentagonal then it is called a pentagonal prism. A pentagonal-prism has:

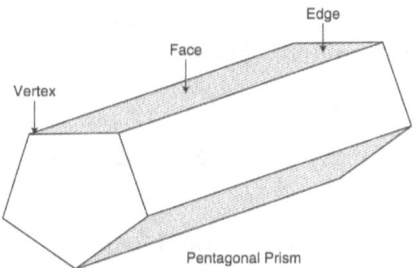

10 vertices, 15 edges and 7 faces.

(V) **HEXAGONAL – PRISM**

A Prism, having its base and top as hexagonal regions is called a hexagonal prism. It has 12 vertices, 18 edges and 8 faces.

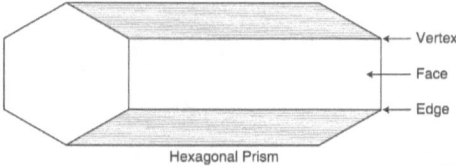

Similarly, prisms having 7 sided, 8 sides and 9 sided bases, are called heptagonal-prism, octagonal-prism and nonagonal-prism respectively.

(VI) **TRIANGULAR – PRISM**

A prism having its base and top as triangles are called triangular prism. Its lateral faces are rectangular regions.

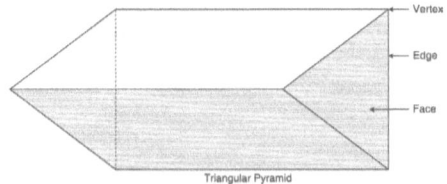

Triangular Pyramid

A triangular prism has: 9 edges, 6 vertices and 5 faces.

REMEMBER

When the lateral faces of a prism are rectangles then it is called as a right prism.

II. PYRAMIDS

Pyramids are also named according to the number of sides of the base. Its lateral faces are triangular regions meeting at a point (vertex).

(I) TRIANGULAR PYRAMID

If the base of a pyramid is a triangle then it is called a triangular pyramid.

It has 6 edges, 4 vertices and 4 faces.

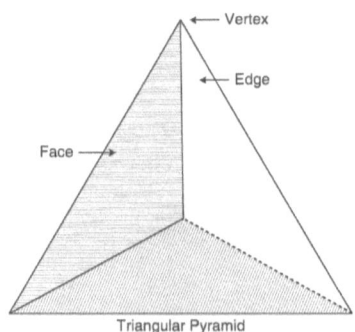

Triangular Pyramid

IMPORTANT:

If all the faces of a triangular-pyramid are equiangular-triangles then it is called a TETRAHEDRON

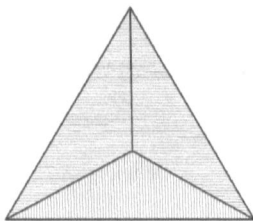

(II) Square Pyramid

A Pyramid, whose base is a square is called a square pyramid. It has 8 edges, 5 vertices and 5 faces.

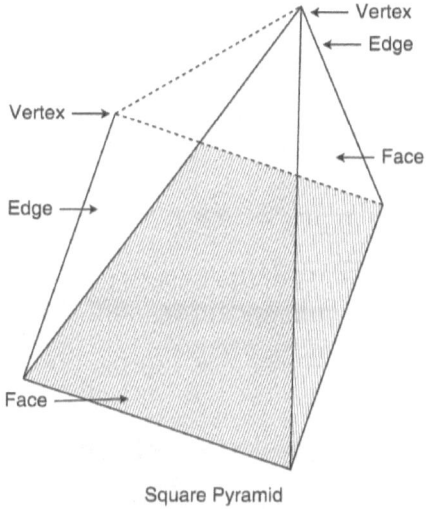

Square Pyramid

(III) Rectangular Pyramid

If the base of a pyramid is a rectangle, then it is called a rectangular-pyramid. It also has 8 edges, 5 vertices and 5 faces.

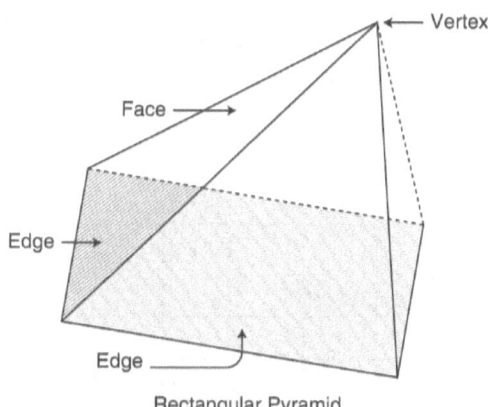

Rectangular Pyramid

(IV) Pentagonal Pyramid

A pyramid having its base as pentagon is called a pentagonal pyramid.

A pentagonal pyramid has: 10 edges, 6 faces and 6 vertices.

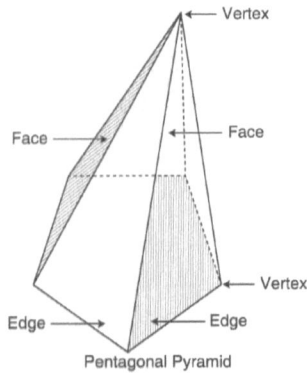

Pentagonal Pyramid

Similarly, if the sides of the base are 6, 7 and 8 then the pyramids are called as hexagonal-pyramid, heptagonal-pyramid and Octagonal-pyramid respectively.

NOTE

I. We have discussed above only about polyhedrons.

II. A Polyhedron is said to be regular if its faces are identical regular polygons and the same number of faces meet at each vertex. For example: a cube, a tetrahedron.

Let us sum up the 'number of edges', 'faces' and vertices of various prisms and pyramids:

SOLID	NUMBER OF SIDES OF THE BASE	NUMBER OF		
		FACES	VERTICES	EDGES
Cuboid (Rectangular Prism)	4	6	8	12
Cube (Rectangle Prism)	4	6	8	12
Triangular-Prism	3	5	6	9
Pentagonal-Prism	5	7	10	15
Hexagonal-Prism	6	8	12	18
Triangular-Pyramid	3	4	4	6
Rectangular-Pyramid	4	5	5	8
Pentagonal-Pyramid	5	6	6	10
Octagonal Pyramid	8	9	9	16

From the above chart we observe that the number of faces, vertices and edges of prisms and pyramids depend upon the number of sides of their bases:

Solid	Number of Faces	Number of Vertices	Number of Edges
Prism	2 + [Number of sides of the base]	2 × [Number of sides of the base]	3 × [Number of sides of the base]
Pyramid	1 + [Number of sides of the base]	1 + [Number of sides of the base]	2 × [Number of sides of the base]

NON-POLYHEDRON PRISM

A *Cylinder* is a prism, because its base and top are identical faces. But it is not a polyhedron (because its base is not a polygon.

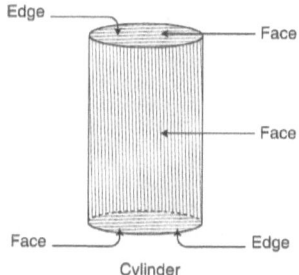
Cylinder

A cylinder has 3 faces (1 curved and 2 flat faces) and 2 edges.

It has no corners.

NON-POLYHEDRON PYRAMID

A *cone* is a pyramid with circular base. Its faces are curved. It has 2 faces, 1 edge and 1 vertex. Since its base is not a polygon, therefore it is not a polyhedron.

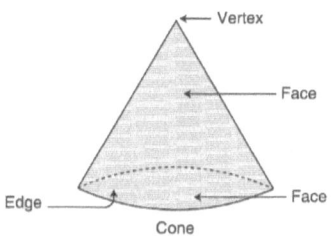
Cone

Sphere

A sphere is a solid. It has only 1 face which is curved.

It has no edges and no corners.

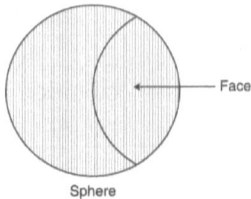
Sphere

Net of a 3D Shape

A net of a 3-dimensional shape is the outline of its faces such that by joining them the solid shape is obtained.

Therefore, the net is a Skelton (outline in 2D) which when folded, a 3D shape is obtained.

Let us join together the following 6 identical squares, edge to edge. We get a cube:

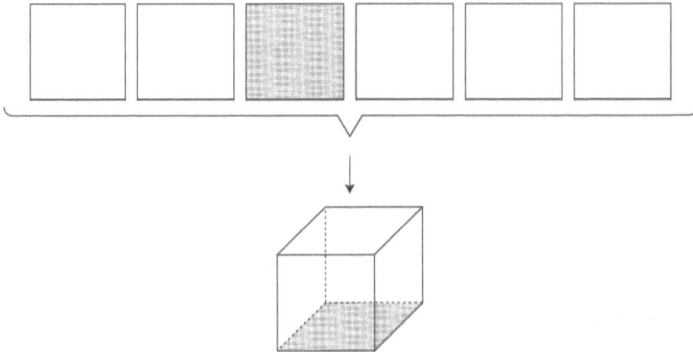

If we need to make a cube from a piece of paper (2D plane), then we make a pattern (net) as given below which when folded along the dotted lines, we get the desired cube.

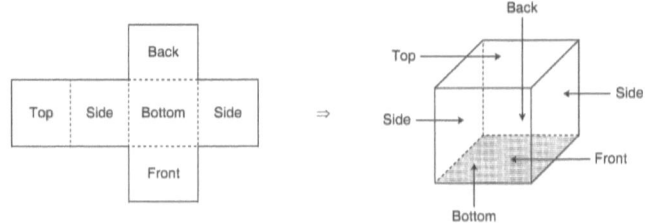

Similarly, nets of some other solids are given below:

SOLID	NET
Cuboid	Back / Side / Base / Side / Front / Top
Cylinder	Top / (rectangle) / Buttom
Cone	(triangle) / (circle)
Prism	
Pyramid	

Exercise - 1

1. Identify the following as prism or pyramid:

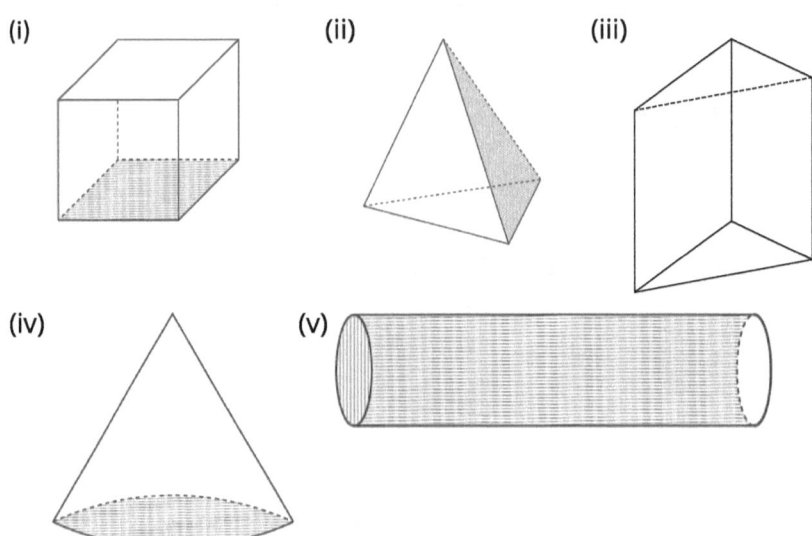

2. Which of the following are not prisms?

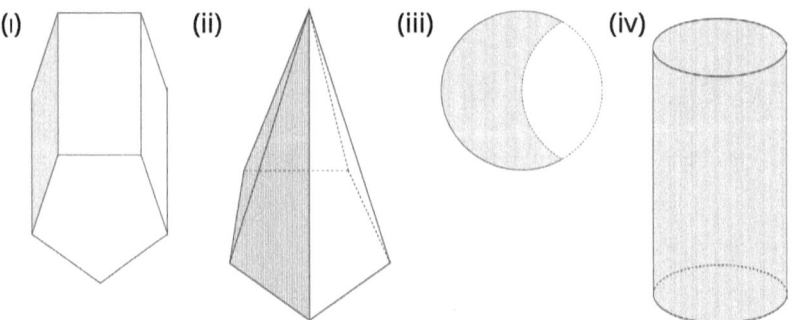

3. Which of the following are not prisms?

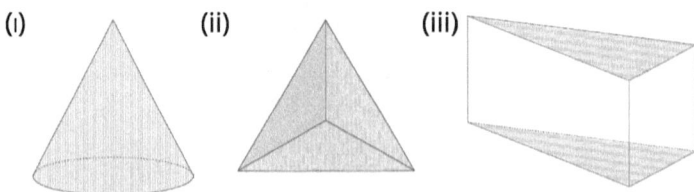

4. The base and top of a solid are identical pentagons. All other faces are rectangular. Name the solid.

5. The base of a solid is a square. Its all the lateral faces are triangles which meet at a common point. Name the solid.

Understanding Elementary Shapes | 305

6. Following is the net of a cube such that the faces are marked with various letters. Identify the face marked with the letter.

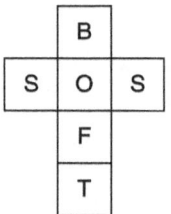

T is opposite to the face marked with which letter when the net is being folded into the shape of a cube?

7. Which of the following nets given below will generate a 'square pyramid'?

(i)    (ii)    (iii)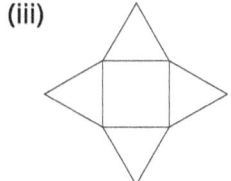

8. One of the following nets, when folded, forms a cuboid. Identify, which net is appropriate?

(i) 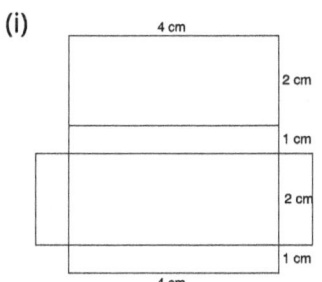   (ii)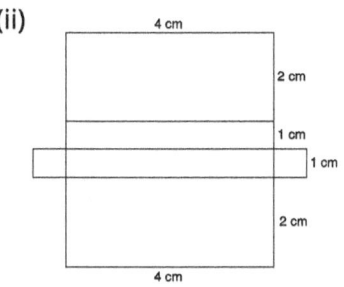

9. Which of the following is the net of triangular-prism?

(i) 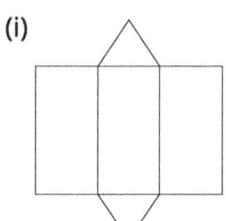   (ii)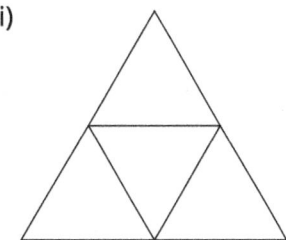

Answers

1. (i) Prism   (ii) Pyramid   (iii) Prism   (iv) Pyramid
   (v) Prism.
2. (ii) and (iii)   3. (i) and (ii)   4. Pentagonal Prism
5. Square-Pyramid
6. 0   7. (iii)   8. (i)   9. (i)

Hots

The measure of an angle 'a' is $\frac{2}{3}$ of a straight angle. The measure of another angle 'b' is $\frac{3}{5}$ of a complete angle. A Third angle 'c' is formed such that c = a + b. (i) Name the type of 'c' (ii) What is the measure of c?

Answer

(i) Reflex angle   (ii) $\angle C = 336$

Miscellaneous Exercise

1. An angle is 40° less than $\frac{2}{3}$ of a complete angle. What type of angle is it?

2. Classify the following triangles as acute angled triangle, right-angled triangle and obtuse angled triangle:

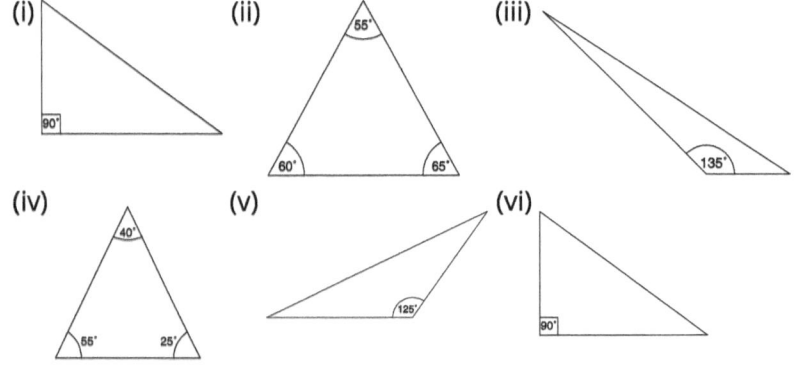

3. Name the polygon:

   (i) That has minimum number of sides.

   (ii) That has each angle equal to 90° and its diagonals bisect at right angles.

   (iii) That has equal diagonals and they bisect each other at right angles.

(iv) That has eight vertices and eight sides.

4. Write the number of faces, number of vertices and number of edges for a:

   (i) Cube   (ii) Cuboid

   (iii) Triangular-Prism (iv) Triangular pyramid.

5. Fill in the blanks:

   (i) The number of sides of a hexagon is _____.

   (ii) A five sided polygon is called as _____.

   (iii) If all sides of a polygon are equal then it is called a _____ polygon.

   (iv) If all angles of a polygon are equal then it is called a _____ polygon.

   (v) If all sides and all angles of a polygon are equal then it is called as _____ polygon.

Answers

1. Reflex angle.
2. acute triangles: (ii) and (iv)
   Right triangles: (i) and (vi)
   Obtuse triangles: (iii) and (v)
3. (i) Triangle   (ii) Square   (iii) Square   (iv) Octagon
4.

	(i)	(ii)	(iii)	(iv)
Faces:	6	6	5	4
Vertices:	8	8	6	4
Edges:	12	12	9	6

5. (i) 6   (ii) Pentagon   (iii) Equilateral   (iv) Equiangular   (v) Regular

Mental Maths

1. $\overline{AB}$ = 6.2 cm, BC = $\frac{1}{31}$ cm and AD = AB + BC. What is the measure of AD?

2. Two angles of a triangle are acute angles. The third angle is a right angle. Is this triangle an acute triangle or a right triangle?

3. Can a square be both equiangular and equilateral quadrilateral?
4. One side of a square ABCD is equal to the side of rhombus shown in the adjoining figure. What is the measure of each side of the square?

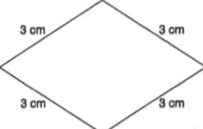

5. Are all rectangles parallelograms?
6. Is a rectangle an equiangular quadrilateral?
7. Is an angle a polygon?
8. Two pairs of adjacent sides are equal in a kite. Is it a parallelogram?
9. What is the smallest number of line-segments such that a polygon is formed?
10. Three equiangular triangular faces are joined together to form a pyramid. What is this pyramid is called?

Answers

1. 6.4 cm   2. Right   3. Yes   4. 3 cm   5. Yes
6. Yes   7. No   8. No   9. 3   10. Tetrahedron.

Multi Choice Questions

1. $\triangle$ ABC is a right angled triangle and $\angle B = 90°$. Which type of the following is the angle $\angle C$?

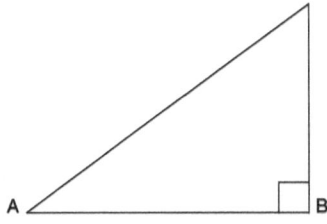

(i) Acute angle (ii) obtuse angle
(iii) Reflex angle (iv) Straight angle

2. Which of the following is correct statement?

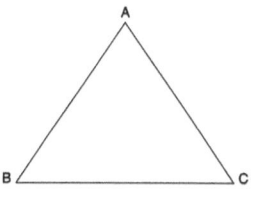

(i) A, B, C lie in the exterior of Δ ABC

(ii) A, B, C lie in the interior of Δ ABC

(iii) A, B, C lie in the triangular region

(iv) A, B, C lie in the interior as well as in its exterior.

3. Which of the following is a correct statement for the isosceles triangle ABC?

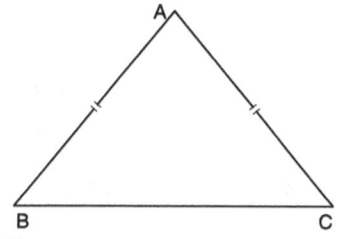

(i) $\overline{AB} = \overline{BC} = \overline{AC}$ (ii) $\overline{AB} = \overline{BC}$

(iii) $\overline{BC} = \overline{AC}$ (iv) $\overline{AB} = \overline{AC}$

4. Which of the following polygon is formed by 8 sides?

   (i) Quadrilateral   (ii) pentagon

   (iii) Octagon   (iv) Hexagon

5. Which of the following is not having a pair of opposite equal sides?

   (i) a kite   (ii) a trapezium

   (iii) an isosceles trapezium   (iv) rhombus

6. Diagonals of which of the following do not bisect at right angles?

   (i) Square   (ii) rhombus

   (iii) Kite   (iv) non of these

7. A cube is formed by folding net. The face marked by the letter B will be opposite to the face marked with which of the following letter?

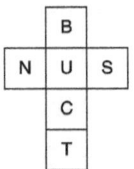

(i) T (ii) U (iii) N (iv) C

8. By which of the following an equilateral triangular pyramid is called?

   (i) Tetrahedron (ii) Prism

   (iii) Square Prism (iv) Square pyramid

9. Which of the following is the number of vertex (vertices) for a cone?

   (i) One (ii) two (iii) Three (iv) four

10. Which of the following is the number of edges of a cuboid?

    (i) 6 (ii) 8 (iii) 12 (iv) 4

Answers

1. (i)   2. (iii)   3. (iv)   4. (iii)   5. (i)

6. (iv) 7. (iv)   8. (i)   9. (i)   10. (iii)

Worksheet

1. Write each of the following angles in the appropriate column:

   70°, 10°, 120°, 300°, 189°

   181°, 62°, 45°, 0°, 180°

   360°, 100°, 200°, 41°, 90°

Zero angle	Acute angle	Right angle	Obtuse angle	Straight angles	Reflex angles	Complete angles

2. Write scalene △ or equilateral △ or isosceles △ for each of the following triangles:

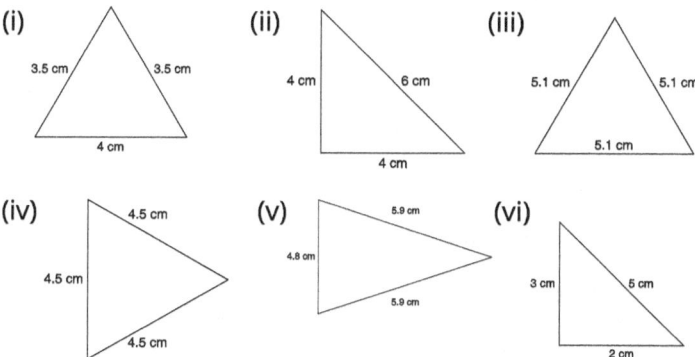

3. Name the parallelogram:
   (i) That has adjacent sides equal and measure of each angle equal to 90°
   (ii) _____ is a pyramid whose all faces are equilateral
   (iii) That has adjacent sides unequal and measure of each angle is equal to 90°
   (iv) That has adjacent sides equal but none of its angle is equal to 90°

4. Name the quadrilateral:
   (i) That has only one pair of opposite sides as parallel.
   (ii) That has no pair of opposite sides parallel but two pairs of adjacent sides are equal.

5. Complete the following table:

Number of sides	3	4	5	6	7	8
Name of the polygon	....	....	....	....	....	....

6. Write convex polygon or concave polygon for each of the following:

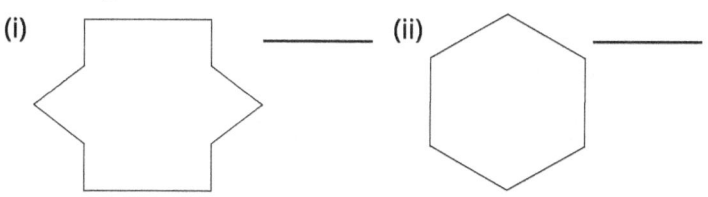

(iii) _____ (iv) _____

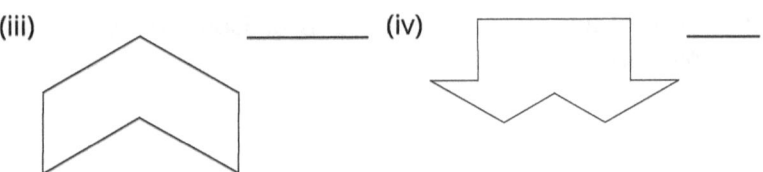

7. Complete the following table:

	Solid Shape	Number of sides of the base	Number of Faces	Vertices	Edges
(i)	Pentagonal Prism				
(ii)	Octagonal Pyramid				
(iii)	Rectangular Prism				
(iv)	Triangular Pyramid				
(v)	Rectangular Pyramid				
(vi)	Pentagonal Pyramid				
(vii)	Hexagonal Pyramid				
(viii)	Triangular Prism				

8. Write the number of faces, edges and Vertices for:
   (i) Cylinder   (ii) Cone   (iii) Sphere

9. Which of the adjoining is an appropriate net of a cone?
   (i)   (ii)

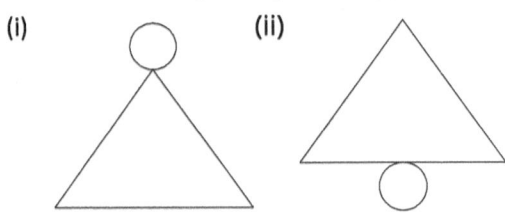

10. Fill in the blanks
    (i) If four straight lines pass through a point O, then O is called the point of _____

(ii) Three of more points, lying on the same line, are called _____ points.

(iii) The measure of a straight angle is _____.

(iv) A triangle, that has two sides equal is called as _____ triangle.

(v) All rectangles are parallelograms but all _____ are not rectangles.

(vi) In a trapezium the parallel sides are called its bases whereas the non-parallel sides are its_____.

(vii) A _____ is also called a regular quadrilateral.

(viii) A _____ is a quadrilateral which has two pairs of equal adjacent sides but unequal opposite sides.

(ix) _____ triangle is a regular polygon with minimum number of sides.

ANSWERS

1. Acute : 70°, 10°, 62°, 45°, 41°
   obtuse : 120°, 100°
   Straight: 180°,
   Reflex: 181°, 189°, 300°, 200°
   Zero: 0°
   Right: 90°
   Complete: 360°

2. (i) Isosceles Δ   (ii) Isosceles Δ   (iii) Equilateral Δ
   (iv) Equilateral Δ   (v) Scalene Δ   (vi) Scalene Δ

3. (i) Square   (ii) Tetrahedron   (iii) Rectangle
   (iv) Rhombus

4. (i) Trapezium   (ii) Kites

5. Triangle, Quadrilateral, Pentagon, hexagon, Septagon, Octagon.

6. (i) Convex   (ii) Concave   (iii) Concave

7.

	Sides	Faces	Vertices	Edges
(i)	5	7	10	15
(ii)	8	9	9	16
(iii)	4	6	8	12
(iv)	3	4	4	6
(v)	4	5	5	8
(vi)	5	6	6	10
(vii)	6	8	12	18
(viii)	3	5	6	9

8.

	(i)	(ii)	(iii)
Faces:	3	2	1
Edges:	2	1	0
Vertices	0	0	0

9. (ii)

10. (i) Concurrence (ii) Collinear (iii) 180°
 (iv) Isosceles (v) Parallelograms (vi) Legs
 (vii) Square (viii) Kite (ix) Equilateral

Chapter 11

Symmetry

[Reflection]

Introduction

In our day to day life, we come across many objects which are beautiful because they have symmetry; *Symmetry refers to the exact match in shape and size between two halves of an object.* The world is full of such beautiful objects. Some of such objects are natural and some of them are manmade as shown in the following examples:

The leaves, butterfly, human body etc., and the temples, churches, gurudwaras and mosques, etc., have symmetry, We see that such objects or figures have evenly balanced proportion. Therefore, such objects are said to the symmetrical.

Most of the beautiful and famous monuments and architectural wonders possess their attraction due to symmetry.

If we draw a vertical line at the middle, the position on either side of the line are identical.

If a MIRROR is placed along this line at the middle, the half part of the figure reflects through the mirror creating the remaining identical half. Thus, the line where the mirror is placed divides the figure into two identical parts. These parts are of the same size and every particular part on one side of the line will have its identical shape at the same distance on the other side. Therefore, this property is called mirror symmetry or mirror-image symmetry. The figures possessing this property are called symmetrical. There are two kinds of symmetry point symmetry and line symmetry.

LINE SYMMETRY

The above shape is equal in size on both sides of dotted line i.e. this shape is symmetrical about the dotted line. If we fold the figure about the dotted line we could get two equal shapes which will overlap each other, as shown in the adjoining figure. Therefore, we can say that one half of the shape in the mirror image of the other half. Thus, the dotted line is the *axis of symmetry.*

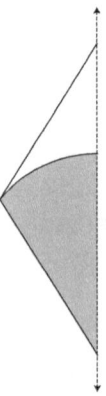

Let us take another example of figure 8, which is when folded on its axis of symmetry, the two halves will match exactly.

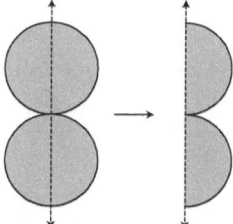

Thus, *the line about which a figure is symmetrical is called the line of symmetry or axis of symmetry.*

NOTE

In a symmetrical figure, each point of the figure has a corresponding point, and the axis of symmetry is the perpendicular bisector of every segment that connects the two corresponding points of the symmetrical figure.

Figures may have *more than one* axis of symmetry. The line of symmetry can be vertical or horizontal (or even slanting also). For example:

A square is symmetrical about the perpendicular bisectors of opposite sides.

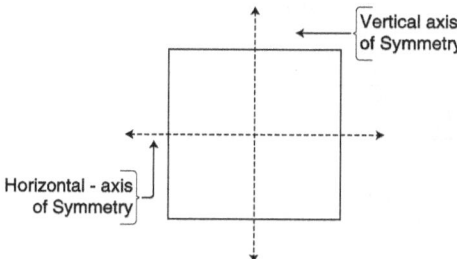

The square is symmetrical also about each of its diagonals (which are slanting lines of axis).

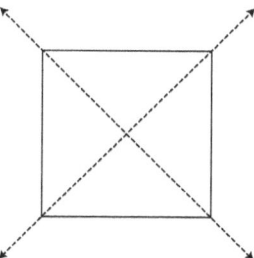

NOTE

A figure may have one, two, three, more or no line of symmetry.

EXAMPLE – 1

Draw the line of symmetry of each of the following shapes.

(i) 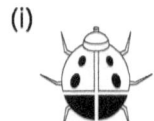 (ii) 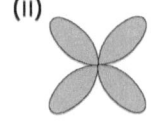 (iii) 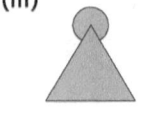 (iv) 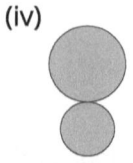 (v)

SOLUTION: The line of symmetry is drawn by a dotted line:

(i) 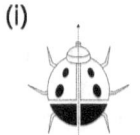 (ii) 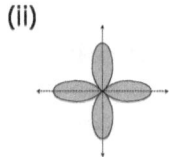 (iii) 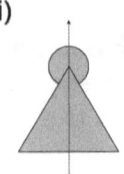 (iv) 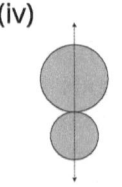 (v)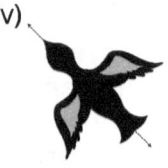

EXAMPLE – 2

Complete each of the following figures to make them symmetrical about the dotted line.

(i) 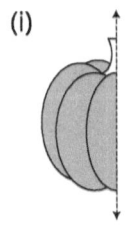 (ii)  (iii) 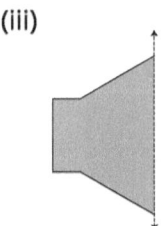 (iv) 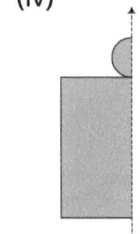 (v)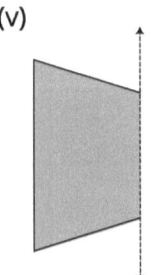

SOLUTION

The completed shapes are given below:

(i) 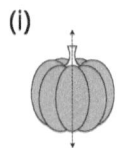 (ii) 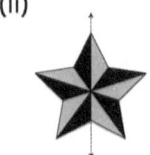 (iii) 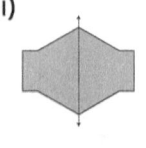 (iv) 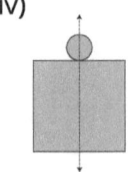 (v)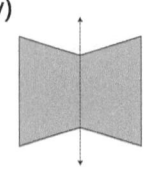

SYMMETRY OF SOME GEOMETRICAL FIGURES

Geometrical figures may have line or lines of symmetry. For example:

Figure	Line or Lines of Symmetry	Number of Lines of Symmetry	Position of line (or lines) of symmetry.
Line 	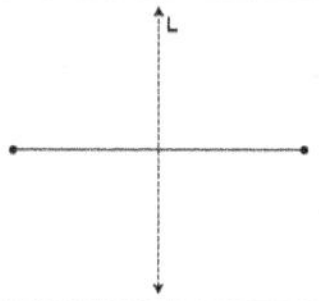	1	The perpendicular bisector of the line is the only line of symmetry
Angle 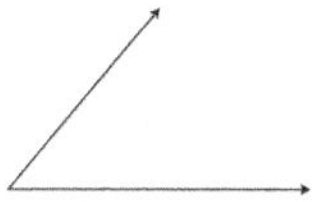	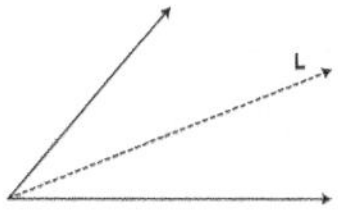	1	The bisector of the angle is the only line of symmetry of the angle
Semi – Circle 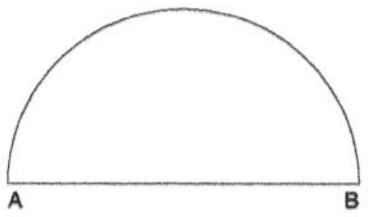	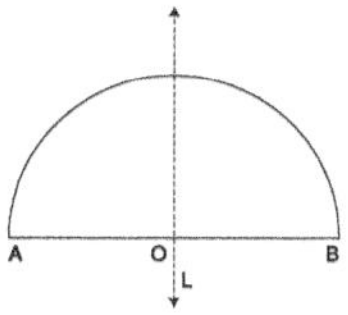	1	The perpendicular bisector of the diameter joining the end points of the semi – circle, is the only line of symmetry of the semi circle.
Isosceles Triangle 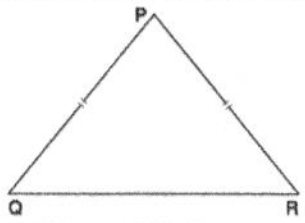	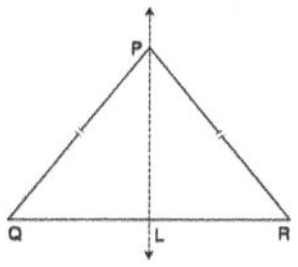	1	The perpendicular bisector of the base is the only line of symmetry of an isosceles triangle
Rectangle 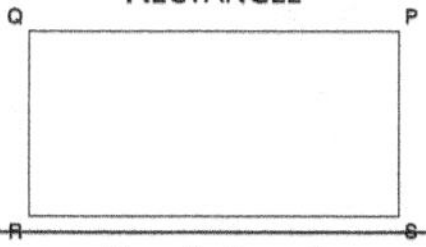	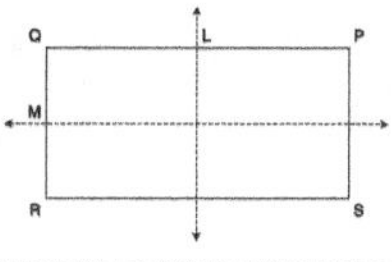	2	The perpendicular bisectors of the two adjacent sides are the lines of symmetry of a rectangle.
Equilateral 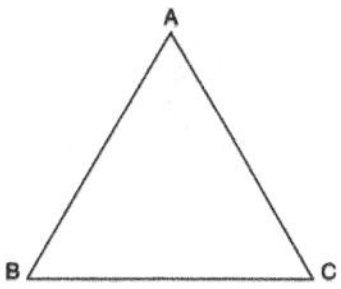	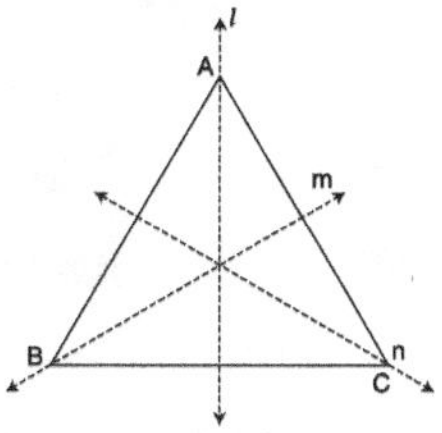	3	The perpendicular bisector of each side is a line of symmetry of an equilateral triangle. In the figure, l, m and n are the line of symmetry.
Square 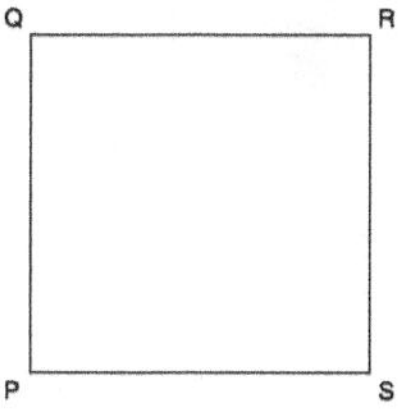	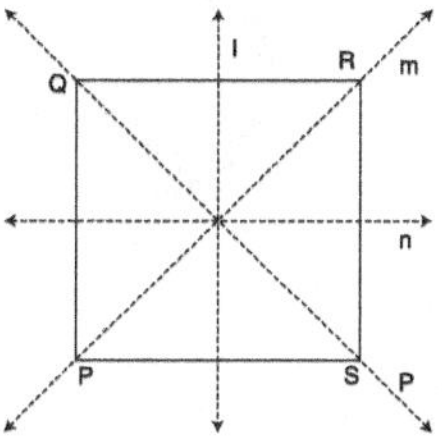	4	The perpendicular bisectors of two adjacent sides and the two diagonals are the four lines of symmetry of a square. In the figure l, m, n and p are the lines of symmetry.

DID YOU KNOW?

Circle is a very special plane figure. Its boundary does not have any line segment, but it has an infinite numbers of lines of symmetry, as shown in the following figure. All the dotted lines and many more are lines of symmetry of the circle.

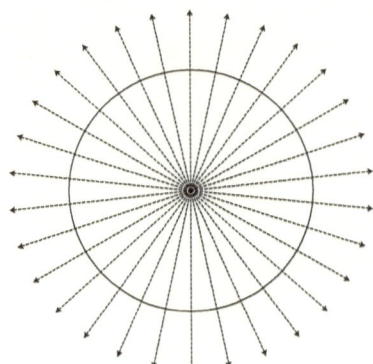

SYMMETRY OF ENGLISH ALPHABET

Some capital letters English alphabet are about a line as shown below by the dotted lines.

Following alphabet have VERTICAL LINE of symmetry:

Some alphabet have HORIZONTAL LINE of symmetry:

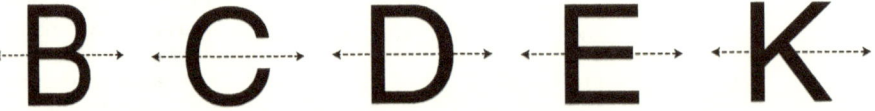

Some alphabet have HORIZONTAL AS WELL AS THE VERTICAL LINES of symmetry:

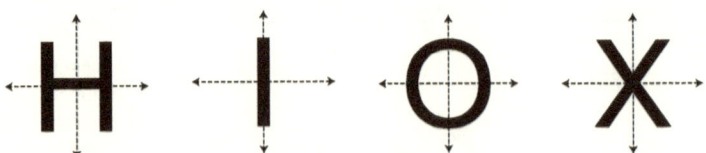

NON – SYMMETRICAL FIGURES

Some of the objects or figures have no lines of symmetry. Such figures are called *Non – symmetrical or (asymmetrical)*.

All irregular shapes are asymmetrical.

For example:

A scalene triangle.

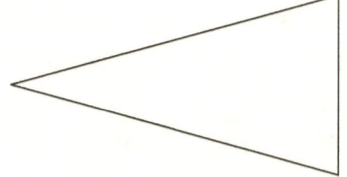

Irregular alphabet:

FGJLNPQRSZ

Irregular shapes:

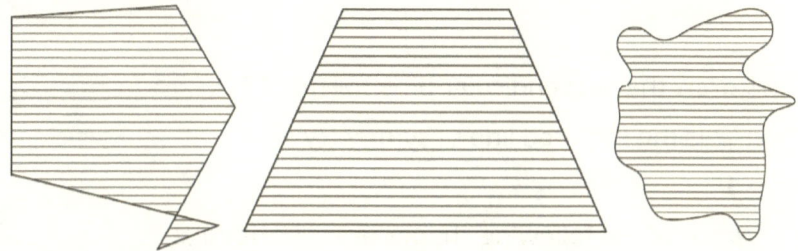

SYMMETRIC POINTS

(I) TO CONSTRUCT A POINT SYMMETRIC TO A GIVEN POINT, WITH RESPECT TO A LINE.

Let 'p' be a line and a point 'A' in the same plane. Draw A x⊥ p which meets p at o.

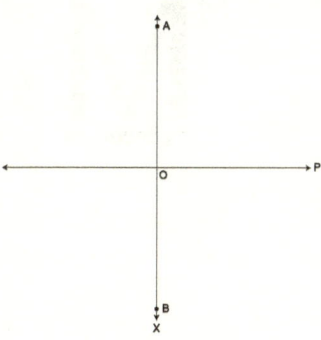

Mark a point B on A × such that $\overline{AO} = \overline{OB}$

Then the point 'B' and point 'A' are symmetric points with respt to p.

OR

The two points 'A' and 'B' are said to be *symmetric points* with respect to the line 'p', if the line 'p' is the perpendicular bisector of the line segment. $\overline{AB}$.

(ii) To construct a line of symmetry such that two points are symmetric with respect to it.

Let P and Q be points in a plane. Join P and Q

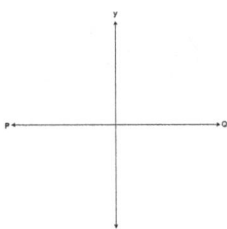

Now, draw the perpendicular bisector (p) of PQ.

Then 'p' is the required line of symmetry.

EXERCISE – 1

1. Which of the following shapes are symmetrical

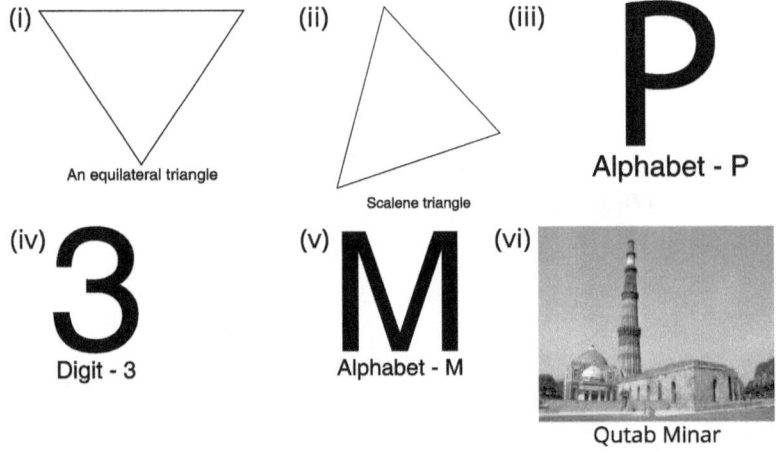

(vii)

2. Complete the following design or shape in the following:

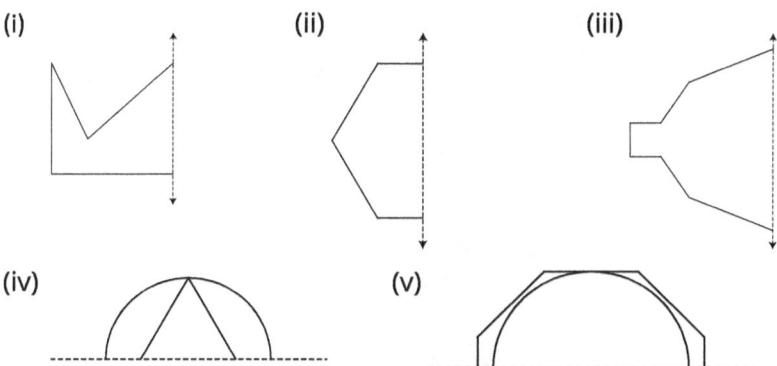

3. Identify the symmetric figures:
   (i) A triangle with angles 60°, 60°, 60°   (ii) A square
   (iii) A circle
   (iv) A triangle with sides 3 cm, 4.5 cm and 5 cm.
   (v) A equilateral with sides as 4 cm, 5 cm, 6.5 cm and 2.5 cm.

4. How many lines of symmetry are there in a
   (i) A rectangle   (ii) A square   (iii) An isosceles triangle

5. Which of the following alphabet have horizontal line of symmetry?
   A, B, K, M and T

6. Which of the following alphabet have vertical line of symmetry?
   D, E, V, W, and Y

7. Which of the following alphabet have both (horizontal and vertical) lines of symmetry?
   K, X, H, C, T and I.

8. If the points A and B are symmetric points with respect to the line 'l' then name segment $\overline{AB}$ in respect of 'l'

9. Draw a line-segment AB = 6 cm. Draw the line about which the points P and Q are symmetric points.

10. Which of the following digits are symmetrical?
    0, 1, 2, 3, 4, 5, 7, 8, and 9

Answers

1. (i) (iv) (v) (vi) and (vii)   3. (i) (ii) and (iii)
4. (i) 2 (ii) 4   (iii) 1   5. B and K   6. V, W, and Y
7. X, H and I   10. 0, 1, 3 and 8

REFLECTION

We know that an image seen in a mirror is obtained after a reflection in the mirror. The reflected image in the mirror is behind the mirror as the object is in front.

In the adjoined figure, let line m be a mirror and A' be the image of A then the 'line m' is the perpendicular bisector of the line AA'.

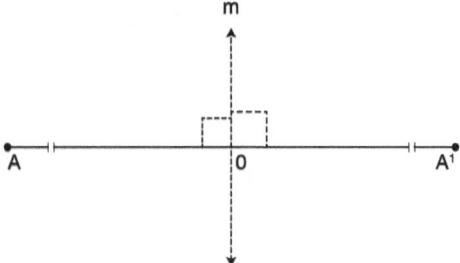

* The 'line m' is called the *mirror line* (or **reflection law**)

Thus, the reflection (or image) of point A in a line – m is a point A' such that 'line - m' is the perpendicular bisector of the line segment AA'.

Here, the 'line m' is called *'axis of reflection'*

Note

If the point 'A' lies on the line - m then the image of A is itself then the point A is called an invariant point with respect to the line m.

REFLECTION OF A POINT

A) **IN A LINE**

To find the reflection of a point p in a line AB, we draw pm, the perpendicular bisector of $\overleftrightarrow{AB}$ (using set squares or ruler + compasses). Produce pm to p', such that p'm = pm. Then the point p' is the reflection (image) of the point p in the line AB.

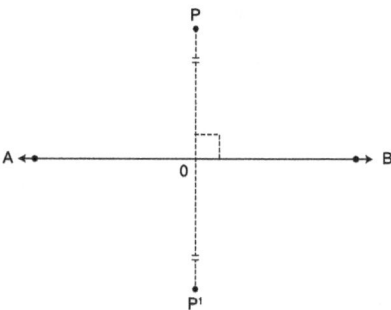

Note

In a plane when two perpendicular lines intersect then these perpendicular lines are called *axes*. In the following figure XOX' is called x- axis and YOY' is called Y-axis. The point of intersection 'O' is called origin and all distances are measured from 'O'. The position of any point is expressed by two points taken together in a bracket. We call them coordinates of the given point.

B) **In the x – axis**

Suppose the given point is p(x, y) in the co-ordinate plane. To find its reflection in the x – axis, draw PM perpendicular to the x – axis and produce it to a point p' such that MP' = MP.

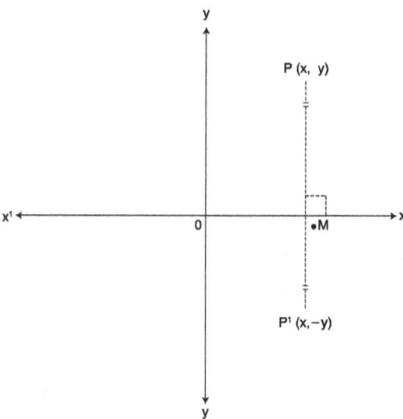

Thus, P' is the *reflection,* (or image) of the print P in the x-axis. In the adjoining diagram, the co-ordinates of the point P' are (x, −y)

Rules

Following rules will help us to find the position of the image of point in the co ordinate planes when it is reflected in the x-axis:

(i) Keep the x co-ordinate as such

(ii) Change the sign of the y-coordinate.

EXAMPLES

GIVEN POINT	REFLECTION IN THE X - AXIS
(2, 3)	2, −3
−4, −2	−4, 2
(5, 0)	(5, 0)

c) **IN THE Y-AXIS**

To find the reflection of a point P in the y-axis, let us suppose the given point as p(x, y). From P, draw pm, the perpendicular to the y-axis and produce it to a point p' such that

MP' = MP

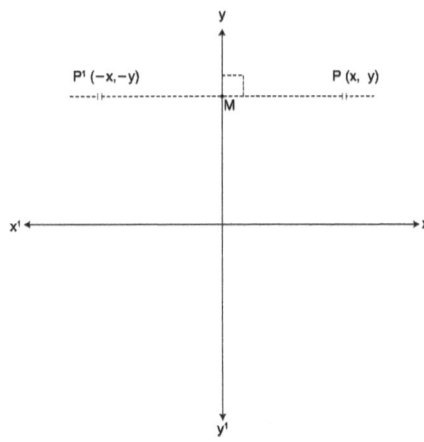

Then P' is the reflection (image) of the point P in the y-axis. From figure, the co-ordinates of P' are (−x, y)

RULES

Following rules help us to find the position of the image of a point when it is reflected in the y-axis.

(i) Keep the y-axis as such.

(ii) Change the sign of the x-coordinate

EXAMPLES

POINT	IMAGE IN THE Y-AXIS
(3, 2)	(−3, 2)
(−4, −3)	(4, −3)
(0, 5)	0, 5

D) IN THE ORIGIN

To find the reflection of a point p(x, y) in the origin, the signs of both of its co-ordinates change, i.e. the image of p(x, y) is p'(−x, −y) as shown in the adjoining diagram.

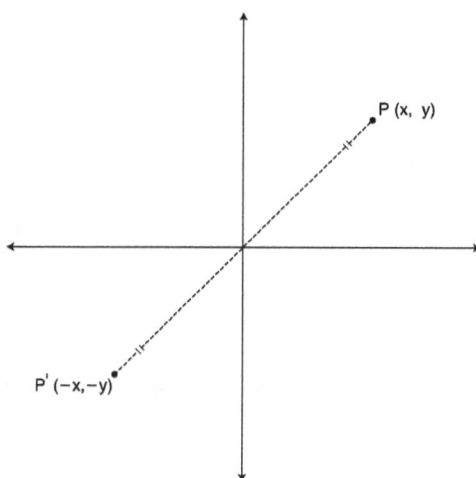

EXAMPLES

Point	Reflection (or image) in the origin
(3, 5)	(−3, −5)
(−5, −4)	(5, 4)
(3, 0)	(−3, 0)
(0, −7)	(0, 7)

NOTE

The reflection (image) of the point (0, 0) when it is reflected in x-axis, y-axis and in origin is (0, 0)

Example – 1

Draw an angle of 90°. Draw its line of symmetry.

Solution

Draw a ray $\overrightarrow{OB}$. Construct $\angle BAC = 90°$. Draw XY, the angular bisector of $\angle BAC$

The line XY is the required line of symmetry of $\angle BAC$.

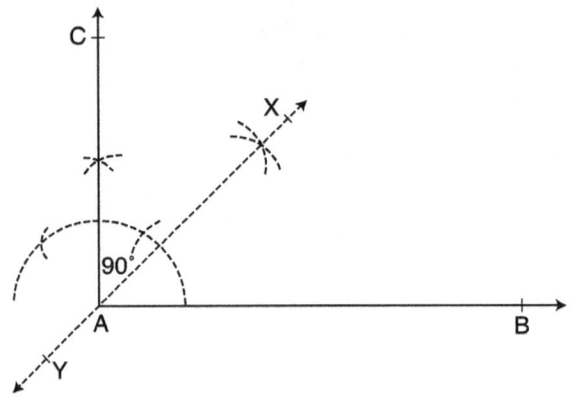

Example – 2

Construct a square of side 6 cm. How many lines of symmetry does a square have? Show them.

Solution

Construct a square whose side in 6 cm as shown in the adjoining figure. A square has 4 lines of symmetry:

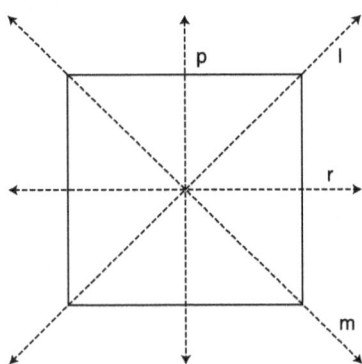

(i) Its 2 diagonals (l and m)

(ii) 2 perpendicular bisectors of the opposite side (p and q)

Example – 3

How many lines of linear symmetry are there of

(i) A isosceles triangle (ii) A scalene triangle

(iii) A circle (iv) The Letter X (v) The digit 8

Solution

(i) An isosceles triangle has only one line of symmetry which is the bisector of the angle included between its equal sides.

(ii) A scalene triangle has no line of symmetry

(iii) A circle has an infinite number of lines of symmetry.

(iv) The letter X has two lines of symmetry: one horizontal and other vertical line of symmetry.

(v) The digit 8 has two lines of symmetry: one horizontal and other vertical line symmetry.

Example – 4

The vertices of a Ä ABC are: A(1, 4), B(3, 1) and C(5, 2). It is reflected in the x-axis as Ä A'B'C'. Draw the reflected Ä A'B'C' on co-ordinate axis. Also write the vertices of Ä A'B'C'.

Solution

By plotting the points A(1, 4), B(3, 1) and C(5, 2) on the co-ordinate axes.

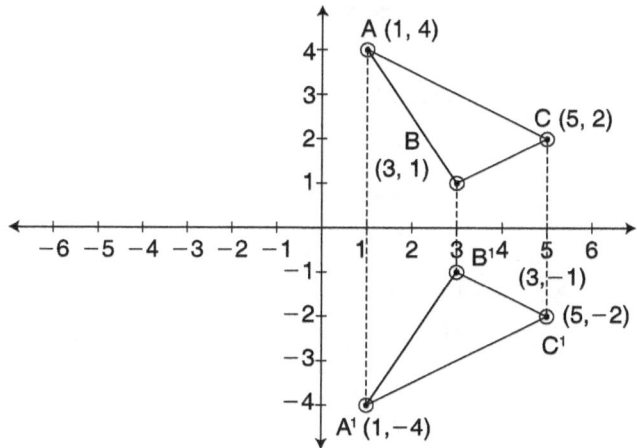

∵ The Ä ABC is reflected in the x-axis as Ä A'B'C'.

Example – 5

Write the co-ordinates of the given points when they are reflected as desired:

(i) P(0, 8) in x-axis (ii) Q(–8, 6) in y-axis

(iii) R(–5, 2) in origin (iv) S(0, 0) in x-axis

(v) T(–2, –4) in y-axis

Solution

Given Point		Reflected in		Co-ordinates of Images
(i) P(0, 8)	→	in x-axis	→	P' (0, –8)
(ii) Q(–8, –6)	→	in y-axis	→	Q' (8, –6)
(iii) R(–5, 2)	→	in origin	→	R' (5, –2)
(iv) S(0, 0)	→	in x-axis	→	S' (0, 0)
(v) T(–2, –4)	→	in y-axis	→	T' (2, –4)

Exercise – 1

1. Write the number of lines of linear symmetry in the following figures:

    (i) Trapezium (ii) Arrowhead (iii) Kite

    (iv) Rhombus (v) Isosceles Ä (vi) A line segment

Write down the co-ordinates of the following points under reflection in x-axis:

2. (i) (4, –2) (ii) (–2, 5) (iii) (–2, –7)

3. (i) (5, 0) (ii) (0, 0) (iii) (0, 2)

Write down the co-ordinates of the following points under reflection in y-axis:

4. (i) (7, 6) (ii) (0, 4) (iii) (3, 0)

5. (i) (–3, 6) (ii) (4, –2) (iii) (–1, –5)

Write down the co-ordinates of the following points under reflection in origin:

6. (i) (6, 2) (ii) (3, 8) (iii) (4, 5)

7. (i) (–2, 1) (ii) (–3, –1) (iii) (0, 0)

8. Draw a line AB, where A(2, 1) and B(5, 4). Reflect in x-axis as (A', B') and write down the co-ordinates of A' and B'.

9. Draw a line PQ, whose co-ordinates P(−3, 2) and Q(3, 4). Reflect PQ in the y-axis as (P'Q'). Also write down the co-ordinates of P' and Q'.

10. Draw the ABC, where A(3, 2), B(6, 2) and C(−6, −4). Reflect the triangle in the x-axis (as A'B'C'). Write down the co-ordinates of A', B' and C'.

11. Draw the Ä ABC, where A (4, −2) B (3, 6) and C(−3, −5). Reflect the triangle in the y-axis (as A'B'C'). Write down the co-ordinates of A', B' and C'

12. The point P (2, −5) is reflected in origin to point P'. Find the co-ordinates of P'.

13. Each of the points A(0, 0), B(5, 0), C(−5, 0), D(0, −5) and E(−9, −2) are reflected in x-axis to points A', B', C', D' and E' respectively. Write the co-ordinates of each of the image points A', B', C', D' and E'.

ANSWERS

1. (i) one (ii) one (iii) one (iv) Two
   (v) one (vi) one
2. (i) (4, 2); (ii) (−2, −5); (iii) (−2, 7)
3. (i) (5, 0) (ii) (0, 0) (iii) (0, −2)
4. (i) (−7, 6) (ii) (0, 4) (iii) (−3, 0)
5. (i) (3, 6) (ii) (−4, −2) (iii) (1, −5)
6. (i) (−6, −2) (ii) (−3, −8) (iii) (−4, −5)
7. (i) (2, −1) (ii) (3, 1) (iii) (0, 0)
8. A'(2, −1); B'(5, −4)   9. P' (3, 2); Q' (−3, 4)
10. A'(3, −2); B'(6, −2); C'(−6, 4)   11. A'(−4, −2); B'(−3, 6); C'(3, −5)
12. P' (−2, 5)
13. A'(0, 0); B'(5, 0); C'(−5, 0); D'(0, 5); E'(−9, 2)

HOTS

Draw lines of symmetry of:

ANSWERS

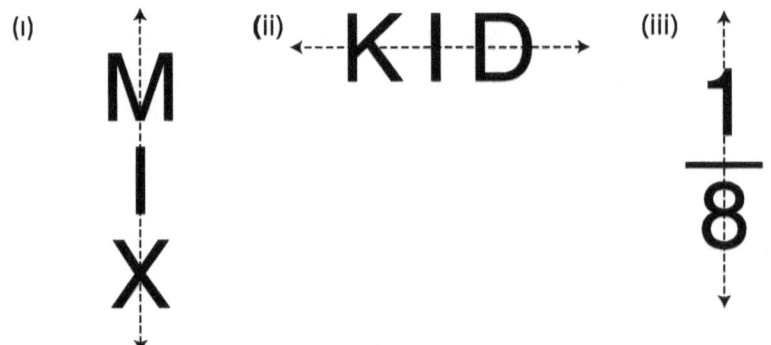

MENTAL MATHS

1. Name the plane geometrical figure which has an infinite number of lines of symmetry.
2. Which triangle has only one line of symmetry?
3. Name the triangle which has maximum number of lines among the triangles.
4. Is a rectangle symmetrical about its diagonals?
5. Is a square symmetrical about its diagonals?
6. Which of the following plane shapes has more lines of symmetry?

    (i) A square (ii) A circle

7. Which of the following digits has a line of symmetry?

    3, 5, 6 and 8

8. Digits '1' and '0' are symmetrical. Is '10' a symmetrical number?
9. Which of X and Y has 2 lines of symmetry?
10. Does the alphabet J has a line of symmetry?
11. What is the reflection of (0, 0) in x-axis?

Answers

1. Circle   2. Isosceles triangle   3. Equilateral
4. No   5. Yes   6. Circle
7. 3, and 8   8. Yes   9. 'X'
10. No   11. (0, 0)

MULTIPL CHOICE QUESTIONS

1. Which of the following is the number of lines of symmetry in an equilateral triangle?

   (i) 1   (ii) 2   (iii) 3   (iv) None of these

2. Which of the following is the number of lines of symmetry in a digit 3?

   (i) 1   (ii) 2   (iii) 0   (iv) None of these

3. Which of the following is the number of lines of symmetry in 8?

   (i) 0   (ii) 1   (iii) 2   (iv) 3

4. Which of the following digits does not have any line of symmetry?

   (i) 0   (ii) 1   (iii) 2   (iv) 3

5. Which of the following alphabet does not have any line of symmetry?

   (i) A   (ii) O   (iii) T   (iv) Z

6. Which of the following is true? The alphabet W has:

   (i) A horizontal line of symmetry

   (ii) A vertical line of symmetry

   (iii) Both the vertical and horizontal lines of symmetry

   (iv) Neither horizontal nor vertical line of symmetry

7. Which of the following is the number of symmetry in the letter M?

   (i) One   (ii) Two   (iii) Three   (iv) Zero

8. Which of the following alphabet has two lines of symmetry?
   (i) K (ii) D (iii) H (iv) W

9. How many lines of symmetry can an angles have?
   (i) Only one (ii) Two (iii) Three (iv) None of these

10. Which of the following shapes have four lines of symmetry?
    (i) A line segment (ii) A square
    (iii) An equilateral triangle (iv) A rectangle

11. Which of the following is the reflection of (3, 5) in the origin:
    (i) (−3, 5) (ii) (−3, 0) (iii) (−3, −5) (iv) (3, 5)

ANSWERS

1. (iii) 2. (i) 3. (iii) 4. (iii)
5. (iv) 6. (ii) 7. (i) 8. (iii)
9. (i) 10. (ii) 11. (iii)

WORK SHEET

1. **Match the COLUMN – (A) with COLUMN – B:**

COLUMN A [Shape]	COLUMN B [Number of lines of symmetry]
(i) (equilateral triangle)	(a) Infinite
(ii) (rectangle)	(b) 0
(iii) (scalene triangle)	(c) 4

Column A [Shape]	Column B [Number of lines of symmetry]
(iv) ○	(d) 3
(v) □	(e) 2

2. **Write 'True' or 'false' from the following**

   (i) 'T' has a horizontal line of symmetry

   (ii) '2' has 1 line of symmetry

   (iii) 0 has an infinite number of lines of symmetry

   (iv) An isosceles Δ has 1 line of symmetry.

   (v) A rectangle has 4 lines of symmetry.

3. **Fill in the blanks:**

   (i) The letters H and I have _____ lines of symmetry each.

   (ii) A scalene Δ has _____ lines of symmetry.

   (iii) The digit 9 has _____ lines of symmetry.

   (iv) A line segment has _____ lines of symmetry.

   (v) A circle has _____ lines of symmetry.

4. List the alphabet which do not have any line of symmetry.
5. List the alphabet which have at least one line of symmetry.
6. List the digits which have 2 lines of symmetry?
7. List the alphabets which have 2 lines of symmetry.

8. List the alphabets which have only a horizontal line of symmetry.
9. List the digits which can have more than two lines of symmetry.
10. Which type of triangle do not have any line of symmetry?
11. Write the reflection of the following points in the origin:

    (i) (0, −5)   (ii) (−3, 0)   (iii) (−2, −1)   (iv) (4, 5)

Answers

1. (i) → (d)   (ii) → (e)   (iii) → (b)   (iv) → (a)
   (v) → (c)

2. (i) False   (ii) False   (iii) True   (iv) True
   (v) False

3. (i) two   (ii) zero   (iii) zero
   (iv) one   (v) an infinite number of

4. F, G, J, L, N, P, Q, R, S, and Z.
5. A, B, C, D, E, H, I, K, M, O, T, U, V, W, X, and Y
6. 0, 1 and 8   7. H, 1 0, and X
8. B, C, D, E, and K,   9. 0
10. Scalene
11. (i) (0, 5)   (ii) (3, 0)   (iii) (2, 1)   (iv) (−4, −5)

Chapter 12
Geometrical Constructions

Uses of Geometrical Instruments in Constructions

Geometrical constructions are made by using some special tools called "Geometrical Instruments." Some of such instruments are:

Ruler (or Scale)

In geometry box, there is a graduated strip with parallel edges. It is called a *ruler* (or scale). It is also called a *straightage*. It is graduated into centimetres (cm) and millimetres (mm) on one side and on the other.

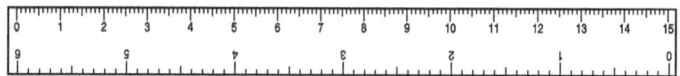

Scale

Other side the divisions are in inches as shown in the above diagram. A ruler is used to draw line-segments and to measure line segments.

Divider

The diagram shown in the adjoining figure is of a divider.

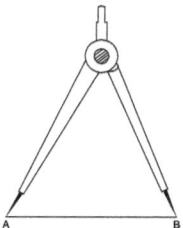

It has two arms with pointed ends. These arms are movable about a point, so the distance between them can be adjusted.

A divider is used to compare lengths.

COMPASS

Like a divider, the compass also has two arms but one of its arms can hold a pencil. The pencil can be fastened with a screw as shown in the figure. It is used to draw circles, to draw angles of specific measures. It is also used to mark off equal lengths.

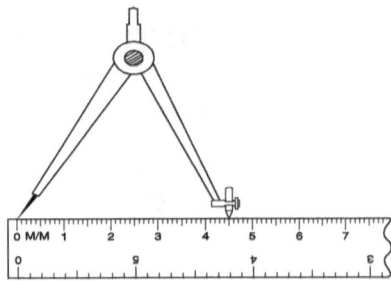

PROTRACTOR

We have already studied about a protractor, as an instrument to measure given angles or to draw angles of specific measures.

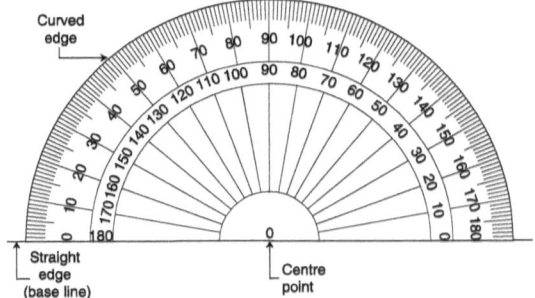

SET SQUARES

The diagram shown below in the figure are set squares. These are two triangular plates of hard plastic and taken together are called set-squares.

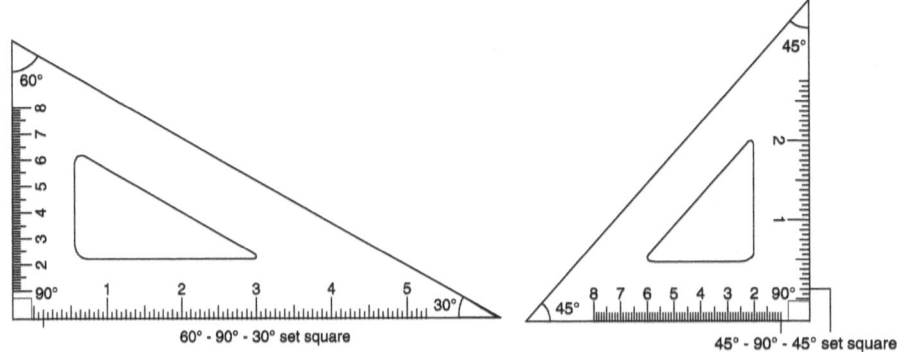

The set square containing 60°, 90° and 30° is called 60°– 90°– 30° set square.

The set square containing 45°, 90°and 45° is called 45°– 90°– 45° set square.

The edges containing 90° (in each set-square) are graduated in "centimetres and millimetres" on one edge and in "inches" on the other edge.

The set squares are used: to construct specific angles (60°, 30°, 45°, 75°, 105°, 90°), to measure lengths of given line-segments, to draw line segments of given lengths, to draw parallel lines, to draw perpendicular lines.

Constructions using Ruler and Pair of Compasses

1. To draw an Angle equal to a Given Angle

Let an angle ∠AOB is given whose measure is not known, as shown in the adjoining figure. To draw an angle equal to ∠AOB, we follow the following *steps*:

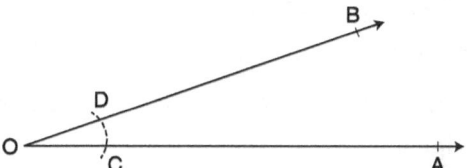

(i) A a point C, draw a line segment CR of any suitable length.

(ii) With centre O and a suitable radius draw an arc which intersects OA and OB at C and D respectively.

(iii) With Q as centre and keeping the radius same as in step-II, draw an arc to cut QP at T.

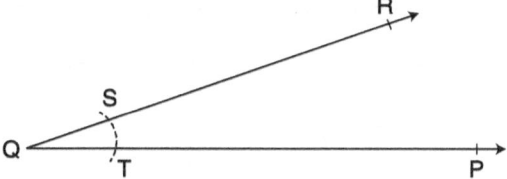

(iv) Now open the compass equal to CD, and then with centre at T draw an arc to cut the earlier arc at S.

(v) Join QS and produce it to R.

Thus, so obtained ∠PQR is equal to the given ∠AOB. i.e. ∠PQR = ∠AOB

2. To Bisect a Given Angle

Let the given angle be $\angle AOB$. To bisect it we follow these *steps*:

(i) With O as centre and a suitable radius, draw an arc to cut the arms OA and OB at C and D respectively.

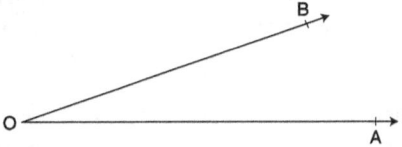

(ii) With centres C and D, draw arcs of same radii (more than $\frac{1}{2}$ CD) to cut each other at E.

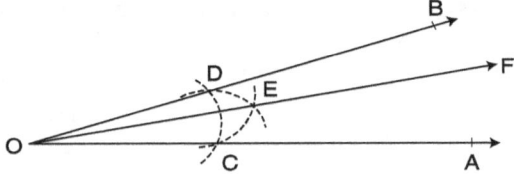

(iii) Join O and E and produce it to F

The line OF, so obtained, is the bisector of $\angle AOB$.

Since, a bisector divides an angle into two equal parts.

∴ $\angle AOF = \angle BOF$.

Construction of Angles of Given Measurement

[Such as 60°, 30°, 15°, 90°, 45°, 120°, 135°, 75°, 150° and 165°]

3. To Construct an Angle of 60°

Steps

(i) Draw a line-segment on of a convenient length.

(ii) With centre O and radius of suitable size, draw an arc to cut OA at C

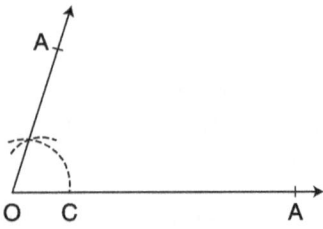

NOTE

60° is the simplest angle from the construction point of view

(iii) With 'C' as centre and keeping the radius same, draw another arc to cut the previous arc at D.

(iv) Join OD and produce it to B.

Thus, $\underline{|AOB} = 60°$

4. TO CONSTRUCT AN ANGLE OF 30°

STEPS

(i) First draw an angle of $\underline{|AOB} = 60°$ as above.

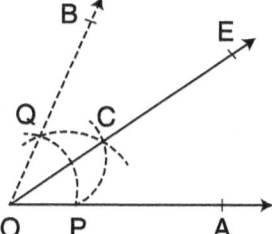

NOTE

$30° = \dfrac{1}{2} (60°)$

(ii) Bisect the angle $\underline{|AOB}$ as explained above (construction 2), by OE.

Thus, $\underline{|AOE} = 30°$

Also, $\underline{|BOE} = 30°$

5. TO CONSTRUCT AN ANGLE OF 90°

STEPS

(i) Draw a line segment OA.

(ii) With centre O and a suitable radius draw an arc to cut OA at B.

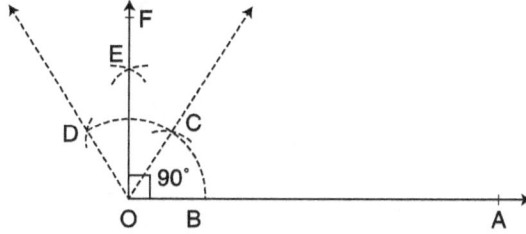

NOTE

$90° = 60° + \dfrac{1}{2}(60°)$

$\Rightarrow 90° = 60° + 30°$

(iii) With B as centre and keeping the radius same, cut the previous arc at C and D, such that $\angle AOC = 60°$ and $\angle COD = 60°$

(iv) Bisect $\angle COD$ and draw its bisector as OE or OF as shown in the figure.

Thus, $\angle AOF = 90°$

ANOTHER METHOD

(i) Draw a line segment AB and mark a point O on it.

(ii) Open the compasses to a suitable width and with centre O, draw an arc to cut AB at two points D and C.

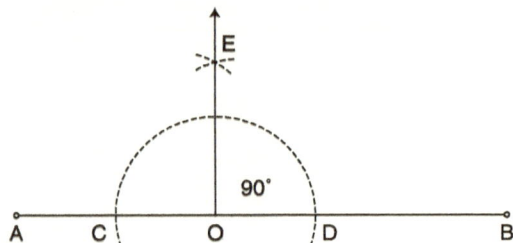

NOTE

(iii) By taking radius greater than half of CD and centre C, draw an arc.

(iv) Keeping the radius same and centre at D, draw another arc to intersect the previous arc at E.

(v) Join OE and produce. Then $\angle AOE = 90°$ (also, $\angle BOE = 90°$)

6. TO CONSTRUCT AN ANGLE OF 45°

STEPS

(i) Draw a line segment OA.

(ii) At O, make $\angle AOB = 90°$ as explained above.

(iii) Draw OE, the bisector of $\angle AOB$

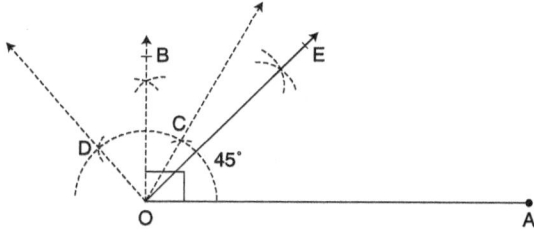

NOTE

$$45° = \frac{1}{2}(90°)$$

Thus, $\angle AOE = 45°$

7. Construction of an Angle of 135°

STEPS

(i) Draw a line segment PA and mark a point O on it.

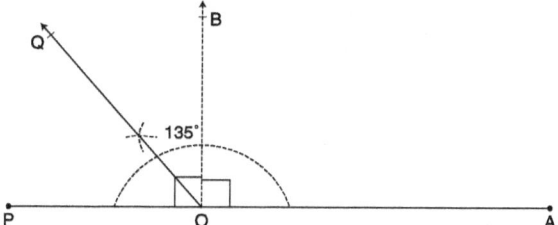

NOTE

$135° = 90° + \frac{1}{2}(90°) = 90° + 45°$

(ii) At O. draw $\angle AOB = 90°$, such that $\angle POB$ is also $= 90°$.

(iii) Bisect $\angle POB$ and draw its bisector OQ.

Thus $\angle AOQ = 135°$.

8. To Construct an Angle of 120°

STEPS

(i) Draw a line segment OA.

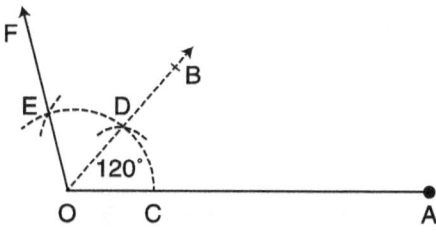

NOTE

120° = 60° + 60°

(ii) With centre O and suitable radius draw an arc at C.

(iii) With centre C and keeping the same radius, mark two points D and E on the previous arc.

(iv) Join OE and produce it to F.

Thus ∠AOF = 120°.

9. To construct an angle of 75°

STEPS

(i) Draw $\overline{PA}$ and mark a point O on it.

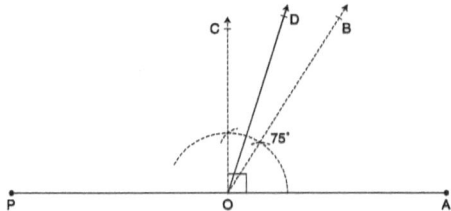

NOTE

75° = 60° + 15°

(i) At o draw OB such that ∠AOB = 60°

(ii) Construct ∠AOC = 90° such that ∠BOC = 30°

(iii) Draw bisector (OD) of ∠BOC, such that ∠BOD = 15°.

Thus, ∠AOD = ∠AOB + ∠BOD = 60° + 15°

i.e. ∠AOD = 75°

10. To construct an angle of 150°

STEPS

(i) Draw a line segment PQ and mark a point O on it.

(ii) Construct ∠QOA = 120° as explained above.

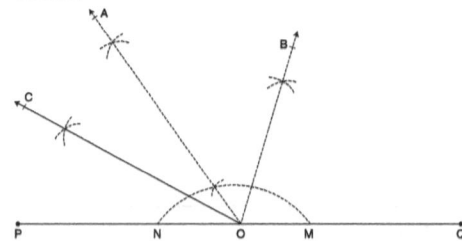

Note

150° = 120° + 30°

    (iii) Obviously $\angle POA$ = 60°. Bisect $\angle POA$ by drawing its bisector $\overline{OC}$, such that
$$\angle AOC = \frac{1}{2} (POA) = \frac{1}{2} (60°) = 30°$$

    (iv) Taking $\angle AOC$ together with $\angle QOA$, we
get $\angle QOC = 120° + 30° = 150°$

Thus, $\angle QOC = 150°$

11. To Construct an Angle of 165°

Steps

    (i) Construct $\angle QOC = 150°$ as explained above.

    (ii) Since $\angle POC = 30°$. So draw OD, the bisector of $\angle POC$, such that
$$\angle COD = \frac{1}{2} \angle POC$$
$$= \frac{1}{2} (30°) \quad = 15°$$

    (iii) Now $\angle QOC$ taken together with $\angle COD$, we get $\angle QOD$ such that
$\angle QOD = \angle QOC + \angle COD = 150° + 15° = 165°$

Thus, $\angle QOD = 165°$

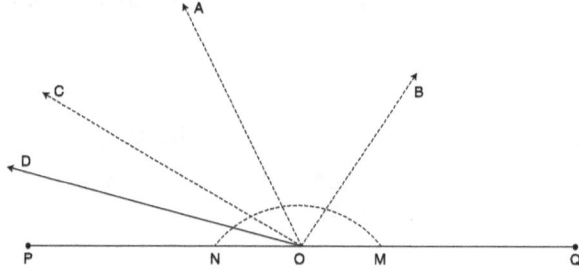

Note

$165° = 120° + \frac{1}{2}(60°) + \frac{1}{2}(30°)$

    = 120° + 30° + 15°

    = 165°

CONSTRUCTION OF PERPENDICULAR TO A LINE

1. To draw perpendicular bisector of a given Line segment

STEPS

(i) Draw the given line segment AB

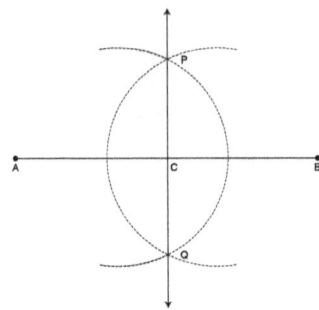

(ii) Open compasses such that the opening is greater than $\left(\frac{1}{2}AB\right)$.

(iii) Keeping the opening same and with centres A and B draw arcs on both sides of AB.

(iv) Let these arcs cut each other at P and Q.

(v) Join PQ which intersects AB at C such that

AC = BC,   also |AOP = 90°

Thus, the line segment PQ is the perpendicular bisector of AB as it bisects it at C and is also perpendicular to it.

2. To draw perpendicular to a given line from a given point outside the line.

STEPS

(i) Take the given line AB and mark a point C, lying outside it.

(ii) With centre C and suitable radius, draw an arc which intersects AB at two distinct points P and Q.

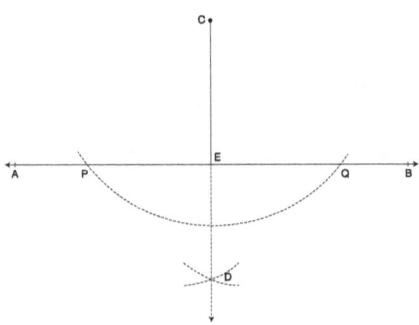

(i) With P and Q as centres, draw two arcs of equal radii such that they cut each other at point D, on the side other than C.

(ii) Join C and D such that it (CD) cuts AB at E.

Thus, CE is the required perpendicular on the given line AB from the exterior point C.

3. To draw a perpendicular to a line at a given point on the line.

STEPS

(i) Take the given line AB and mark the given point C on it.

(ii) Taking C as centre and any suitable radius, draw two arcs cutting AB at points P and Q

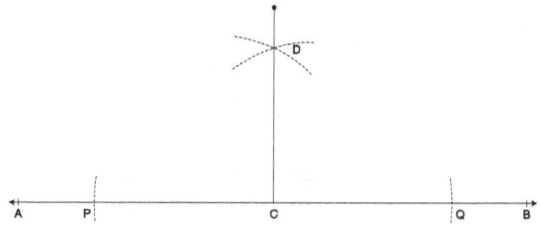

(iii) With centres P and Q and same suitable radius draw arcs to cut each other at D

(iv) Join DC.

Thus, DC is the required perpendicular to the line AB at a point C on it.

Here, $\lfloor ACD = 90°$ and $\lfloor BCD = 90°$.

TO DRAW A *CIRCLE*, USING A COMPASS

CIRCLE

A circle is a simple closed curve all of whose points are at the same distance from a fixed point inside it.

CENTRE OF A CIRCLE

The fixed point, O (as shown in the figure) is called the centre of the circle.

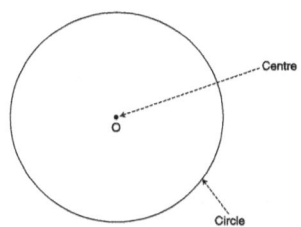

To construct a circle using compass, we follow the following steps:

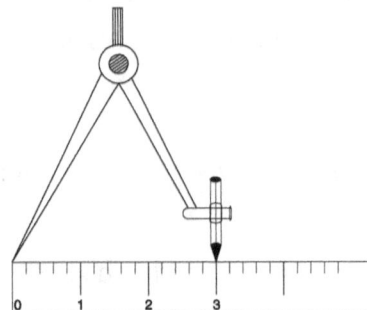

(i) Fix a sharp pencil in the pencil-holding arm of the compass firmly. If need arises, tighten the pencil with the help of a screw.

(ii) Take a point O on the plane of the paper.

(iii) Rest the metal tip of the compass at O and open the arms of the compass.

(iv) Hold the head of the compass firmly between the thumb and the fore-finger such that the pencil end touches the paper.

(v) Now move the pencil point around so that it traces a circle, as shown in the following figure:

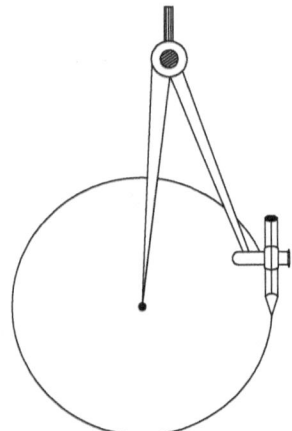

EXAMPLE

Draw a circle of radius 4 cm.

SOLUTION

(i) To open the compass for radius 4 cm, place the steel point on O of the scale and open the pencil point up to 4 cm.

(ii) Mark a point O on the paper. This point is taken as the centre of the circle.

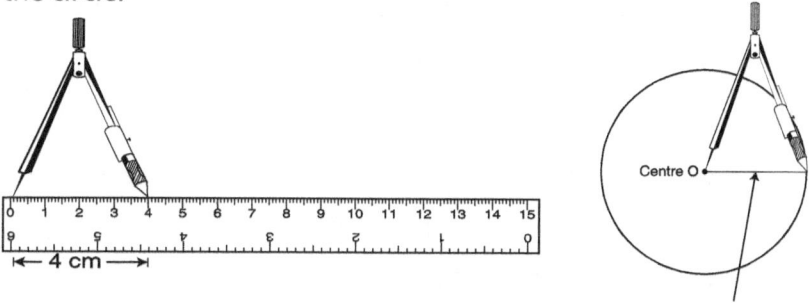

(iii) Place the steel point at O, the centre of the circle.
(iv) Holding the paper properly, turn the pencil leg of the compass slowly and smoothly around to draw the circle.

The figure so obtained is the required circle of 4 cm radius.

Exercise – 1

1. Copy the angles given below:

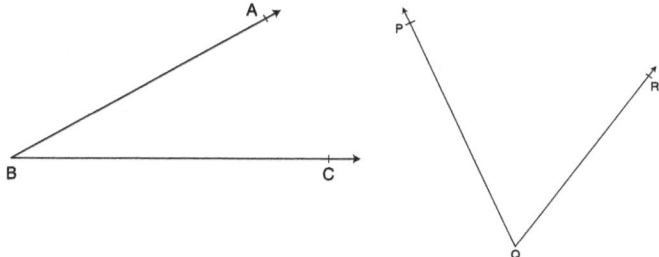

Construct the following angles using ruler and compasses:

2. 45° 3. 75° 4. 105°
5. 150° 6. 90° 7. 15°

8. Draw a line segment PQ = 6.5 cm. Construct ∠PQR = 60°. Draw the bisector of ∠PQR and then measure each part so obtained.

9. Draw a line segment AB = 8 cm. Mark a point P outside it. Draw PQ ⊥ AB.

10. Take a line segment MN = 10.5 cm. Mark a point O on MN. Draw OP ⊥ MN.

11. Draw $\overline{AB}$ = 9.4 cm. Draw the perpendicular bisector of $\overline{AB}$. Measure each part of AB, so obtained.

12. Draw circles of the following radii:

    (i) 3.4 cm   (ii) 5 cm   (iii) 4.2 cm

    (iv) 5.4 cm   (v) 6.3 cm

WORKSHEET

1. **Write 'True' or 'False' for each of the following:**

    (i) A protractor is a triangular disc.

    (ii) A protractor has two scales.

    (iii) One set-square has angles at its corners as 30°, 45° and 90°.

    (iv) A compass has two arms. Its one hand has a pointer whereas the other hand can hold a pencil.

    (v) A ruler and a set square can be used to measure lines, measure line-segments and to draw line-segments.

2. **Fill in the blanks:**

    (i) A ruler is graduated into centimetres on one side and on the other side the division are in _____.

    (ii) A _____ has both of its hands as pointed ends.

    (iii) The scale marks on a _____ are written in clockwise as well as anticlockwise direction.

    (iv) The midpoint of the base line of a protractor is called its _____.

    (v) Usually, the two types of _____ are 45°– 90°– 45° and 30°– 90°–60°.

4. Draw a line segment $\overline{AB}$. Mark a point P on it. Draw a perpendicular to $\overline{AB}$ through P, using:

    (i) ruler and compasses

    (ii) set squares.

5. Mark a point 'Q' outside a line segment $\overline{AB}$. Draw a perpendicular to $\overline{AB}$ from Q, using:

    (i) ruler and compasses,

    (ii) set-squares.

6. Draw a line segment $\overline{PQ}$ = 10.5 cm. Construct a perpendicular bisector of $\overline{PQ}$.
7. Using ruler and compasses construct an angle of 75° and then bisect it.
8. Draw an angle of 120° and then divide it into four equal parts.
9. Using set squares, construct an angle of 105°.
10. From the same centre O, draw circles with radii:

    3 cm, 4 cm, 5 cm, and 5.5 cm.
11. Draw a line segment $\overline{PQ}$ and mark a point S outside it. Using set squares, draw a line RS through S, such that:

    RS ☐ PQ

Answers

2. (i) inches   (ii) divider   (iii) protractor

   (iv) centre   (v) set-squares.

CHAPTER 13
MENSURATION

INTRODUCTION

To build a wall around a plot, we need to know the total length of the wall around the plot. This length of the wall along the boundary given by total length the boundary. In many situations in our daily life we have to get the total length of a boundary of a plane figure. This total length of boundary is called the perimeter.

Thus,

The PERIMETER of a close plane simple figure is the total length of the boundary.

For example:

The perimeter of the figure (i), or (shape) given along side is ($\overline{AB}+\overline{BC}+\overline{CD}+\overline{DE}+\overline{EA}$).

(i)

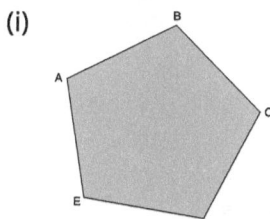

Whereas the amount of surface enclosed by the boundary is called its AREA. The area of the figure ABCDE is shown by the shaded region.

The area and perimeter are concerned with the closed shapes only. So we do not speak about the area or perimeter of adjoining figures.

(i)   (ii)

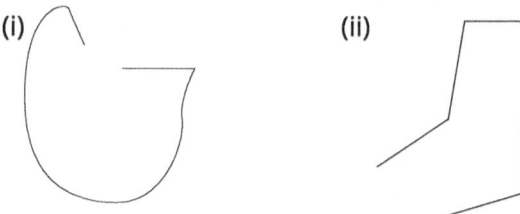

Perimeter of Plane Figures

Every *closed* plane shape is bounded by line segments or curves. The *Total length of the boundary of a closed figure is called its perimeter.*

Since, the perimeter is a length, its unit of measurement is the same as that of length. Let us recall the measure of length:

10mm = 1cm

10cm = 1dm

10dm = 1m   mm, cm, dm, m, hm and km are abbreviations respectively for millimetre; Centimetre, decimetre, metre, hectometre and kilometre.

10dam = 1hm

10hm = 1km

For example:

1. Perimeter of △ ABC

    = sum of the length of the line segments forming the triangle ABC

    = $\overline{AB} + \overline{BC} + \overline{CA}$

    Thus

    The perimeter of a triangle

    = The sum of the lengths of the three sides

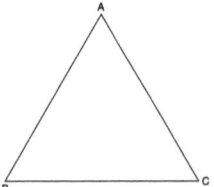

2. Perimeter of quadrilateral ABCD

    = The sum of the lengths of the line-segments forming the quadrilateral ABCD

    = $\overline{AB} + \overline{BC} + \overline{CD} + \overline{DE}$

    The Perimeter of a quadrilateral = The sum of its four sides.

Example - 1

Find the perimeter of the following figures:

(i)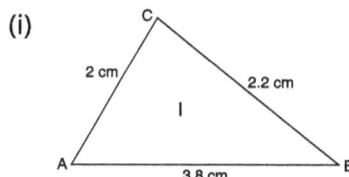

(ii)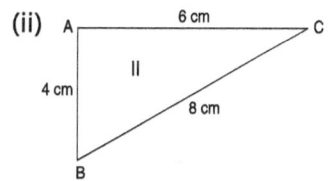

(iii)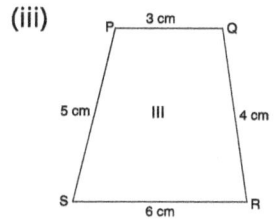

(iv)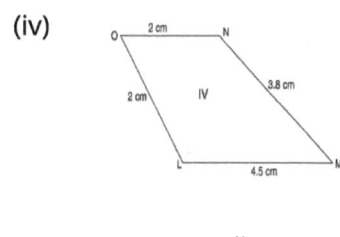

Solution

(i) Perimeter of Figure-I = AB + BC + CA
   = 3.8 cm + 2.2 cm + 2.0 cm
   = 8.0 cm

(ii) Perimeter of Figure-II = AB + BC + CA
   = 4.0 cm + 8.0 cm + 6.0 cm
   = 18.0 cm

(iii) Perimeter of Figure-III = PQ + QR + RS + SP
   = 3.0 cm + 4.0 cm + 6.0 cm + 5.0 cm
   = 18.0 cm

(iv) Perimeter of Figure-IV = LM + MN + NO + OL
   = 4.5 cm + 3.8 cm + 2 cm + 2.0 cm
   = 12.3 cm

Perimeter of a Rectangle

Let us consider a rectangle PQRS having its length = 6.2 cm and breadth = 4.4 cm.

$$\therefore PQ = 6.2 \text{ cm and } QR = 4.4 \text{ cm}$$

We know that opposite sides of a rectangle are equal.

∴ PQ = RS and PS = QR

∴ PQ = RS = 6.2 cm

PS = QR = 4.4 cm

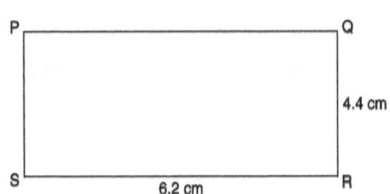

Now, Perimeter of rectangle PQRS

= PQ + RS + QR + SP

= 6.2 cm + 6.2 cm + 4.4 cm + 4.4 cm = 21.2 cm

We notice that, PQ + RS = 2(PQ)

[∵ PQ = RS]

and QR + SP = 2(QR)

[∵ QR = SP]

∴ Perimeter of rectangle = 2PQ + 2QR

= 2(PQ + QR)

= 2(Length + Breadth)

Thus, The perimeter of a rectangle = 2 (Length + Breadth)

If we represent perimeter by p,

Length by l and Breadth by b,

Then $P = 2(l + b)$

and $l = \dfrac{p}{2} - b$ $b = \dfrac{p}{2} - l$

For example if length and breadth of a rectangle are 10.5 cm and 8.7 cm respectively, then its perimeter (P) is given by:

$P = [l + b] \times 2$

$= [10.5 \text{ cm} + 8.7 \text{ cm}] \times 2$

$= [19.2 \text{ cm}] \times 2 = 38.4 \text{ cm}$

Did You Know?

The pattern on a chess board is an 8 × 8 grid and has 64 squares

Perimeter of a Square

We know that in a square the length of its four sides are equal. Let it be 'a'

Then Perimeter of the square

$\qquad$ = Sum of length of four sides

$\qquad$ = AB + BC + CD + DA

$\qquad$ = a + a + a + a = 4a

$\qquad$ = 4 (length of a side)

Thus, Perimeter of a square = 4 × (length of a side)

Let Perimeter of the square be P then

P = 4 × a $\qquad \Rightarrow a = \dfrac{P}{4}$

Example - 2

The length and breadth of a rectangular field are 150 m and 100 m respectively. Find the perimeter of the rectangle. Also, find the cost of fencing the field at the rate Rs 120 per meter.

Solution

Length of the field (l) = 150 m

Length of the field (b) = 100 m

Since perimeter of a rectangle = 2(l + b) and the given field is rectangular

∴ Perimeter of the field (P) = 2(150 m + 100 m) = 2(250 m) = 500 m

Thus, the perimeter of the field is 500 m

Cost of Fencing

Cost of 1 metre of fencing $\qquad$ = Rs 120

∴ cost of 500 metre of fencing = Rs 500 × Rs 120

$\qquad$ = Rs (500 × 120) = Rs 60,000

Example - 3

If the perimeter of a rectangle. Whose length is 9.8 m, is 34 m, then find the breadth of the rectangle.

Solution

Here P = 34 m and l = 9.8 m

$$\because b = \frac{p}{2} - l$$

$$\therefore \text{Breadth (b)} = \frac{34}{2}m - 9.8 \, m$$

$$= 17 \, m - 9.8 \, m = 7.2 \, m$$

Thus, the required breadth of the rectangle is 7.2 m

Example - 4

The length and breadth of a rectangle are in the ratio 5:3. If its perimeter is 320 cm then find its length and breadth.

Solution

Since, Length: Breadth = 5:3

i.e. if length (l) = 5x then breadth (b) = 3x

$\therefore$ Perimeter = 2(l + b) = 2(5x + 3x)

$$= 2(8x) = 16x$$

Since, the perimeter of the rectangle = 320 cm

$\therefore$ 16x = 320

$$\Rightarrow x = \frac{320}{16} = 20$$

$\therefore$ Length of the rectangle = 5x

$$= 5 \times 20 \, cm = 100 \, cm$$

Breadth of the rectangle = 3x

$$= 3 \times 20 \, cm = 60 \, cm$$

Example - 5

If a side of a square is 7.2 cm. Find its perimeter.

Solution

Side of the square 'a' = 7.2 cm

∴ Perimeter of the square = 4 × side

= 4 × 'a

= 4 × 7.2cm = 28.8 cm

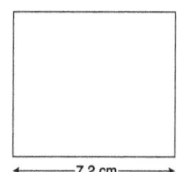

Example - 6

The perimeter of a square of side 6.3 cm is equal to the perimeter of a rectangle whose length is 8.5 cm. Find the breadth of the rectangle.

Solution

∵ Side of the square 'a' = 6.3 cm

∴ Perimeter of the square = 4a = 4 × 6.3 cm

Since, the perimeter of a rectangle = 2(l + b)

∴ 2(l + b) = 4 × 6.3

⇒ 2(8.5 + b) = 4 × 6.3 | ∵ l = 8.5 cm

⇒ 8.5 + b = $\frac{4 \times 6.3}{2}$ = 2 × 6.3 = 12.6

⇒ b = 12.6 − 8.5 = 4.1

Thus, breadth of the rectangle = 4.1 cm

Perimeter of Some other Regular Polygons

Polygon

A simple closed plane figure bounded by line-segment only.

Perimeter of a polygon

It is the sum of the length of all line segments bounding it.

Regular Polygon

If the length of each side of a polygon is same then, it is called a regular polygon.

The simplest polygon is a triangle which is formed by 3 line-segments. The regular polygon of three sides is given a special name i.e. *"Equilateral-triangle."* The Regular polygon of four sides is called a *"Square."* All other polygons are called as: A regular polygon of 5 sides, A regular polygon of 6 sides and so on.

Perimeter of a Regular Polygon = (Number of sides) × Length of a side.

Example of Perimeter of some Regular Polygons:

Figure	Number of Sides	Length of a side	Perimeter of the regular polygon
(triangle, 6 cm)	3	6 cm	3 × [Length of a side] = 3 × 6 cm = 18 cm
(rectangle/square, 5 cm)	4	5 cm	4 × [Length of a side] = 4 × 5 cm = 20 cm
(pentagon, 4.2 cm)	5	4.2 cm	5 × [Length of a side] = 5 × 4.2 cm = 21 cm
(hexagon, 3.4 cm)	6	3.4 cm	6 × [Length of a side] = 6 × 3.4 cm = 20.4 cm
(octagon, 2.5 cm)	8	2.5 cm	8 × [Length of a side] = 8 × 2.5 cm = 20 cm

Exercise - 1

1. Find the Perimeter of following closed shapes:

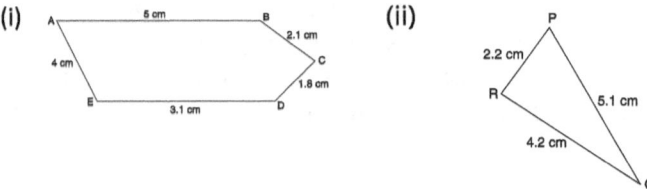

(i) A–B = 5 cm, B–C = 2.1 cm, C–D = 1.8 cm, D–E = 3.1 cm, E–A = 4 cm

(ii) P–Q = 5.1 cm, Q–R = 4.2 cm, R–P = 2.2 cm

(ii) 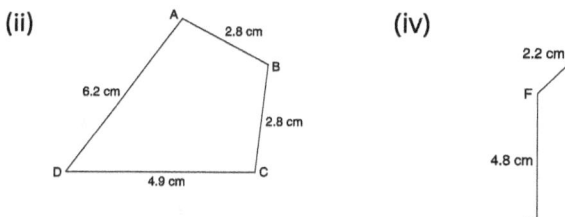 (iv)

2. Find the perimeter of the following plane closed shapes:

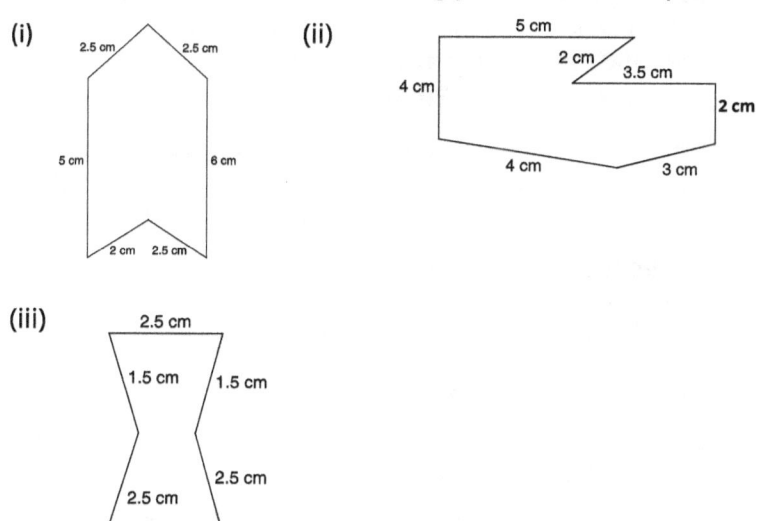

3. Find the perimeter of the following regular polygons:

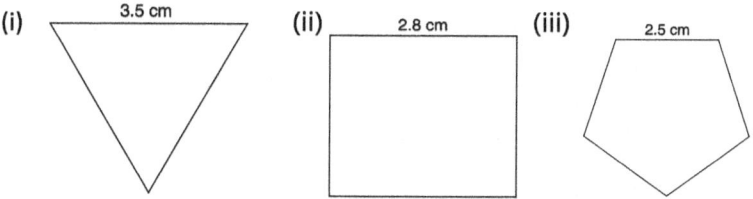

4. Find the perimeter of the regular polygons whose number of sides and length of a side are give:

Number of Sides	Length of a Sides
(i) 6	3.6 cm
(ii) 7	2.7 cm
(iii) 8	1.4 cm
(i) 9	2.2 cm

5. The perimeter of an equilateral triangle is 16.5 cm. Find its side.
6. The perimeter of a square is 34.4 cm. Find its side.
7. What is the length of a regular octagon if its perimeter is 36 cm.
8. A park is in the form of a regular pentagon. If its one side is 10.5 m. Then find the length of the barbed wire required to fence it by 3 rounds of the barbed wire?
9. Ratna has a plot whose shape is a regular hexagon. A wall is to be erected around it. Find the cost of erection of the wall at the rate of Rs. 350 per metre leaving an entrance 8 m wide, if each side of the plot is measuring 20 m.
10. The cost of fencing a square field of side 150m is Rs 27000. What will be cost of fencing of another rectangular field, at the same rate, whose length and breadth are 250 m and 150 m respectively?
11. The length and breadth of a rectangle are in the ratio 5:2. If its perimeter is same as the perimeter of a regular pentagon of side 5.6 cm, find the length and breadth of the rectangle.
12. If the side of a square is 22.5 cm. Find:
    (i) The length of a rectangle whose perimeter is equal to that of the square and breadth is equal to 20 cm
    (ii) side of an equilateral triangle whose perimeter is equal to that of the square.

ANSWERS

1. (i) 16 cm   (ii) 12.5 cm   (iii) 16.7 cm   (iv) 17.3 cm
2. (i) 20.5 cm   (ii) 23.5 cm   (iii) 14 cm
3. (i) 10.5 cm   (ii) 11.2 cm   (iii) 12.5 cm
4. (i) 21.6 cm   (ii) 18.9 cm   (iii) 11.2 cm   (iv) 19.8 cm
5. 5.5 cm   6. 8.6 cm   7. 4.5 cm   8. 157.5 m
9. Rs. 39200   10. Rs. 36000   11. 10 cm, 4 cm
12. (i) 25 cm   (ii) 30 cm

AREA OF PLANE FIGURES

Every closed plane figure encloses certain amount of the surface. In the following diagram, the shaded parts are the amount of surface of this plane paper enclosed by the boundaries of these figures. Just by observation, we can say

(i) Fig-A   (ii) Fig-B

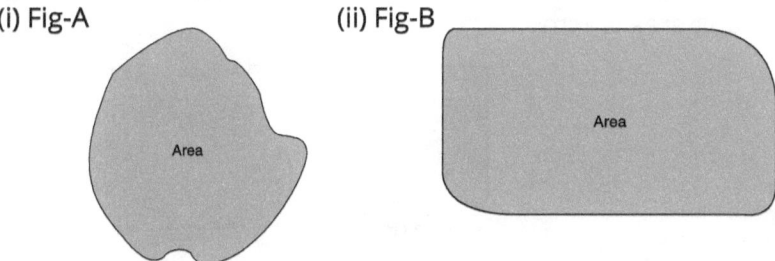

That fig B has enclosed bigger part of surface than fig A

The amount of surface of a plane enclosed by a closed figure is called its area.

We know that every closed figure drawn on a plane surface, has two regions: Interior-region and the exterior region.

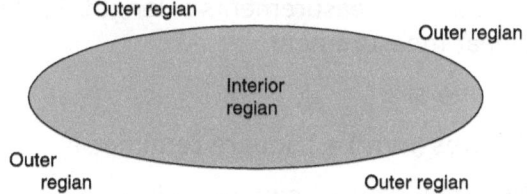

REMEMBER

The boundary of a closed curve is always a part of the interior region

The area of a shape is the measure of the interior-region of the shape.

Measuring the area of a closed plane figure means to compare it with a 'unit area' For our convenience, we have adopted the 'unit' of area as the amount of surface enclosed by a square of side 1 unit i.e. 1 cm or 1 m etc.

If we choose the side of a unit square as 1 m then the unit of area is 1 m^2

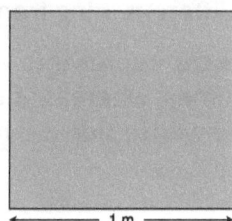

NOTE

1 m^2 is the (abbreviated) short form of '1 square metre.'

∴ 1 metre × 1 metre = 1 square metre

If we choose the side of a unit square as 1 cm (for smaller measurements), then the unit area is 1cm²

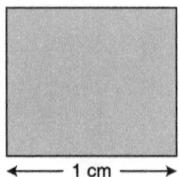

NOTE

1cm² is the (abbreviated) form of 1 square centimetre.

∴ 1 centimetre × 1 centimetre = 1 square centimetre

UNITS OF AREA

Area, being the product of measurements of length. So, it is expressed as square units of linear measurement.

Some measure of area are:

100 square millimetres (mm²) = 1 square centimetre

100 square centimetres (cm²) = 1 square decimetre

100 square decimetres (dm²) = 1 square metre

100 square metres (m²) = 1 square decametre

100 square decametres (dam²) = 1 square hectometre

100 square hectometres (hm²) = 1 square kilometre

100 square kilometres (km²) = 1 square myriametre

NOTE

'**1 square metre**' is quite a different thing from '**1 metre square**.'

For example, '**3 square metre**' represents an amount of area which contains 1 square metre 3 times, whereas '**3 metre square**' represents area of a square plane shape whose side is 3 metres long, i.e 3x3 square metres or 9 square metres.

AREA OF RECTANGLE AND SQUARE

Let us take three rectangle of different sizes and make a chart as given below:

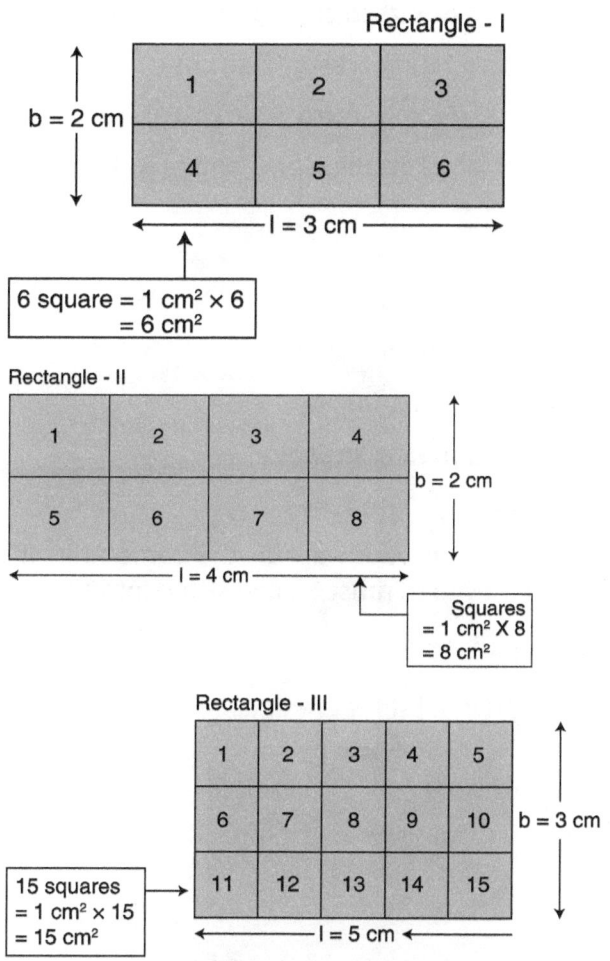

Rectangle	Length	Breadth	Number of Squares of Area = 1 cm²	Area
I	3 cm	2 cm	6	6 cm²
II	4 cm	2 cm	8	8 cm²
III	5 cm	3 cm	15	15 cm²

From the above table, we get

Area of rectangle I   = 6 squares = 6 sq.cm

                           = (3 × 2) sq.cm = Length (l) × Breadth (b)

Area of rectangle II   = 8 squares = 8 sq.cm

Area of rectangle III
= (4 × 2) sq.cm = Length (l) × Breadth (b)
= 15 squares = 15 sq.cm
= (5 × 3) sq.cm = Length (l) × Breadth (b)

It follows that if length of a rectangle be 'l' and breadth 'b' then its area = length × breadth.

Thus,

Area of Rectangle = Length × Breadth

Hence, A = l × b, where A means 'area'

l means 'length'

b means 'breadth'

From, A = l × b, we get $l = \dfrac{A}{b}$ and $b = \dfrac{A}{l}$

Provided the length and breadth are in the same unit. If they are in different units then they first, must be converted into a common unit.

AREA OF A SQUARE

In a square, the length of all sides are equal.

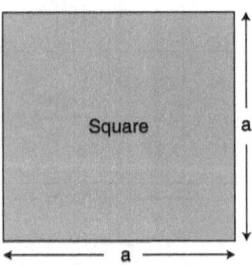

Let each side be 'a'

∴ Length = breadth = side = a

i.e. l = b = a

∴ l × b = a × a = a^2

i.e. Area of a square = (Length of a side)^2

Thus, A = a^2 | where 'A' means 'area' and 'a' means side of square

1 square decameter = 100 square metres = 1 are

1 square hectometer = 1000 square metres = 1 hectare

EXAMPLE

The length of a rectangle is 1.5 metres and breadth is 100 cm. Find its area.

SOLUTION

The units of Length and breadth are different. We convert them in common unit 'metre', using

  100 cm = 1 m

  l = 1.5 m

  b = 1.0 m

Since A = l × b

$\Rightarrow$ A = 1.5 m × 1.0 m = 1.5 m²

Thus, the required area of the rectangle = 1.5 m²

EXAMPLE – 8

If area of a rectangle whose length is 14.5 cm is 140.65 cm². Find its breadth.

SOLUTION

∵ Length of the rectangle (l) = 14.5 cm

Let breadth of the rectangle be 'b' cm

∴ [Area of the rectangle] = length × breadth

$\Rightarrow$ 140.65 = 14.5 × b

$\Rightarrow$ b $= \dfrac{140.65}{14.5} = \dfrac{14065}{100} \times \dfrac{10}{145} = \dfrac{97}{10} = 9.7$

Thus, the breadth of the rectangle = 9.7 cm.

EXAMPLE - 9

If the length and breadth of a rectangle are in the ratio 4:3 and its area is 4800 cm², then find the length and breadth of the rectangle.

SOLUTION

Since the ratio of length to breadth of a rectangle is 4:3

Let Length (l) = 4x and breadth (b) = 3x

$\therefore 4x \times 3x$ = Area of the rectangle

$12x^2 = 4800 \text{ cm}^2$

$x^2 = \dfrac{4800}{12} = \dfrac{\cancel{4800}^{400}}{\cancel{12}_{1}} = 400$

$x = \sqrt{400} = 20$

$\therefore$ Length = (4 × 20) cm = 80 cm

Breadth = (3 × 20) cm

EXAMPLE - 10

The side of a square is 8.1 cm. Find its area.

SOLUTION

Here, side of the square (a) = 8.1 cm

$\because$ Area of a square = a × a

$\therefore$ Area of the given square = 8.1 × 8.1 cm²

$= \dfrac{81 \times 81}{100} \text{cm}^2 = \dfrac{6561}{100} \text{cm}^2 = 65.61 \text{ cm}^2$

Thus, The required area of the square = 65.61 cm²

EXAMPLE - 11

Find the area of a square, whose side is equal to the length of a rectangle. If the breadth and area of rectangle are 13 cm and 195 cm² respectively.

SOLUTION

Area of the rectangle = 195 cm²

Let the length of the rectangle be 'l' cm

Since, breadth of the rectangle (b) = 13 cm

$\therefore l = \dfrac{A}{b} = \dfrac{195}{13} = 15 \text{ cm}$

$\because$ Length of the rectangle = side of the square

$\therefore$ 15 cm = side of the square.

$\Rightarrow$ a = 15 cm

Now Area of square = a × a

= 15 cm × 15 cm = 225 cm²

Thus, the required area of the square = 225 cm²

EXAMPLE - 12

The side of a square plot is 14m. If the cost of its levelling Rs. 98, 00 then find the rate of levelling the plot.

SOLUTION

∵ Side of the square(a)   = 14 m

∴ Area of the plot = a × a

   = 14 × 14 m²

   = 196 m²

Let the rate of levelling be Rs R per metre².

∴ Cost of levelling = Rs R × 196

But the cost of levelling is Rs. 9800.

⇒ R × 196 = 9800

$$\Rightarrow R = \frac{9800}{196} = \frac{\cancel{9800}^{\,50}}{\cancel{196}_{\,1}} = 50$$

∴ Rate per m² = Rs 50

EXAMPLE - 13

A rectangle is 25 m by 9 m. A square has the same area as the rectangle. Find the perimeter of the square.

SOLUTION

l = 25 m and b = 9 m

∴ Area (A) of rectangle   = 25 m × 9 m

Since [Area of square]   = [Area of rectangle]

⇒ Area of square   = 25 × 9 = 5² × 3² = (5 × 3)²

⇒ Side of the square   = 15 m

⇒ Perimeter of the square = 4 × side

   4 × 15 = 60 m

Exercise - 2

1. Find the area of a square of side:
   (i) 4.5 cm   (ii) 7.2 cm   (iii) 8.1 cm
   (iv) 5.4 cm   (v) 2.7 cm   (vi) 9.9 cm

2. Find the area of a rectangle of
   (i) Length = 3.5 cm and Breadth = 2.1 cm
   (ii) Length = 15 cm and Breadth = 10.8 cm
   (iii) Length = 20 cm and Breadth = 18 cm
   (iv) Length = 12 cm and Breadth = 10.5 cm
   (v) Length = 11.2 cm and Breadth = 10 cm

3. Find the area of a square whose perimeter:
   (i) 76 cm   (ii) 100 cm   (iii) 42 cm
   (iv) 27.2 cm   (v) 75.2 cm

4. Find the area of a rectangle of:
   (i) Perimeter = 126 cm and length = 40 cm
   (ii) Perimeter = 230 cm and length = 70 cm
   (iii) Perimeter = 316 cm and length = 100 cm
   (iv) Perimeter = 102 cm and length = 35 cm
   (v) Perimeter = 624 cm and length = 190 cm

5. A square of side 15.0 cm and a rectangle of length 25 cm have equal area. Find the breadth of the rectangle.

6. A square and a rectangle have equal area. If the length and breadth of the rectangle are 36 cm and 25 cm respectively, then find the side of the square.

7. The length of a rectangular field is 36 m and its breadth is $\frac{2}{3}$ of its length. Find the cost of leveling the field at the rate of Rs 15 per square meter.

8. The breadth of a rectangle field is $\frac{2}{5}$ of its length. If its length is 60 m, find the cost of fencing it at the rate of Rs.5 per metre.

9. Determine the cost of ploughing a tract of land 50 m long and 30 m wide at the rate of Rs. 10 per sq.m. Also find the cost of fencing it all around at the rate of Rs. 15 per metre.

10. The perimeter of a square is 80 cm. A rectangle of length 40 cm has the same area as that of the square. Find the breadth of the rectangle.

11. The length and breadth of a rectangle are 50 cm and 8 cm respectively. Find the perimeter of a square of area equal to that of the rectangle.

12. A square and rectangle have equal area. If a side of a square is 24 cm and breadth of the rectangle is 18 cm, find the length of the rectangle.

Answers

1. (i) 20.25 cm²   (ii) 51.84 cm²   (iii) 65.61 cm²
   (v) 2916 cm²   (iv) 7.29 cm²   (v) 98.01 cm²
2. (i) 7.35 cm²   (ii) 162 cm²   (iii) 360 cm²
   (iv) 126 cm²   (v) 112 cm²
3. (i) 361 cm²   (ii) 625 cm²   (iii) 110.25 cm²
   (iv) 46.24 cm²   (v) 353.44 cm²
4. (i) 920 cm²   (ii) 3150 cm²   (iii) 5800 cm²
   (iv) 560 cm²   (v) 23180 cm²
5. 9 cm   6. 30 cm   7. Rs. 12960   8. Rs. 840
9. Rs. 15000; Rs. 2400   10. 10 cm   11. 80 cm
12. 32 cm

Area of Irregular Plane Figures

For finding area of irregular plane shapes, we use squared paper. We can find approximate area. We use following three steps:

(i) Count the number of complete squares (say m) enclosed by the figure.

(ii) Count the number of squares whose more than half parts are enclosed by the figure (say n). Each such part of the square is treated as one complete square. We ignore those squares whose less than half parts are enclosed.

(iii) Count the number of squares whose exactly half parts are enclosed (say p). Then the area of the given figure = $(m + n + \frac{1}{2}p)$.

EXAMPLE - 13

Find the area of the following plane figure such that each square is a unit square:

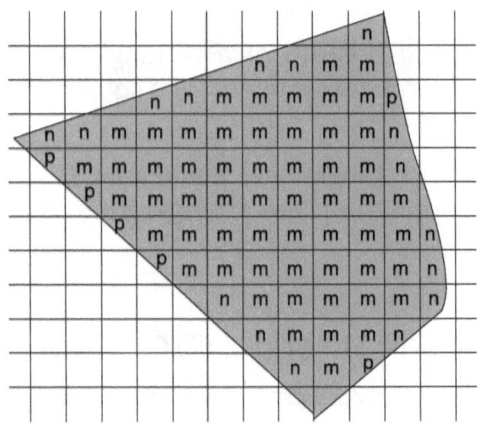

SOLUTION

Number of complete unit squares (m) = 57

Number of squares whose more than half are enclosed (n) = 16

Number of squares, whose exactly half part are enclose (p) = 6

∴ Area of the plane shape = $(m + n + \frac{1}{2}p)$ square units

= $(57 + 16 + \frac{1}{2}6)$ sq.units

= (57 + 16 + 3) sq.units

= 76 sq.units.

DID YOU KNOW?

Circle has the shortest perimeter of all plane shapes with the same area.

EXERCISE - 3

Find the area of the following plane shapes by counting the number squares enclosed by then, where each square is a unit square.

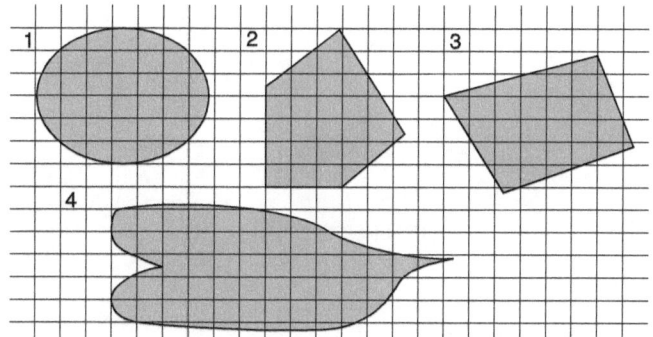

MISCELLANEOUS EXERCISE

1. Find the perimeter of following plane figures:

   (i)

   (ii)

   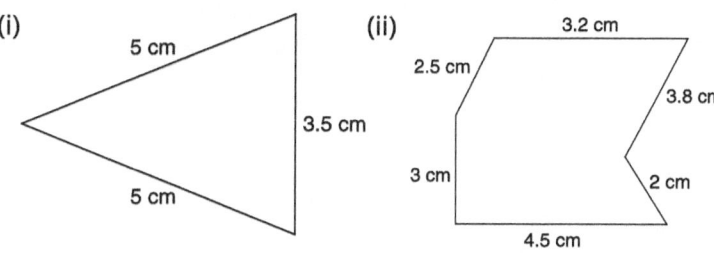

   (iii)

   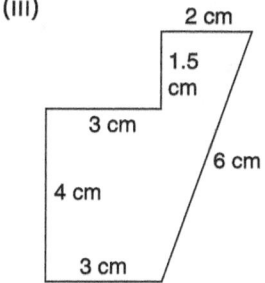

2. Find the perimeter of an equilateral triangle of side 4.5 cm
3. Find the perimeter a square whose side is 5.5 cm
4. Find the perimeter of a regular polygon of 9 sides each of its side 4.2 cm
5. Find the perimeter a hexagon whose one side is 3.5 cm
6. Find the perimeter of an octagon whose side is 2.8 cm

   Find the perimeter of the following plane figures.

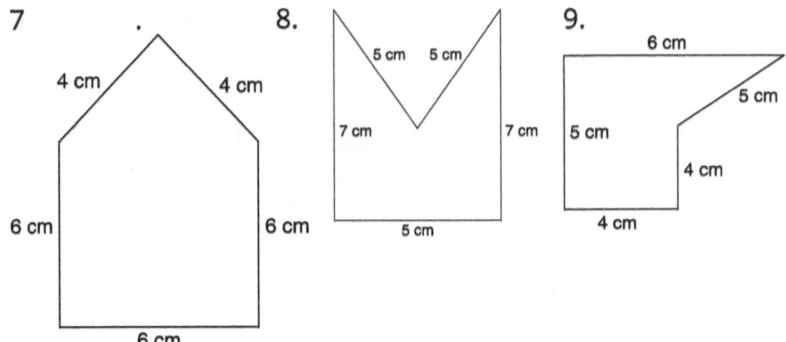

10. Find the perimeter of a rectangle whose length and breadth are 10.5 cm and 7.2 cm respectively.

11. Find the cost of fencing a square plot of side 12 m at the rate of Rs 8 per meter

Find the area of:

12. A rectangle having length = 12 cm Breadth = 9.5 cm

13. A square of side 8.3 cm

14. The area of a rectangular field is equal to the area of a square field whose perimeter is 80 m. If length of the field (rectangular) is 25 m, find the breadth of the rectangular field.

15. A rectangular field is 150 m × 100 m. Find the cost of ploughing it twice at the rate of Rs. 2 per sq.m.

ANSWERS

1. (i) 13.5 cm	(ii) 19.0 cm	(iii) 19.5 cm
2. 13.5 cm	3. 22 cm	4. 37.8 cm
5. 21 cm	6. 22.4 cm	7. 26 cm
8. 29 cm	9. 24 cm	10. 35.4 cm
11. Rs. 384	12. 114 cm²	13. 68.89 cm²
14. 16 m	15. Rs. 60000.	

HOTS

1. Find the perimeter of the following figure (ABCDEFGHIJK)

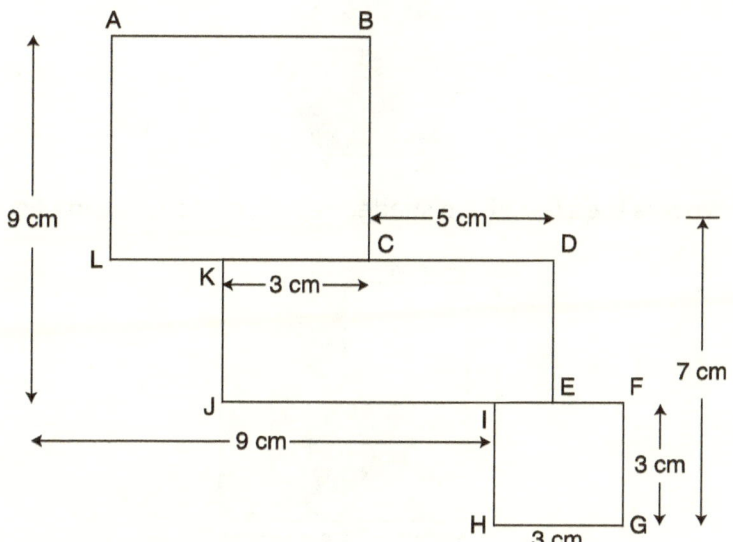

2. Find the area of the above figure (ABCDEFGHIJKL):

ANSWERS

1. 48 cm   2. 66 cm²

MENTAL MATHS

1. What is the length of a side of a square such that its area and perimeter are numerically the same?
2. A square and a rectangle have same area. Whose perimeter is more?
3. The side of a pentagon is 5 cm. What is its perimeter?
4. ABCDE is a regular pentagon. If its surface area is 90 cm² them what is the area of △ ABE?

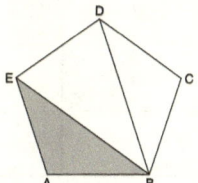

5. ABCDEF is a regular hexagon. If AB = 4 cm then what is the perimeter of ABCDEF?

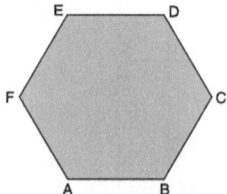

6. What is the area of the shaded region in the following figure?

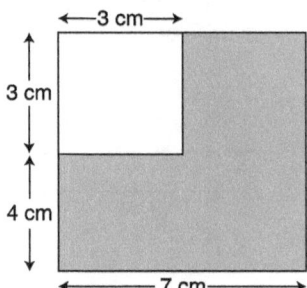

7. What is the perimeter of the figure (adjoining)?

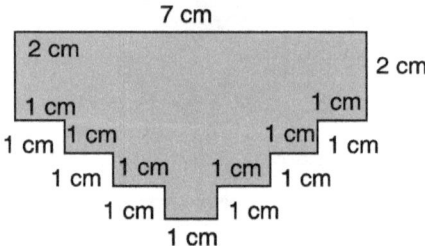

8. The perimeter of a square is 4 cm. What its area?

9. What is the area of the shaded region in the following figure? (All blocks are identical)

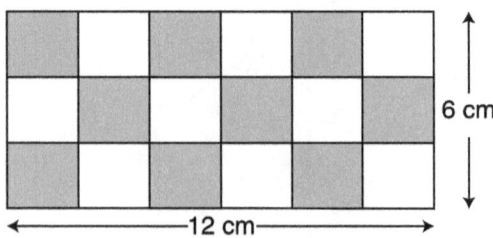

10. In the above figure what is the perimeter of all the shaded blocks such that they are identical?

Answers

1. 4 units   2. Rectangle   3. 25 cm   4. 30 cm²   5. 24 cm
6. 40 cm²   7. 24 cm   8. 1 cm²   9. 36 cm²   10. 72 cm

Multi Choice Questions

1. If perimeter of a square is 8 cm then which of the following is its area?

   (i) 2 cm²   (ii) 4 cm²   (iii) 8 cm²   (iv) 64 cm²

2. If the area of a square is 25 cm² then which of the following is its perimeter?

   (i) 10 cm   (ii) 25 cm   (iii) 20 cm   (iv) 50 cm

3. If the length and area of a rectangle are 2 cm and 2 cm² respectively. Which of the following is its breadth?

   (i) 1 cm   (ii) 2 cm   (iii) $\frac{1}{2}$ cm   (iv) $\frac{1}{4}$ cm

4. If the breadth and area of a rectangle are 1 cm and 2 cm² respectively. Which of the following is its perimeter?

   (i) 3 cm   (ii) 4 cm   (iii) 5 cm   (iv) 6 cm

5. The length and breadth of a rectangle are 9 cm and 4 cm respectively. Which of the following is the perimeter of the square whose area equal to that of the rectangle?

   (i) 26 cm   (ii) 24 cm   (iii) 18 cm   (iv) 16 cm

6. The perimeter of a pentagon is 50 cm. which of the following is the length of its side?

   (i) 5 cm   (ii) 15 cm   (iii) 25 cm   (iv) 10 cm

7. The area of a square of side 10 cm is equal to the area of a rectangle of length 25 cm. which of the following is the breadth of the rectangle?

   (i) 2 cm   (ii) $12\frac{1}{2}$ cm   (iii) 4 cm   (iv) 10 cm

8. The area of a rectangle of length 8 cm is 48 cm². What is the length of the 4 cm broad rectangle whose perimeter is equal to that of the above rectangle?

   (i) 8 cm   (ii) 6 cm   (iii) 10 cm   (iv) 14 cm

9. Which of the following is the perimeter of a square of area 4 cm²?

   (i) 2 cm   (ii) 4 cm   (iii) 8 cm   (iv) 16 cm

10. Which of the following is the length of a 2 m broad rectangle whose perimeter is equal to the perimeter of a square of side 3 cm

   (i) 4 cm (ii) 3 cm (iii) 2 cm (iv) 6 cm

ANSWERS

1. (ii)   2. (iii)   3. (i)   4. (iv)   5. (ii)
6. (iv)   7. (iii)   8. (iii)   9. (iii)   10. (i)

WORKSHEET

1. **Fill in the blanks:**

   (i) The _____ is the total length of the boundary of closed plane simple figure.

   (ii) The _____ is the total amount of the surface enclosed by the boundary of a plane figure.

   (iii) The curve itself is the _____ of the interior region.

   (iv) The perimeter of a rectangle = _____.

   (v) The perimeter of a square = 4 × _____.

   (vi) The perimeter of a regular pentagon = _____ × (length of the side of the pentagon)

   (vii) The perimeter of an equilateral triangle = _____.

   (viii) The area of a square = _____.

   (ix) The area of a rectangle = _____.

   (x) A square is a rectangle whose _____.

2. **Write true or false for each of the following:**

   (i) The area of a square of side 'a' is = 4 × a.

   (ii) Perimeter of a rectangle = length + breadth.

   (iii) The area of a rectangle whose length = l and breadth = b is given by (l × b) sq.units.

   (iv) If the side of an octagon is 6 cm then its perimeter is 48 cm.

   (v) 1 sq.metre = 100 cm

   (vi) The perimeter and area of a square are numerically equal if its side is 4 cm.

(vii) Two different rectangles can have equal area.

(viii) The perimeter of two different rectangles can be same.

(ix) The perimeter of a square and a rectangle having equal areas are always equal.

(x) If the perimeter of a regular polygon of 10 sides is 100 cm then length of its side is 10 cm

3. Find the number of tiles required to cover the floor of a room of size 10 m × 6 m if the size of each tile is $\frac{1}{10}$m × $\frac{1}{2}$m

4. Find the total length of a fence-wire to make 3 rounds of it around the rectangular park shown in the figure:

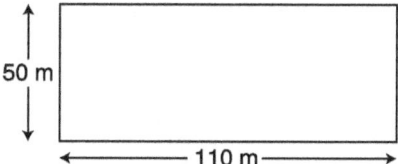

5. Match the column A (shape) and column B (perimeter)

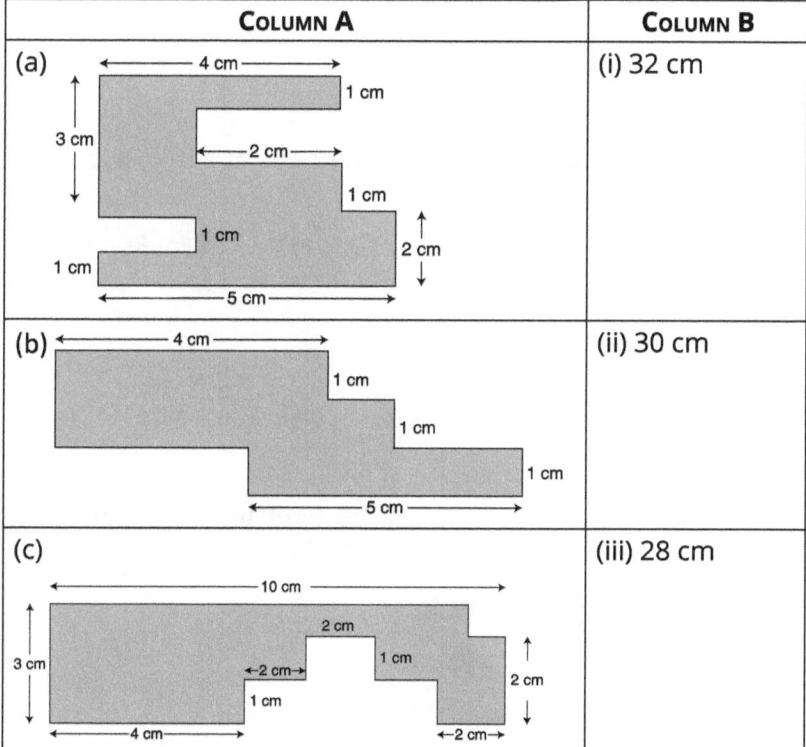

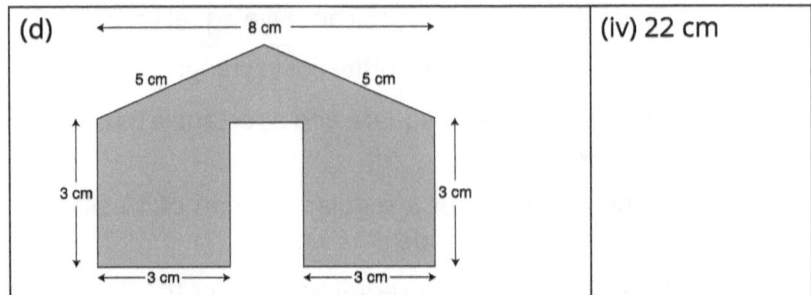

(d) (iv) 22 cm

6. What is the cost of leveling a rectangular plot of land 800 m long and 200 m wide at the rate of Rs. 10 per 100 sq.m?

7. 5 square flower beds are dug on a lawn as shown in the figure. If side of each bed is 1 m then find:

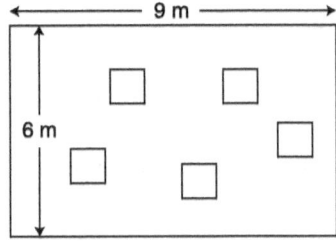

(i) The cost of levelling the remaining area of the lawn at the rate of Rs. 4 per sq.m

(ii) The cost of fencing all the beds at the rate of Rs.10.50 per metre.

(iii) The cost of fencing the park at the rate of Rs. 10.50 per meter leaving an entrance 8 metre wide.

Answers

1. (i) Perimeter (ii) area (iii) boundary (iv) 2(l + b)
   (v) Side (vi) 5 (vii) 3 × side (viii) side × side
   (ix) (l × b) (x) l = b

2. (i) False (ii) False (iii) True (iv) True
   (v) False (vi) True (vii) True (viii) True
   (ix) False (x) True.

3. 1200   4. 960 m

5. (a) → (iii)   (b) → (iv)   (c) → (i)   (d) → (ii)

6. Rs. 16000   7. (i) Rs. 196   (ii) Rs. 210   (iii) Rs. 231

CHAPTER 14
DATA HANDLING

INTRODUCTION

In most of our day to day activities we use numerical information. While watching a cricket-match on T.V. We get lot of interesting facts from the scorekeeper and statistician, such as: Shekhar's average score in last two years, India's performance in last world cup, etc.

DATA

Each number which gives some information is data.

For example, in a cricket test match Rohit played for 5 overs. Overwise his score was as under:

Over	1st	2nd	3rd	4th	5th
Score	8	10	2	25	12

Such data is collected and used in many way by various agencies, government departments, educational institutions and different companies etc. data can be about many activities and actions. Data can be about education, population, taxes, food, finance, agriculture, defence, elections etc. Infact in every field of life data is used. We get lot information from such data. The modern society is essentially data-oriented.

STATISTICS

The branch of mathematics that is mainly concerned with *COLLECTION, CLASSIFICATION, TABULATION* and *PRESENTATION* of data and the conclusions that are drawn from it is called *'statistics'*.

REPRESENTATION OF DATA

Representation of data is done mainly in two ways:
  (i) Tabular Representation
  (ii) Graphical Representation.

Let us collect information about favourite ice-cream of ten students of your class. We can record it as under:

1. Sawita	Chocolate	2. Chhavi	Chocolate
3. Reena	Strawberry	4. Vashav	Strawberry
5. Kiran	Strawberry	6. Nandini	Black current
7. Harish	Black current	8. Mani	Vanilla
9. Gautam	Vanilla	10. Bablu	Mint

If we have to collect the above information about 100 students then it would be very difficult to write names of all the students and write name of the ice-cream again and again. If we want to know that how many students like strawberry or vanilla, etc, then we are not interested in the names of students. We simply need to know the number of students, who like a particular ice-cream.

A quick way is to ask each student about their favourite ice-cream and mark a 'standing-line' against that ice-cream. We call these standing lines as **'tally marks'**. These tally marks are written in the following manner:

We write

   '|' for 1 student against the ice cream

   '||' for 2 students against the ice cream

   '|||' for 3 students against the ice cream

   '||||' for 4 students against the ice cream

   ‚ЖЖГ , for 5 students against the ice cream

   ‚ЖЖГ | for 6 against the ice cream

   ‚ЖЖГ ||' for 7 against the ice cream

   ‚ЖЖГ |||' for 8 against the ice cream

   ‚ЖЖГ ||||' for 9 against the ice cream

   ‚ЖЖГ ЖЖГ , for 10 against the ice cream

   ‚ЖЖГ ЖЖГ |' for 11 against the ice cream

and so on.

If there are 18 students against a particular ice cream, then

    ЖЖГ ЖЖГ ЖЖГ ||| Represent 18

i.e. we make groups of five tally marks. Thus, above data can be represented (using tally marks) as under:

Ice cream	Tally marks	Number of students
Chocolate	||	2
Black current	||	2
Strawberry	|||	3
Mint	|	1
Vanilla	||	2
Total		10

EXAMPLE

Following are the marks (out of 10) obtained by 40 students of class VI in a class test

    1, 5, 6, 2, 3, 5, 2, 4, 1, 5   6, 5, 5, 4, 2, 5, 3, 2, 2, 1

    2, 5, 3, 4, 5, 1, 5, 5, 6, 1   6, 1, 4, 5, 1, 3, 5, 6, 6, 3

Using tally marks, prepare the table for the above data. What score is obtained by the maximum number of students.

SOLUTION

Crossing out the marks one by one and marking a tally side by side, we get the following table:

~~1~~ ~~5~~ ~~6~~ ~~2~~ ~~3~~ ~~5~~ ~~2~~ ~~4~~ ~~1~~ ~~5~~   ~~6~~ ~~5~~ ~~5~~ ~~4~~ ~~2~~ ~~5~~ ~~3~~ ~~2~~ ~~2~~ ~~1~~

~~2~~ ~~5~~ ~~3~~ ~~4~~ ~~5~~ ~~1~~ ~~5~~ ~~5~~ ~~6~~ ~~1~~   ~~6~~ ~~1~~ ~~4~~ ~~5~~ ~~1~~ ~~3~~ ~~5~~ ~~6~~ ~~6~~ ~~3~~

MARKS	TALLY MARKS	NUMBER OF STUDENTS
1									7
2								6
3							5
4						4
5														12
6								6
Total		40

From the table, we observe that maximum students (i.e. 12 students) have obtained 5 marks.

PICTOGRAPHS

Representation of data by using symbols and pictures is called *pictograph*. We use pictures or symbols to denote numerical data. Let us have some examples.

EXAMPLE - 1

Number of students enrolled in various classes in a school are:

Class – VI	Class – VII	Class – VIII	Class – IX	Class – X
25	30	35	20	15

Represent the data using a pictograph.

SOLUTION

Let us choose a symbol which is equal to a certain quantity or number.

Here, let the symbol 👤 = 5 students

∴ 25 students = 👤👤👤👤👤

30 students = 👤👤👤👤👤👤

35 students = 👤👤👤👤👤👤👤

20 students = 👤👤👤👤 and 15 students = 👤👤👤

NOTE

The symbol selected to represent a certain number or quantity is called the *key* of the pictograph.

Thus, the pictograph showing the number of students enrolled in various classes is:

Class	Number of students enrolled
vi	👤 👤 👤 👤 👤
vii	👤 👤 👤 👤 👤 👤
viii	👤 👤 👤 👤 👤 👤 👤
ix	👤 👤 👤 👤
x	👤 👤 👤

Advantages of a pictograph

It is eye catching. It leaves a more lasting impression on the minds of the observer. Information is obtained from a pictograph at glance.

Drawbacks of a pictograph

Representation of data through pictograph is difficult and time-consuming sometimes it fails to represent fractional or very large number of data.

BAR GRAPH [or column Graph]

Thin rectangles of same width are used to represent the data. The bars (Thin rectangles) are also called columns. The bars drawn can be vertical, or horizontal but *Their breadth must be equal.*

The given example can be represented by a column graph as:

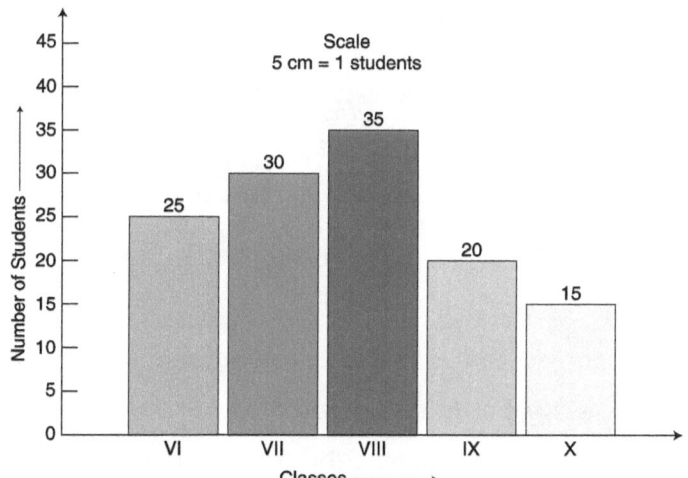

To draw a bar graph, we make use of mutually perpendicular axes.

We can use a graph paper in drawing the column graphs in an easier manner. In a bar-graph, it must be clearly mentioned on both the axes what is being represented. Here we have taken the various 'class' along the x-axis and corresponding number of students along the y-axis. The height of the column represent the numerical value of the data (**Here, the number of students**)

Remember

Vertical bars are called columns.

INTERPRETATION OR READING A BAR GRAPH

By reading a bar graph, we can extract the information and can easily make interpretation of the bar graph. Let us take the following example:

EXAMPLE: Read the following graph (bar graph) and answer the questions:

Favourite Subject for 100 Students

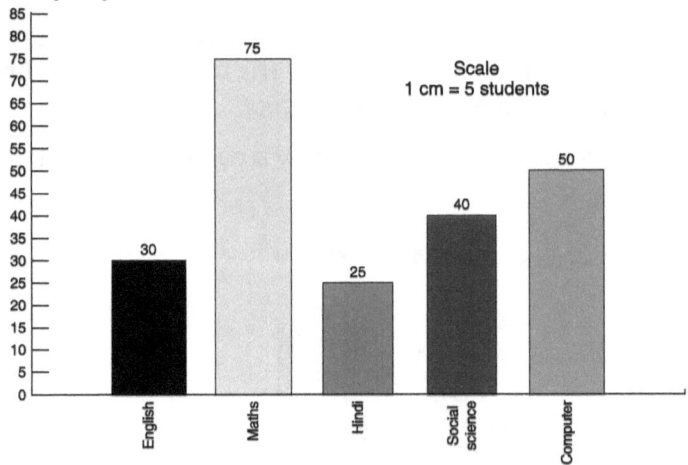

Note

More than one subject can be favourite to a student

(i) What does the bar-graph represent?

(ii) Which is the most popular subject as per this graph?

(iii) What is the number of students liking English?

(iv) What fraction of total students like computer?

(v) Which subject is least popular?

(vi) What percent of students like social-sciences?

Solution

(i) The column graph represents the subjects which are popular among 100 students.

(ii) The most popular subject is mathematics

(iii) 30 students like English the most.

(iv) Number of students liking computer = 50

Total number of students under observation = 100

∴ Required fraction = $\frac{50}{100} = \frac{1}{2}$

(v) Hindi is the least popular subject

(vi) Number of students liking social science the most = 40

∴ Required percentage = 40

or

40 % students like social sciences.

EXERCISE

1. Find the number of students liking various ice creams from the following table:

ICE CREAMS	TALLY MARKS	NUMBER OF STUDENTS
Strawberry	𝐼𝐼𝐼𝐼 𝐼𝐼𝐼𝐼 𝐼	—
Chocolate	𝐼𝐼𝐼𝐼 𝐼𝐼𝐼𝐼 𝐼𝐼𝐼𝐼	—
Vanilla	𝐼𝐼𝐼𝐼 𝐼𝐼𝐼𝐼 𝐼𝐼𝐼𝐼	—
Black current	𝐼𝐼𝐼	—
Mint	𝐼𝐼𝐼𝐼 𝐼𝐼	—

2. 40 students of a class obtained the following marks in a class test (out of 10). Using tally marks arrange these marks in a table:

1, 9, 3, 4, 7, 7, 4, 8, 3, 5, 8, 5, 6, 6,

7, 2, 4, 6, 6, 7, 2, 9, 5, 1, 8, 6, 6,

5, 6, 9, 4, 4, 5, 5, 4, 4, 2, 8, 6, 7,

What is the score of maximum number of students?

3. Following pictograph shows the number of mobile phones sold by a shop on first 4 days a week:

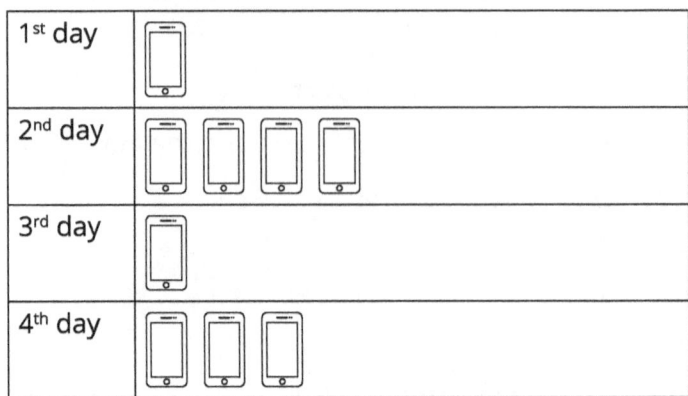

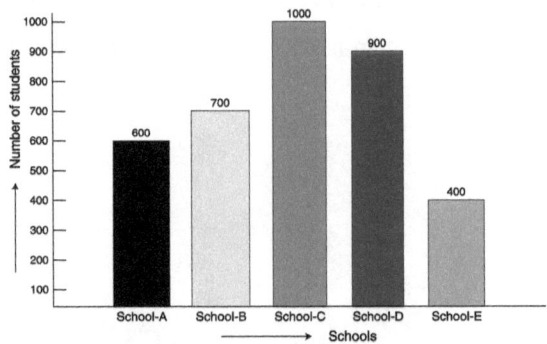 If represents 10 cell phones, then answer the following:

(i) How many cell phones are sold on first four days?

(ii) Which day the sale of Mobile phones was maximum?

(iii) Which day the sale of Mobile phones was minimum?

(iv) What is the sale of Mobile phones on first two days?

(v) What is the sale of Mobile phones on the last two days?

4. Number of students who remained absent from the school on various days of a week are given below. Represent the data using a pictograph:

Week days	Monday	Tuesday	Wednesday	Thursday	Friday	Saturday
Number of students	5	3	10	8	4	2

5. Following bar graph represents the number of total students in 5 school of a town.

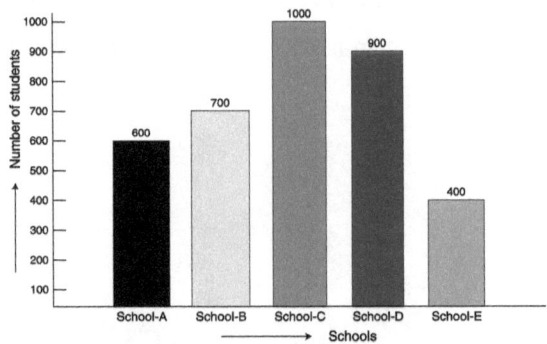

Read the column graph and answer the following question:

(i) What is the strength of school D?

(iii) Which school has highest strength?

(vi) Which school has minimum number of students?

(v) What is the total strength of all the four schools of the town?

6. Following table shows the number of students admitted to sections 'A', 'B', 'C' and 'D' of class VI of a school:

Section	Section A	Section B	Section C	Section D
Number of students admitted	42	40	50	48

Represent the above data by a pictograph

7. The temperature of a patient was recorded as under by a nurse in a hospital:

Time	6 am	8 am	10 am	12 noon	2 pm
Temperature	38°c	40°c	38°c	40°	37°

Draw a bar-graph for the above data.

8. The number of visitors in a Bal mela on 14th November in a school were as under:

Time	During 1st hour	During 2nd hour	During 3rd hour	During 4th hour	During 5th hour
Number of visitors	210	400	620	550	100

Represent the above data by a bar-graph.

9. The following table shows the percentage of buyers five different brands of mobile phones:

Brand	Apple	Blackberry	Samsung	Motorola	Vivo
Percentage of Buyers	15%	5%	35%	20%	25%

Draw a bar graph to represent the above data.

Answers

1. Strawberry = 11, chocolate = 14, Vanilla = 15, black current = 3, Mint = 7

2.

Marks	1	2	3	4	5	6	7	8	9
Tally marks	II	III	II	𝌷 II	𝌷 I	𝌷 III	𝌷	IIII	III
Number of Students	2	3	2	7	6	8	5	4	3

Score of maximum students = 6

3. (i) 90 (ii) 2nd day (iii) 1st and 3rd day (iv) 50 (v) 40
5. (i) 900 (ii) School - C (iii) School - E (iv) 3600

Mean and Median

We know that average is single value that represents each of the set of unequal values. For example:

The daily saving of Ratanlal during a week is given as:

Monday: Rs 3300   Tuesday: Rs 2350   Wednesday: Rs 1800
Thursday: Rs 3650   Friday: Rs 3200   Saturday: Rs 2750
Sunday: Rs 2550

The weekly average saving is:

$$\frac{[Rs\ 3300 + Rs\ 2350 + Rs\ 1800 + Rs\ 3650 + Rs\ 3200 + Rs\ 2750 + Rs\ 2550]}{7} = \frac{Rs\ 19600}{7} = Rs\ 2800$$

We can say that average weekly saving of Ratanlal is Rs 2800.

Thus, the average of any number of given values is given by:

$$Average = \frac{Sum\ of\ the\ given\ values}{Number\ of\ values}$$

Example

In a class test 6 students obtained marks (out of 100) as under:

Rajan: 87,   Prashant: 95   Ajanita: 67,   Pramod: 98
Sarang: 48,   Priyush: 61

Find their average marks.

Solution

Sum of all the marks obtained

$$= 87 + 95 + 67 + 98 + 48 + 61 = 456$$

$$\therefore \text{ Average marks} = \frac{\text{Sum of marks of all students}}{\text{Number of students}} = \frac{456}{6} = 76$$

∴ Required average marks are 76.

MEAN

Mean in the statistics is the same average in Arithmetic.

For example, mean of 7, 9, 11 and 13 is given by:

$$\frac{7+9+11+13}{4} = \frac{40}{4} = 10$$

so mean of given data, $= \frac{\text{Sum of the values}}{\text{Number of values}}$

i.e.

Mean of 'n' numbers $= \frac{\text{Sum of the numbers}}{\text{Number of addends}}$

Example

Find the mean of 6, 7, 9, 5 and 8.

Solution

Sum of the given numbers $= 6 + 7 + 9 + 5 + 8 = 35$

Number of addends $= 5$

$$\therefore \text{ Mean} = \frac{\text{Sum of given numbers}}{\text{Number of addends}} = \frac{35}{5} = 7$$

Hence, the mean of the given numbers is 7.

Example

The heights of 7 students in group are: 150 cm, 170 cm, 155 cm, 165 cm, 158 cm, 147 cm and 140 cm.

Find their mean height.

SOLUTION

Sum of the given observations (heights)

= 150 cm + 170 cm + 155 cm + 165 cm + 158 cm + 147 cm + 140 cm

= 1085 cm

Number of observations = 7

∴ Mean height = $\dfrac{\text{Sum of observations}}{\text{Number of observations}} = \dfrac{1085 \text{ cm}}{7} = 155$ cm

Thus, the mean height is 155 cm.

MEDIAN

In a set of data, the median is the value separating the higher half of the given data from the lower half. For example, if the given observations are 2, 1, 6, 2, 4, 3, 1 then arranging them in an ascending order, we have:

$$\underbrace{1, 1, 2,}_{\text{Lower half}} \quad \underset{\uparrow}{2,} \quad \underbrace{3, 4, 6}_{\text{Higher half}}$$

(middle most)

Term

Median

The given observations are 7 and 7 is an odd number

Thus, median of 2, 1, 6, 2, 4, 3 and 1 is 2

Similarly to find the median of 5, 18, 2, 21, 25, 3, 12 and 10, let us arrange them in ascending order, we have:

$$\underbrace{2, 3, 5,}_{\text{Lower half}} \quad \underset{\uparrow}{10, 12,} \quad \underbrace{18, 21, 25}_{\text{Higher half}}$$

(Two Middle most term)

The given observations are 8 and 8 is an even number.

Since, there are two middle most terms,

∴ median = (mean of the two middle most terms) = $\dfrac{10+12}{2} = \dfrac{22}{2} = 11.0$

Thus, median is 11.0

Note

(i) When the number of given observations is odd, then the middle-most term is equal to the median of the data.

(ii) When the number of given observations is even the mean of the values of two middle most terms in the median.

Example

Find the median of the following data:

38, 10, 21, 15, 40, and 25.

Solution

Arranging the given data in ascending order:

10, 15, $\underbrace{21, 25,}_{\text{Two middle most terms}}$ 38, 40

Here two middle most terms are: 21 and 25

∴ Median = mean of 21 and 25

$= \frac{1}{2}(21+25)$

$= \frac{1}{2}(46)$ = 23

Thus, required median is 23

Note

(i) Instead of arranging the given data in ascending order, we can arrange them in descending order; i.e., the descending order of 38, 10, 21, 15, 40 and 25 is:

40, 38, 25, 21, 15, 10

↑

Two middle most terms

(ii) When the number of observations is odd, then there will be only one middle most term, and this term is the median.

(iii) When the number of observations is even then there will be two middle terms, and the average of these two middle-most terms will be the median of the data.

Example

Find the median of the following data:

17, 24, 32, 9, 20, 13, 27, 36 and 15

Solution

Arranging the given data in ascending order, we have:

9, 13, 15, 17, 20, 24, 27, 32, 36

     Middle most term

Here, the middle most term is 20

∴ Median = 20

Exercise

1. Find the mean of:
   (i) 9, 11, 13, 17, 21 and 25 (ii) 15, 10, 6, 26, 10, 3 and 7
   (iii) 21, 23, 20, 20, 6, 2 and 27 (iv) First 5 odd natural numbers.
   (v) First 8 even natural numbers.

2. Find the median of:
   (i) 1, 2, 3, 1, 1, 3, 4 and 6 (ii) 21, 22, 48, 26, 24, 10, 16 and 18
   (iii) 25, 26, 18, 11, 25, 14 and 9 (iv) 1, 3, 11, 9, 7, 5, 3, 1 and 5
   (v) 23, 13, 10, 8, 9, 21, and 15

3. Find the (i) mean and (ii) median of first 6 odd natural numbers.

Answers

1. (i) 16 (ii) 11 (iii) 17 (iv) 5 (v) 9
2. (i) 2.5 (ii) 21.5 (iii) 18 (iv) 5 (v) 13
3. (i) 6 (ii) 6

HOTS

Find the value of x when the mean of 8, 7, x + 10, 10, 13 – x, 6 and 9 is 9. Now, substitute the value of x, so obtained, in the above data and find median.

ANSWER

    x = 2;   median = 9

MISCELLANEOUS EXERCISE

1. Following table shows the number of bicycles manufactured in a factory during the years 2015 to 2018. Represent this data, using a bar-graph, choosing appropriate scales.

YEAR	Number of bicycles manufactured
2015	900
2016	1200
2017	800
2018	1100

   Write *True* or *False*:

   (i) In a bar graph space between consecutive bars must be different.

   (ii) The height of a bar represents a quantity, whereas its width represents nothing.

   (iii) Each number, collected for giving some information, is called data.

   (iv) In a bar-graph, the heights of bars should be same.

   (v) Line graph are generally used to show the change in a quantity.

2. **Fill in the blanks:**

   (i) In a bar graph, the bars drawn can be vertical or horizontal, but their _____ must be equal

   (ii) Data means _____ information

   (iii) There are mainly two types of data representation: (i) _____ and (ii) Graphical representation.

   (iv) In pictograph, the symbol selected is called the _____ of the pictograph.

   (v) In a bar-graph, the vertical rectangles are specially called as _____.

3. Represents the following data, which shows the monthly budget of Mr. Bharat Prasad, using a bar graph:

Item	Education	Clothing	Food	House rent	Power
Amount in Rupees	Rs 5000	Rs 2000	10,000	Rs 20,000	Rs 3000

4. In a class test 40 students of class-VI obtained the following marks (out of 10):

9	1	5	2	6	3	2	4	9	5
2	2	3	3	6	9	4	5	6	6
1	5	2	4	5	5	5	4	5	9
5	3	8	6	3	2	5	5	9	6

Make a table using tally-marks and answer the following:

(i) What score is obtained by maximum number of students.

(ii) Which scores are obtained by equal number of students?

5. Following is a chart of favourite subjects for some students:

Subject	Tally marks
English	|||| |||| |||| |||| |||
Maths	||||	
Computer	||
Social Science	||||
General studies	|||| |||| |||| ||

Answer the following questions:

(i) Which subject is the most favourite?

(ii) Which subject is the least favourite?

(iii) How many students like English the most?

(iv) How many students like the General studies the most?

(vi) How many students were surveyed?

6. 10 students obtained following marks in a class test (out of 100). Find their mean marks.

   70, 48, 68, 57, 75,
   81, 88, 54, 61, 69,

7. If the mean height of 9 students is 145 cm. If the mean height of 8 students is 147 cm then what is the height of 9th student?

8. Find the median of the following numbers:

   41, 18, 26, 10, 12, 42, 48, 12, 5

9. Find the median of first 7 multiples of 3.

10. The price status of green vegetables for a week is given below:

Day	Monday	Tuesday	Wednesday	Thursday	Friday	Saturday	Sunday
Price Per kg	40	35	55	60	60	45	42

Represent the data by a bar-graph.

Answers

1. (i) False   (ii) True   (iii) True   (iv) False   (v) True
2. (i) Width   (ii) Numerical   (iii) Pictograph
   (iv) Key   (v) Column
4.

Marks	1	2	3	4	5	6	9
Tally marks	||	||||	||||	||||	|||| ||||		|||| ||	||||
Number of students	2	6	5	4	11	7	5

   (i) 5   (ii) 3 and 9

5. (i) English   (ii) Computer   (iii) 23   (iv) 17   (vi) 52
6. 67.1   7. 129 cm   8. 18   9. 12

Multiple Choice Questions

1. In a bar graph, when bars are vertical then the graph (representation) is called as:

   (i) Pictograph   (ii) Column-graph
   (iii) Bar graph   (iv) Line graph

2. In a pictograph, the symbol taken to represent a certain amount or numbers is called as:

   (i) Key of the pictograph (ii) Scale of the pictograph

   (iii) Colour of the pictograph (iv) Size of the pictograph.

3. Reading a bar-graph is related with which of the following terms?

   (i) Colouring the bars (ii) Measuring the bars

   (iii) Scaling (iv) Interpretation

4. The above column graph represents the type of game popular in 100 students of a school.

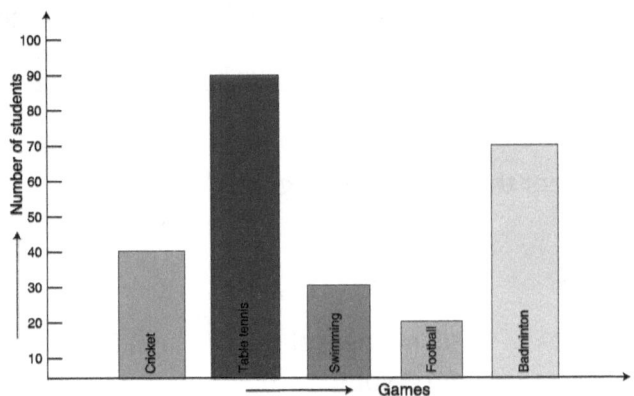

   Which of the following is the fraction of students who like cricket?

   (i) $\dfrac{2}{5}$ (ii) $\dfrac{1}{5}$ (iii) $\dfrac{9}{10}$ (iv) $\dfrac{7}{10}$

5. Following chart gives information about the favourite ice cream of a group of children:

Ice cream	Vanilla	Chocolate	Strawberry	Mint
Number of children	||||	ЖТ ЖТ |||	ЖТ ||||	ЖТ	

   Total number of children in the group is:

   (i) 42 (ii) 37 (iii) 32 (iv) 31

6. The mean of first ten two digit numbers is:

   (i) 14.7 (ii) 14.5 (iii) 14.6 (iv) 14.4

7. The median of 92, 108, 107, 93, 94, 106 and 100 is:

(i) 97   (ii) 100   (iii) 103   (iv) 57.14

ANSWERS

1. (ii)   2. (i)   3. (iv)   4. (i)   5. (iii)   6. (ii)   7. (ii)

WORKSHEET

1. Complete the following table:

MARKS	TALLY MARKS	NUMBER OF STUDENTS
1	|||| |||| |||| ||	—
2	—	9
3	|||| |||| ||||	—
4	—	13
5	|||| |||| |||| ||||	—

2. Following is the data showing the size number of shoes sold on Sunday by a trader. Represent the data by a table using tally-marks:

   8, 7, 9, 10, 6, 5, 8, 8, 9, 9, 10, 8, 7, 5, 5, 7, 7, 9, 10, 5,

   5, 9, 10, 6, 5, 6, 7, 7, 8, 8,

3. The heights of a group of 20 students are given below:

150 cm,	150 cm,	148 cm,	149 cm
150 cm,	151 cm,	151 cm,	148 cm
149 cm,	150 cm,	148 cm,	150 cm
148 cm,	148 cm,	150 cm,	149 cm
151 cm,	149 cm,	150 cm,	148 cm

   Represent the above date by a bar graph

4. Read the following bar graph and answer the questions :

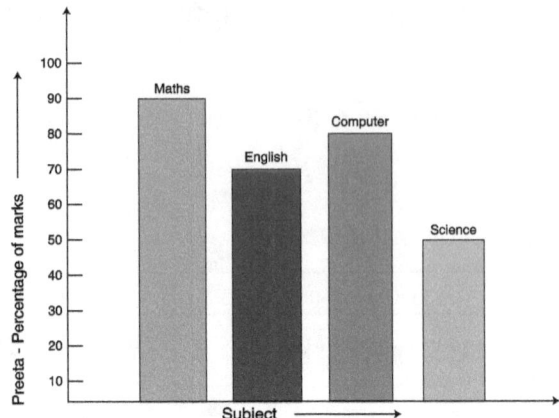

(i) In which subject Preeta got highest marks?

(ii) What is the percentage of marks in English?

(iii) In which subject Preeta got lowest percentage?

(iv) Name the subjects in which Preeta obtained marks more than 50%

(v) Name the subjects in which Preeta obtained marks less than 80%

5. Following table shows the different modes of travel to school for 50 students of a locality:

Mode of travel	Bicycle	School-bus	car	Two wheeler	Rick show
Number of students	12	10	8	15	5

Represent the above data by a bar-graph.

6. Find the mean of marks obtained by 10 students as given below:

Rohit: 68   prabhat: 75   Sahil: 63   Rajani: 65

Laxman: 81   Rohita: 82   Komal: 57   Mukul: 60

Buntee: 58   Sohit: 65

7. Find the mean of: 98, 101, 102, 97, 99, 100, 103, 96, 104.

8. Find the median of the following data: 16, 28, 23, 24, 31, 14, 38.

9. Find the median of the following data: 36, 5, 26, 3, 5, 1, 4, 5, 5, 18.

10. Find the mean and median of the following data:
    (i) 1, 2, 1, 3, 1, 4, 9, 4, 8, 3
    (ii) 2, 2, 5, 5, 1, 1, 3, 3
    (iii) 92, 100, 94, 96, 108, 106, 104
    (iv) First 6 multiples of 2
    (v) First 7 prime numbers

www.ingramcontent.com/pod-product-compliance
Lightning Source LLC
Chambersburg PA
CBHW020722180526
45163CB00001B/69